Dear John,
I hope you enjoy these books. You helped make them possible.
Thank you,
Neal

EXPERIENCE TECHNOLOGY

COMMUNICATION TRANSPORTATION
PRODUCTION BIOTECHNOLOGY

Dr. Stuart Soman
Technology Teacher
West Hempstead Public Schools
West Hempstead, New York

Neal Robert Swernofsky
Technology Teacher
Lincoln Orens School
Island Park, New York

GLENCOE
Macmillan/McGraw-Hill
New York, New York
Columbus, Ohio
Mission Hills, California
Peoria, Illinois

Cover: The Stock Market
Text Design: Design Five
Title Page: Mark Tomalty/Masterfile

Edited by Jody James, Editorial Consultant

Send all inquiries to:
GLENCOE DIVISION
Macmillan/McGraw-Hill
3008 W. Willow Knolls Drive
Peoria, IL 61614

ISBN: 0-02-646945-6 (Student Edition)
ISBN: 0-02-646951-0 (Student Workbook)
ISBN: 0-02-646948-0 (Teacher's Resource Binder)

Printed in the United States of America.

3 4 5 6 7 8 9 10 RRW 96 95 94

Dedication

My love and thanks to my wife, Bonnie, an adroit navigator on the highway of life; my fifteen-year-old "like, awesome" daughter, Amy; my twelve-year-old son, Philip (AKA "the human dynamo"); and my parents, Henry and Ida, who of course have retired to Florida.

—Stu Soman

I wish to thank those people who have helped make my teaching career a complete joy: my wife, Cindi, my daughter, Sarah, and my son, David, who have always been a source of support and encouragement; my school community for the opportunity to grow professionally; my students for challenging my creativity; and my colleagues across the country who share the same concerns for children.

—Neal Swernofsky

Acknowledgments

The publisher wishes to thank the following people, who served as planning consultants for this book.

Mr. Joe Charles "Chuck" Bridge
Industrial Technology Education Teacher
Chisholm Trail Middle School
Round Rock, TX

Ms. Barbara Brock
Industrial Technology Teacher
Friendswood, TX

Mr. Ronnie McQueen
Technology Teacher
Madison Middle School
Abilene, TX

Mr. Alan L. Towler
Technology Education Curriculum Specialist
The University of Texas at Austin
Austin, TX

For his help in planning and developing this program, special thanks to

Dr. Ronald E. Jones
Professor and Coordinator of Technology Teacher Education
Department of Industrial Technology
University of North Texas
Denton, TX

The following have contributed section opening activities:

Sections I, II, and VII
Mr. Neal Swernofsky
Lincoln Orens School
Island Park, NY

Section III
Mr. James A. Hendricks
Desert Hills Middle School
Kennewick, WA

Section IV
Mr. Joseph H. Orr
Eagles Landing Junior High School
McDonough, GA

Table of Contents

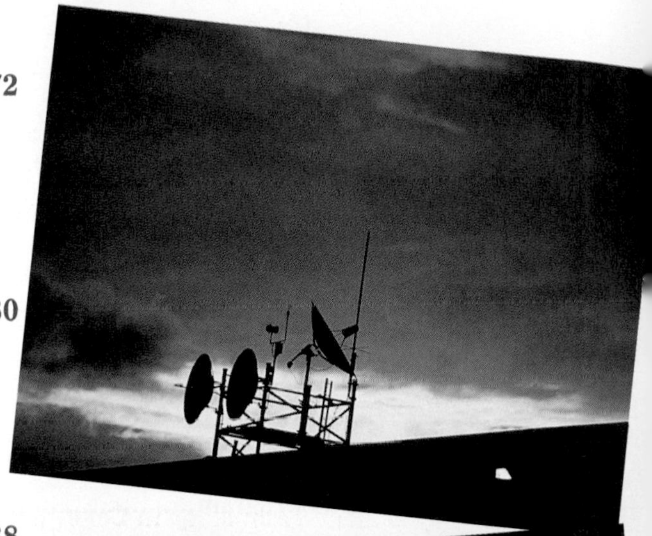

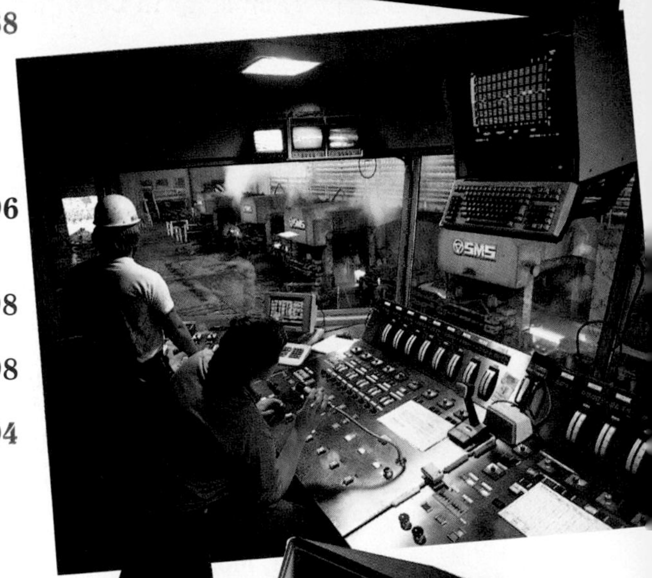

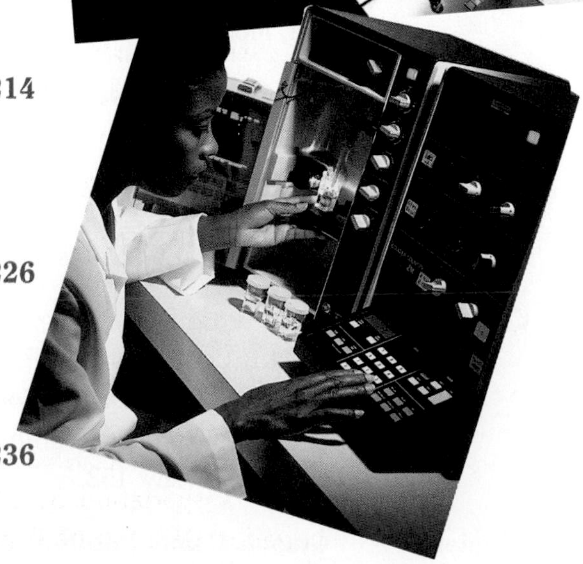

Introduction

"Experience is the best teacher." This statement means that we learn best by actually doing something. The title of this textbook is *Experience Technology: Communication, Production, Transportation, Biotechnology*. Notice the first word in the title—"experience." In this course, you will learn through experience—by "doing." You will not only read, but you will also interact with your classmates during discussions and while performing a wide variety of interesting and challenging activities.

The next word in the title is "technology." What is technology? It is many things to many people. Basically, technology is the study and improvement of our people-made world. It includes information, ideas, actions, and things. Technology is all around us every day. It plays an important role in our lives. As you study this textbook, you will learn about technology in general and the various resources on which it relies.

The last four words in the title are the four families of technology: communication, production (manufacturing and construction), transportation, and biotechnology. In this course, you will see how technology in many forms shapes the world in which we live.

Section Content and Organization

Information in this book is presented in seven sections. Each section presents various aspects of a basic part of technology or a family of technology. For example, Section II discusses the resources of technology and Section IV discusses systems of communication.

Opening Activities

Each section opens with an activity that will give you the opportunity to learn through hands-on experience what the section is basically about. For example, in the activity that introduces you to Section VI, *Production Systems: Construction*, you will design and build a model bridge.

Chapters

Following the section activity are chapters. Each major section consists of two or more chapters. Each chapter discusses a topic related to the basic idea of the section or to a family of technology. For example, in Section II, *Technology Relies on Resources*, Chapter 5 discusses material resources; in Section V, *Production Systems: Manufacturing*, Chapter 19 discusses manufacturing processes.

Section Closings

Each section closes with a feature called "You Can Make a Difference." These are true stories about young people who have used technology in a positive manner to achieve something.

Special Section

The last section of the book is a special section called *Putting It All Together*. In this section, you will see how the four families of technology can be used together to plan a space community. You will "learn by doing," for that's what this textbook is all about.

Chapter Content and Organization

There are 31 chapters in this book. These include the following elements.

Chapter Openings

Each chapter begins with an introduction, followed by a list of objectives and a list of words you will need. The introduction is a brief overview of chapter content. Objectives state specific facts or concepts you will learn. The "words you will need" are terms used in technology. Understanding these will help you understand chapter content. These terms are also printed in **boldface type** and defined within the chapter.

Heads

Each major part of a chapter is introduced by a short title called a "head." These titles are set in different type sizes and colors according to the importance of the information and its relationship to other information in the chapter. The number one heads identify the major concepts presented in the chapter. The number two and three heads identify subdivisions of the major concepts. Examples of heads from Chapter 12 are:

Number 1 head:

Modern Communication

Number 2 head:

Electronic Communication

Number 3 head:

Oral/Electronic Communication

For Discussion

Each major part of a chapter is followed by one or more discussion questions. These give you the opportunity to extend your learning by sharing ideas with your classmates in discussions.

Chapter Closings

Chapter closings include three sections. "Chapter Highlights" list important points to remember from the chapter. "Test Your Knowledge" includes 10 questions that help you identify, express, and remember key concepts in the chapter. "Correlations" are activities that show how technology can be applied in science, math, language arts, and social studies.

Special Activities and Features

This textbook is designed to allow you to learn through experience and apply what you learn to daily life. It includes many special activities and features to help you accomplish these things.

Activities

All chapters contain extension activities and community activities.

Extension Activities

Extension activities are suggestions of things you can do to apply and expand what you learn as you read the text. For example, in Chapter 29, "Power in Transportation," you may design and build a wind-powered land vehicle.

Community Activities

A continuing series of activities in this course provides the opportunity for you and your classmates to work together to design and build a model community. You can apply and expand what you learn in a simulated community setting. For example, in Chapter 7, "Energy Resources," you and your classmates may decide together whether wind, water, or solar energy would be an appropriate supplementary energy source to use in your community. Then you will design and build a model of a device for capturing energy from that source.

Features

Besides activities, each chapter contains two special features: impacts and trivia.

Impact

The ways in which technology affects the world and society are impacts. For example, the "Impact" in Chapter 20, "Biotechnical Systems," will help you recognize the enormous effect that the use of machines has had on agriculture, the industry that supplies our food. At least one impact of technology is identified in each chapter.

Technology Trivia

"Technology Trivia" features present interesting tidbits of information about technology. For example, do you know about how many telephones are in use in the United States? You'll find out in "Technology Trivia" in Chapter 1.

Glossary and Index

At the end of the book are two sections that can be helpful to you as you study—the glossary and the index.

Glossary

If you need to know the meaning of a term, look in the glossary. Terms identified at the beginning of each chapter are listed in alphabetical order and defined in the glossary.

Index

If you would like to know where a certain subject is discussed in this book, look in the index. It will give you the numbers of pages that provide information on that subject.

Experience Technology

As you use this textbook, *Experience Technology: Communication, Production, Transportation, Biotechnology*, you will explore the exciting and dynamic technical fields that make up the broad, general area of technology. You will do this not only by reading, but also by "doing." You will actually "experience" technology.

SECTION ONE

Understanding Everyday Technology

Activity Brief

Combining Science and Technology

PART 1: Here's the Situation

People have always used their creativity to develop products that make their lives easier. When people use information, materials, and tools to develop products that make their lives easier, they are using **technology**. We don't know who invented technology, but it was probably the cave dwellers of prehistoric times. When a cave dweller first used a stick as a club in self-defense, that person used technology. As people learned more about the world around them and how to change it, more technology developed.

Science is the study of the natural world. As people developed new tools and methods to investigate the natural world, more science developed.

Science and technology grew together. Science shaped technology and technology shaped science. As you do this activity and as you read Chapters 1 and 2, you'll learn for yourself how science and technology work together.

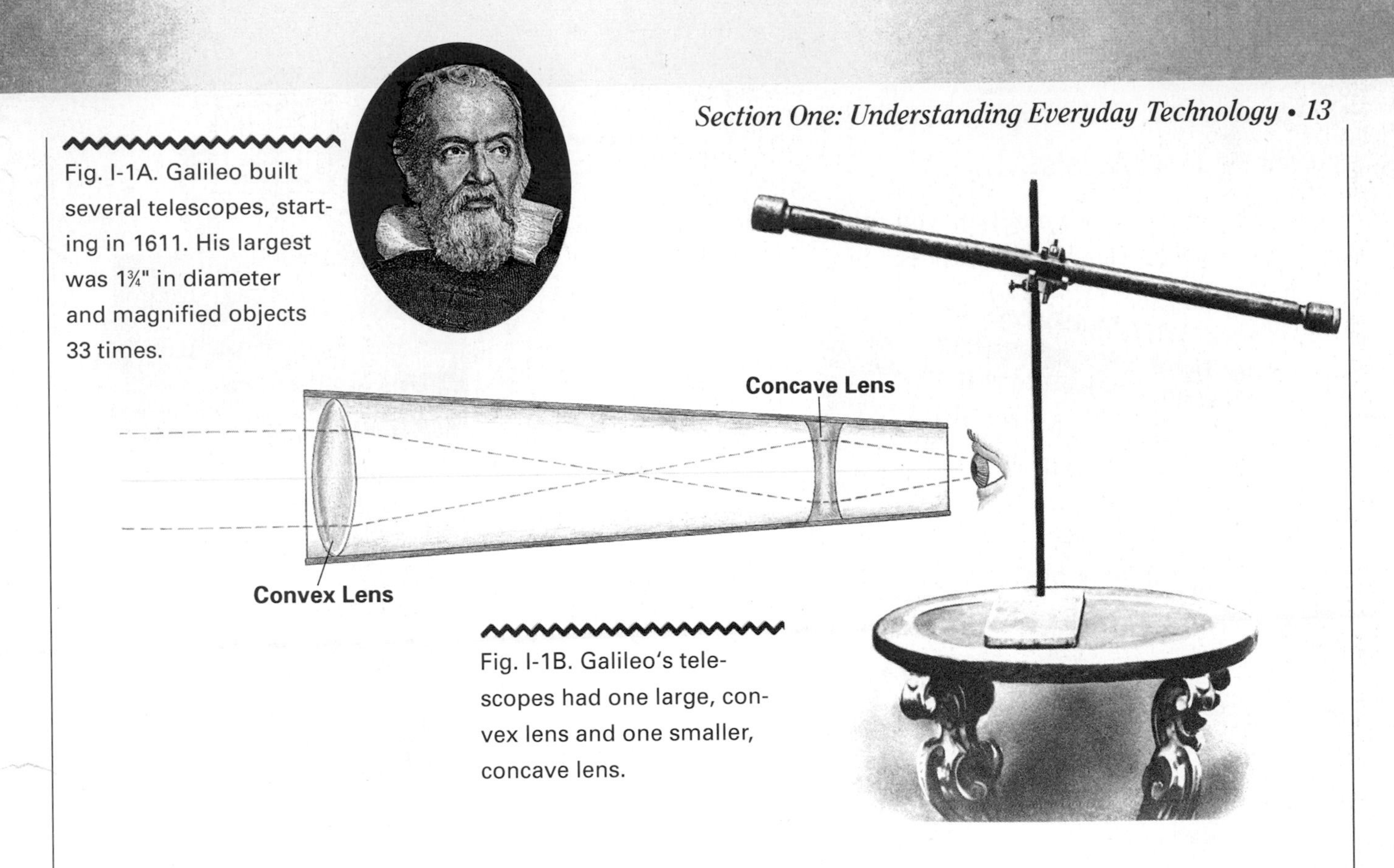

Fig. I-1A. Galileo built several telescopes, starting in 1611. His largest was 1¾" in diameter and magnified objects 33 times.

Fig. I-1B. Galileo's telescopes had one large, convex lens and one smaller, concave lens.

PART 2: Your Challenge..........

Galileo, the famous astronomer from Italy, knew how powerful science and technology could be. In the early 1600s, he used the information he gathered about light and lenses (science) to improve the design of the telescope (technology). Fig. I-1. With his powerful new telescopes, Galileo became the first person to see the mountains of the moon.

In this activity, you will gather information and then you will design and build a device that magnifies, reflects, or bends light to accomplish a task of your choice. Some suggested devices include a telescope, periscope, kaleidoscope, or camera obscura.

PART 3: Specifications and Limits..........

Your device will need to meet certain standards. Read the following specifications and limits carefully before you begin.

1. Your device must be an application of a scientific principle of light, such as:
   - reflection
   - refraction (the bending of light due to a change in its speed as it moves from air to another medium, such as glass)
   - mirror image
   - lens magnification
2. If you use glass mirrors or lenses, you must cover all raw edges.
3. You must hand in the following:
   - all sketches and drawings
   - a log sheet of your work
   - a statement of the scientific principles your device is based on

PART 4: Materials...........

There are many materials you might use to build your device. Here is a list of possibilities.

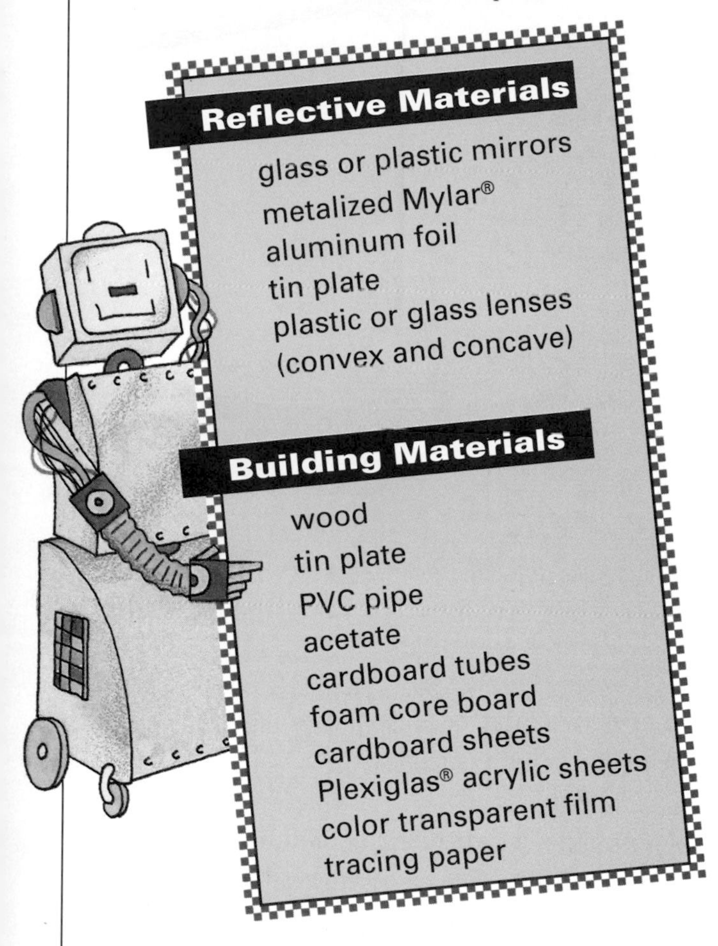

Reflective Materials
glass or plastic mirrors
metalized Mylar®
aluminum foil
tin plate
plastic or glass lenses
(convex and concave)
Building Materials
wood
tin plate
PVC pipe
acetate
cardboard tubes
foam core board
cardboard sheets
Plexiglas® acrylic sheets
color transparent film
tracing paper

Fig. I-2.

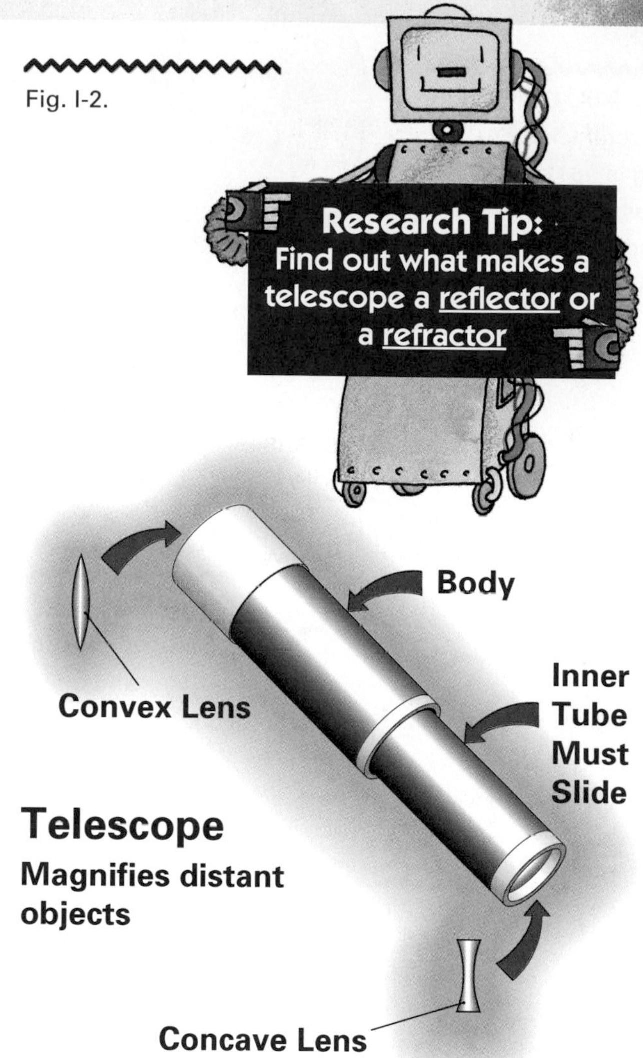

Find out what makes a telescope a reflector or a refractor

Inner Tube Must Slide

Magnifies distant objects

PART 5: Procedure...........

The device you choose and how you build it will be up to you. Still, there are certain steps to follow that will make your work easier.

1. Using your science book or other reference books, investigate some basic principles of light. Choose a principle, such as reflection or refraction, to demonstrate.
2. Select a device to build that will demonstrate the principle you chose. See Figs. I-2 through I-5 for suggestions.
3. Make sketches and drawings showing how you will make your device.
4. Make a materials list of what you will need.
5. Construct your device using the tools, machines, and processes your teacher will demonstrate.

Safety Notes

- Do not look at a strong light source, such as the sun, through a telescope. It could permanently harm your eyes.
- If you use glass lenses or mirrors, handle them with care. Be sure to cover all raw edges.
- As you do this activity, remember to follow all the safety guidelines your teacher has explained to you. Before using any power tools, be sure you understand how to operate them and always get your teacher's permission.

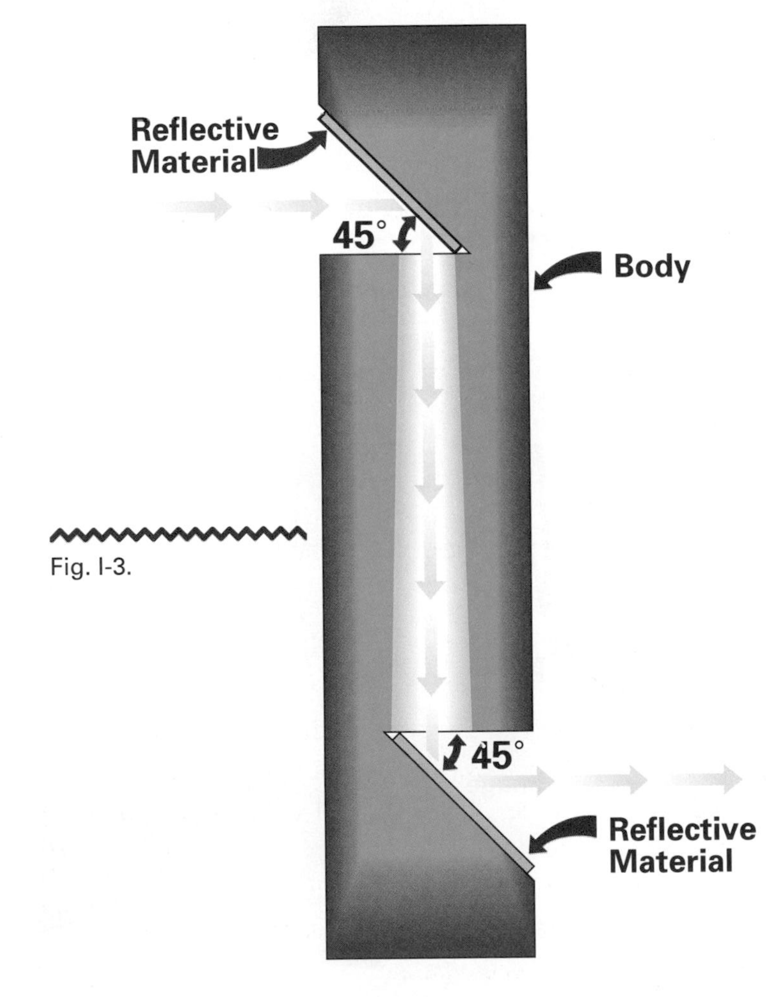

Fig. I-3.

Periscope

Allows people to see objects that are not in a straight line of sight

Used in submarines to see ships on the surface

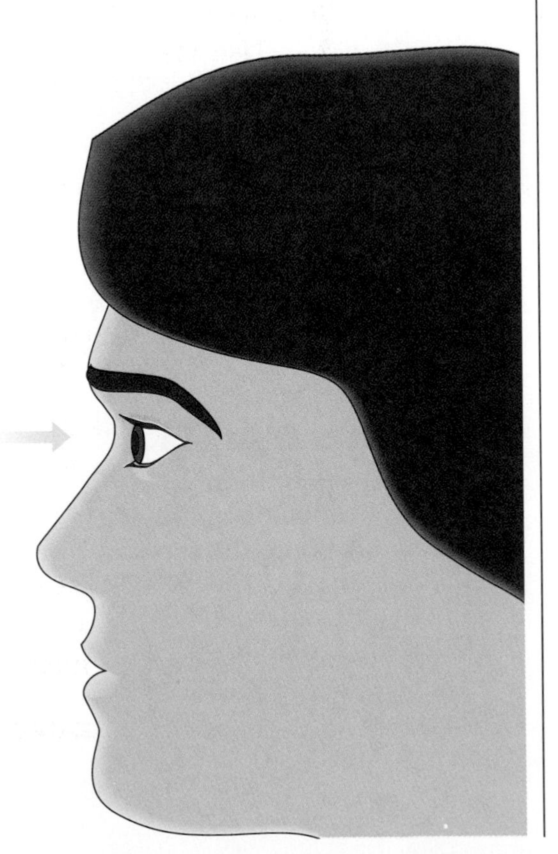

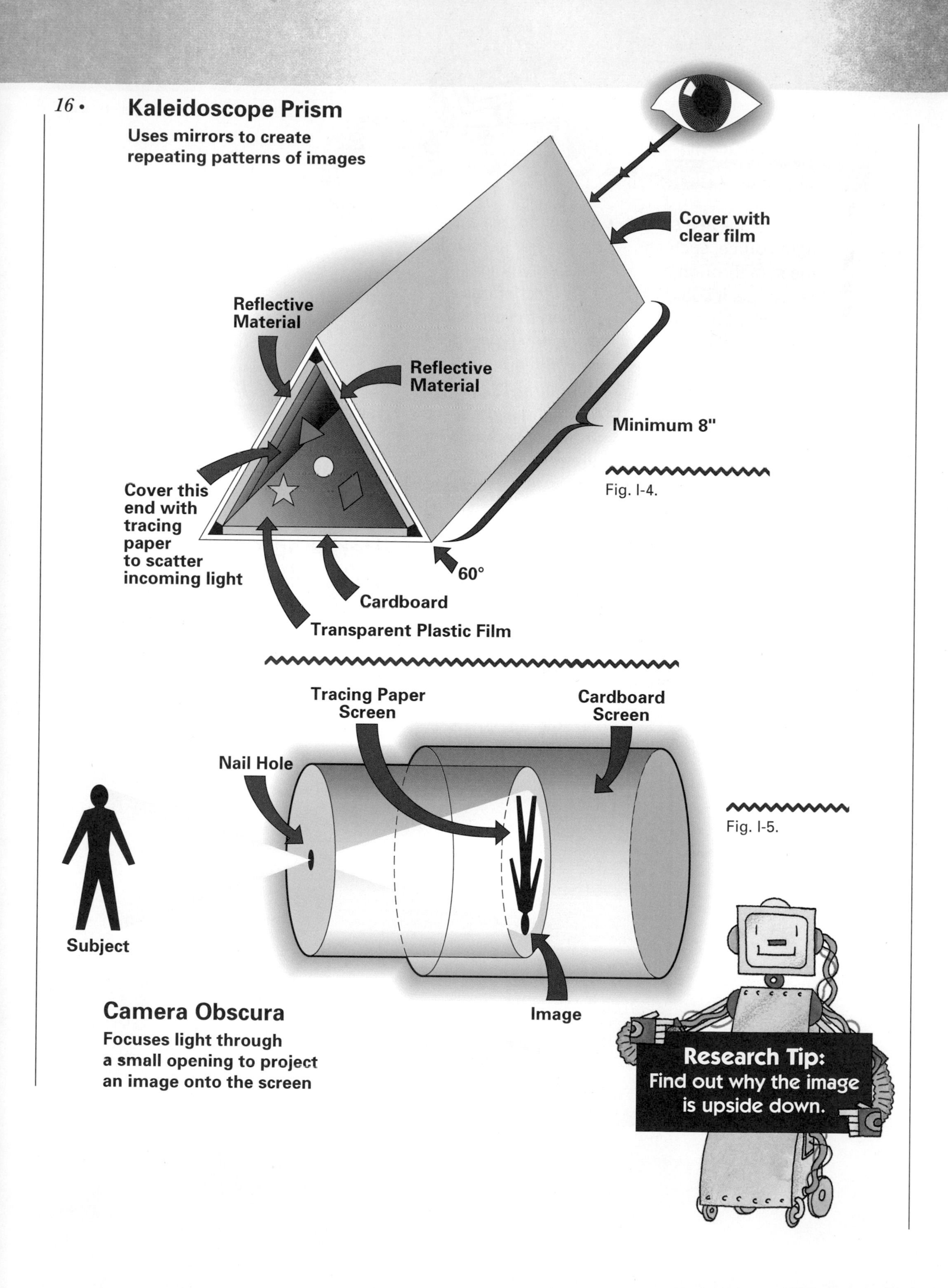
Kaleidoscope Prism
Uses mirrors to create repeating patterns of images
Cover with clear film
Reflective Material
Reflective Material
Minimum 8"
Cover this end with tracing paper to scatter incoming light
60°
Cardboard
Transparent Plastic Film
Tracing Paper Screen
Cardboard Screen
Nail Hole
Subject
Image
Camera Obscura
Focuses light through a small opening to project an image onto the screen
Research Tip:
Find out why the image is upside down.

Fig. I-4.

Fig. I-5.

PART 6: For Additional Help...........

For more help with this activity, look up the following terms. You'll find some of them in this book. (Check the index.) You'll find others in science books, dictionaries, or encyclopedias.

light energy
reflection
refraction
prism
mirror
lenses
concave
convex
optics
telescope
camera
microscope
fiber optics
kaleidoscope
periscope
camera obscura

PART 7: How Well Did You Meet the Challenge?...........

When you've finished building your device, evaluate it. Does it do what it is supposed to do? Ask yourself the following questions. Your teacher and classmates may take part in this evaluation.

1. How does your device use light?
2. Why is your device a good example of technology?
3. How might you improve the image made by your device?

PART 8: Extending Your Experience...........

This activity helps you learn about the connection between science and technology. Science and technology affect us and the world around us in many ways. Think about the following questions and discuss them in class. You'll find more about science and technology in Chapter 1, "What Is Technology?" and Chapter 2, "Technology Brings About Change."

1. How have the following devices made life easier and better?
   - microscope
   - telescope
   - periscope
2. In your opinion, which came first, science or technology? Why?

CHAPTER 1

What is Technology?

Introduction

Our people-made world is a very complex place. Each day we use many machines and systems that can be very hard to understand. How does a microwave oven cook food without a flame? How does a home heating system know when to turn on or turn off? What controls the traffic lights?

People have created these and many other familiar products to make our lives easier and more enjoyable. From alarm clocks that wake us in the morning to toothpaste that cleans our teeth before we go to bed, the products we produce using technology are countless.

After reading this chapter, you should be able to

Define technology.

Discuss how technology has become part of our everyday lives.

Describe the four families of technology.

Describe the technology-science relationship.

Words you will need

technology
communication technology
production technology
transportation technology
biotechnology
science

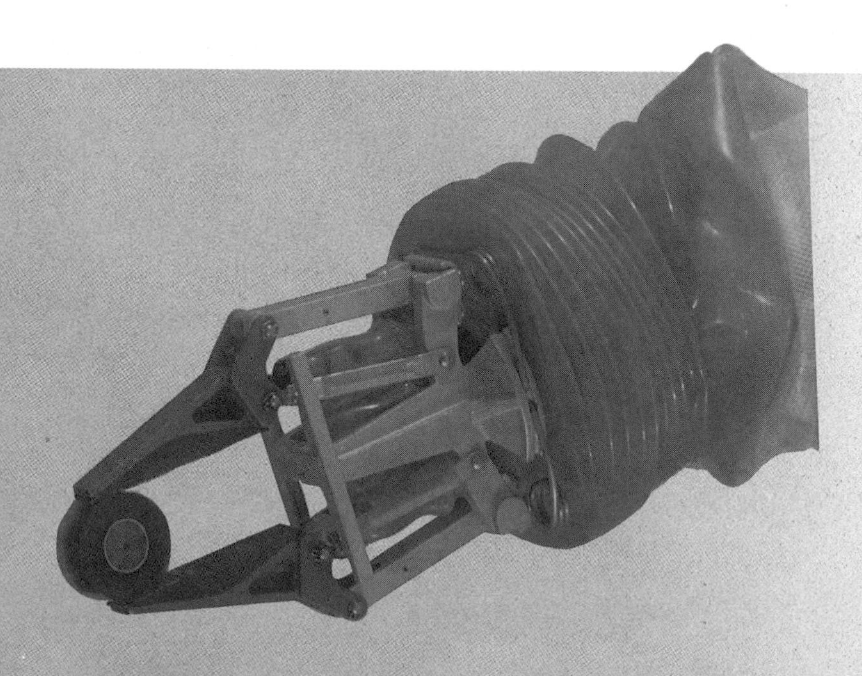

Technology Is All Around Us

When people use knowledge, materials, and tools to create things that meet their wants and needs, they create technology. **Technology** is the means by which we try to improve our people-made world.

Technology is all around us. It provides the home we live in, the food we eat, and the clothes we wear. Builders use technology to construct homes and other structures. Farmers use technology to grow and harvest the food we buy at the supermarket. Fig. 1–1. Stores use technology to keep the food fresh and to calculate the prices of our purchases at the checkout counter. Fig. 1–2. Our clothes are the products of technology, too. Technology was used to make the fabric, cut the pieces, and sew the pieces into garments.

Technology is part of our everyday lives. The names of some products of technology have even become part of our everyday speech: Nintendo, Walkman, Watchman, The Pump.

Fig. 1-2. Computer and laser technology make store checkout faster and easier. A laser beam scans the bar code on the package, then sends information to a computer about the product and its price.

Fig. 1-1. Farmers use technology to help grow and harvest food products. Here a farmer is harvesting corn with a machine called a combine.

The way people live and work in the community is the result of technology. Think about some of the things you do almost every day. How do you get to school? What do you do in your spare time? How do you stay in touch with friends and family? We use technology to do all these activities. In fact, we depend on technology for our way of life.

People need technology, but technology also needs people. People with various jobs and skills keep our technological world running. Bankers, teachers, technicians, scientists, engineers, and hundreds of others help design, produce, and consume (use) the products and services of technology.

Our community is one small part of a technological world. Technology has helped bring the global community closer together. The many countries of the world are at our fingertips, only a phone call away. Satellite and broadcast technology can bring us shows and concerts live from remote parts of the world. Fig. 1–3.

Fig. 1-3. Direct broadcasting by satellite (DBS) brings television programs from around the world directly into our homes.

Technology has made the world community so much smaller that we can have breakfast in New York and lunch in Paris, France. The *Concorde* supersonic transport makes this possible. Fig. 1–4.

From products in our homes to the services that bring our world closer, the advantages of technology make life easier and more enjoyable. Technology has worked its way into almost everything we do. We can say that technology is pervasive. *Pervasive* means "throughout." Technology can be found throughout our people-made world.

Fig. 1-4. Technology has brought the world community closer through high-speed travel. The *Concorde* cruises at an altitude of 50-60,000 feet, traveling at 1,350 miles per hour.

IMPACT

As you can see from these examples, technology changes our world in many ways. The impacts of technology, whether they are social, economic, or environmental, affect us all. Throughout this book, you will learn about these impacts.

▶▶▶ FOR DISCUSSION ◀◀◀

1. What products of technology found in your classroom might not have been found in your parents' classroom?
2. How has technology made your life easier? How has it made your life more complicated?

■ Using pictures cut from magazines and newspapers, make a poster that illustrates *one* of the following statements:
"Technology affects our daily lives."
"Technology makes the world grow smaller."

The Four Families of Technology

There are four families of technology. Each family provides products or services that can help us every day. Fig. 1–5. The four families are:

- communication technology
- production technology
- transportation technology
- biotechnology

TECHNOLOGY TRIVIA

At last count, the United States had a whopping 181,893,000 telephones in use.

The family of technology that helps us gather, store, and share important information is **communication technology**. This family of technology allows us to communicate ideas clearly and quickly across longer distances.

Production technology provides us with the manufactured and constructed products we use each day. Automobiles, furniture, magazines, pens, and even buildings are products of production technology.

People have become very dependent on **transportation technology**. Products, materials, and people can be sent quickly anywhere in the world through a complex system of transportation technologies.

The newest technology family is called *biotechnology*. **Biotechnology** is the use of living cells to help create new products. Genetic engineering is one part of biotechnology. Genetic engineering allows us to change the characteristics of plants and other organisms. Imagine designing a tomato plant that repels worms or a vaccine that protects people from cancer. Biotechnology may make these things possible.

All of the families, or systems, of technology work together. All four families are related. For example, an automobile is an example of transportation technology, but it is produced using the principles of production technology. By using technology from all four families, people can create a much greater variety of useful products.

Fig. 1-5. Each of the four families of technology provides products or services that make our lives better and easier.

▶▶▶ FOR DISCUSSION ◀◀◀

1. Consider the Walkman radio. Which families of technology are involved in its creation and use?
2. Many people are concerned that some products of transportation technology are polluting the environment. Explain how transportation technology can be used to solve this problem.

The Science-Technology Relationship

Many people think science and technology are the same thing. This just isn't true. Science and technology help one another, but they are very different subjects.

Science is the study of our natural world. In science class, we explore nature by learning about biology, physics, chemistry, and earth science. For example, we might observe living microorganisms under a microscope or test chemicals for acidity.

As we have noted, technology is the study of our people-made world. In technology class, we might explore transportation systems, communication systems, and production systems. We might build a rocket and study its flight. We might calculate the altitude it reaches.

Even though science and technology are very different subjects, they often work together. Technology and science are most powerful when they work hand in hand. When science and technology work as teammates, the results are often incredible.

The pacemaker is a good example of how this teamwork can produce a miracle. Your heart beats about 70 times each minute. Each beat sends blood from the heart to all parts of the body.

In science class, you have learned that the body needs oxygen, carried in the blood, to stay alive. Unfortunately, some people suffer from a defect in the natural system that controls the number of times the heart beats. The result is that not enough oxygen gets to the cells in the body.

Doctors and engineers together developed the pacemaker to fix this breakdown in the natural system. The pacemaker is a small machine that is placed under the skin. It controls the number of times per minute the heart beats. The machine sends a tiny electrical shock to the heart muscle each time the heart muscle should pump. Fig. 1–6.

Battery provides electricity, sending it through a wire that connects to the heart.
Another wire leads from the heart back to the battery, completing the electrical circuit.

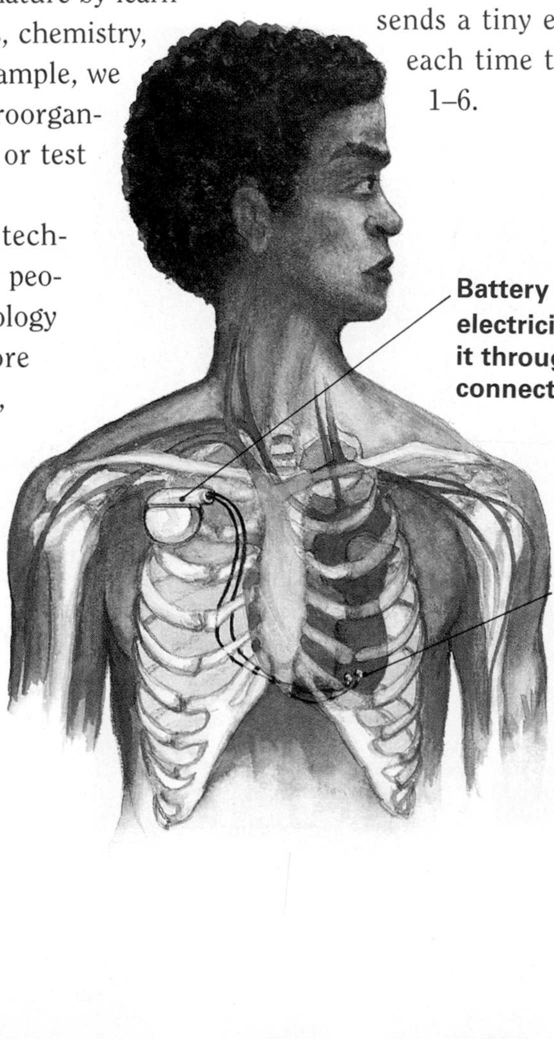

Fig. 1-6. The pacemaker is a product of technology that helps replace a faulty natural system.

■ **Look around your community. Identify systems in each of the four families of technology. Make a list of these systems. Later, as your class plans a model community, you will find this information useful.**

Science provided the information about the natural function of the human body. Technology, using this information, developed a machine that could replace the faulty natural system. The powers of science and technology now keep 500,000 people alive each year with pacemakers.

Science and technology are very different subjects. Together, however, they form a powerful tool to make life easier and better. Together, they improve our lives and surroundings. Fig. 1–7.

▶▶▶ FOR DISCUSSION ◀◀◀

1. The pacemaker is one example of technology and science working together. Name some others.
2. Why are science and technology working together more powerful than either one working alone? Explain.

Fig. 1-7. The connection between science and technology can easily be seen in the space program. Scientists and technologists are working together to develop life support systems for space travel.

THE TECHNOLOGY/SCIENCE CONNECTION

TECHNOLOGY	SCIENCE
Material engineers design new materials to meet new needs.	Chemists help us understand the structure of materials better.
Structural engineers design structures that will support specific loads.	The science of physics explains how forces act on an object under a load.
Technology provides life-support systems for long stays in space.	Physiology is the study of living organisms and their biological systems.
Hydroponic food production will support long stays in space.	Geneticists develop new varieties of plants.

Chapter Highlights

- Technology is people using tools, materials, and information to create products or services that meet our wants and needs.
- The products of technology have become part of our everyday lives. Technology can be found throughout everything we do.
- The families of technology are communication technology, production technology, transportation technology, and biotechnology.
- Science is the study of our natural world, whereas technology is the study of our people-made world.
- Science and technology together make a powerful team that can improve our people-made world.

Test Your Knowledge

1. Define *technology*.
2. What family of technology helps us gather, store, and share information?
3. What family of technology provides most of our everyday products?
4. How has technology brought the world community closer together?
5. If you were studying how the planets revolve around the sun, would you be studying science or technology?
6. Give two examples of how science and technology working together have improved your life or surroundings.
7. Farmers may soon be harvesting plants that have been developed to resist freezing in cold weather. What family of technology will this product come from?
8. Explain the differences between science and technology.
9. List five ways technology has influenced your daily routines.
10. What products in your home are the result of communication, transportation, and production technology? List two products for each of these families of technology.

Correlations

SCIENCE

1. Tomatoes are fragile. Packing them for shipment is hard because they are round. Find out what science and biotechnology have done to create a more cube-shaped tomato.

MATH

1. You read in this chapter that pacemakers regulate heartbeats. If a person's heart beats 70 times per minute, how many times will the heart beat in one 24-hour period?

LANGUAGE ARTS

1. For one day, try not using a television or telephone. How did you spend your time? How did you communicate with friends outside your home? In a paragraph or two, describe your day.

SOCIAL STUDIES

1. Make a list of the brand-name products in your home. Report your findings to the class. Determine the top ten brand names listed in the entire class and find out where these companies have factories.
2. Draw a map of the world and write the company names from Activity 1 in their respective countries.

CHAPTER 2

Technology Brings About Change

Introduction

Technology is a powerful force that brings about changes in the way people live and work. The impacts, or influences, of technology can be positive, negative, or sometimes both. You can feel the impacts of technological change in your home, your community, and the world.

New technologies have changed forever how people communicate, produce products, transport things, and meet their daily needs. Changes due to technology can take place quickly or very gradually. In a mere 75 years, people have gone from dreaming about flying to landing on the moon. In a short 50 years, we have gone from no television to live TV programs viewed around the world by way of satellite systems.

After reading this chapter, you should be able to

Explain when and how technological change began.

Discuss how people satisfy their needs with technology.

Discuss the impacts of the Agricultural Era.

Discuss the impacts of the Industrial Era.

Discuss the impacts of the Information Era.

Words you will need

Agricultural Era
Industrial Revolution
Industrial Era
mass production
factory system
cottage industry
Information Era
service industry
electronic cottage

When Did Technological Change Begin?

Technology began when people first attempted to satisfy their needs by developing tools. Who were these early toolmakers? Archaeologists tell us they were our ancestors, the cave dwellers. Cave dwellers lived during the Stone Age, which began about 1,000,000 B.C.

Early humans had a very difficult life. Finding food, staying warm, and protecting themselves from wild animals amounted to a full-time job. Early technology helped make their harsh lives easier.

Early humans were *nomadic*—they moved from place to place in search of food. They hunted, fished, and picked wild fruits and vegetables to survive.

The first products of technology were probably simple sticks and bones. Cave dwellers used these simple tools to kill animals for food and to defend themselves from enemies. Fig. 2–1.

Can a stick really be considered technology? We have learned that technology is defined as people using tools, materials, and information to satisfy their needs. Sticks, stones, and bones were the only toolmaking materials available. What greater need exists than that of finding food? As primitive as these tools seem today, they were high technology a million years ago.

These simple tools changed how people satisfied their everyday needs. Tools of stone and wood marked the beginning of a process of change that still goes on today.

▶▶▶ FOR DISCUSSION ◀◀◀

1. If you were a cave dweller drawing on a cave wall, what kind of thoughts would you want to express?
2. If you were stranded on an unpopulated island, what would be the first three tools you would make?

■ Using sticks, stones, and string or twine, make a small version of the type of hammer that cave dwellers may have used.

Fig. 2-1. Early humans made tools from materials that were available. They shaped stones, wood, and bones into spears, clubs, and axes. Perhaps the first attempts at technology, these simple tools made life easier and better for our ancestors.

Stone Hand Ax **Stone Ax with Wood Handle**

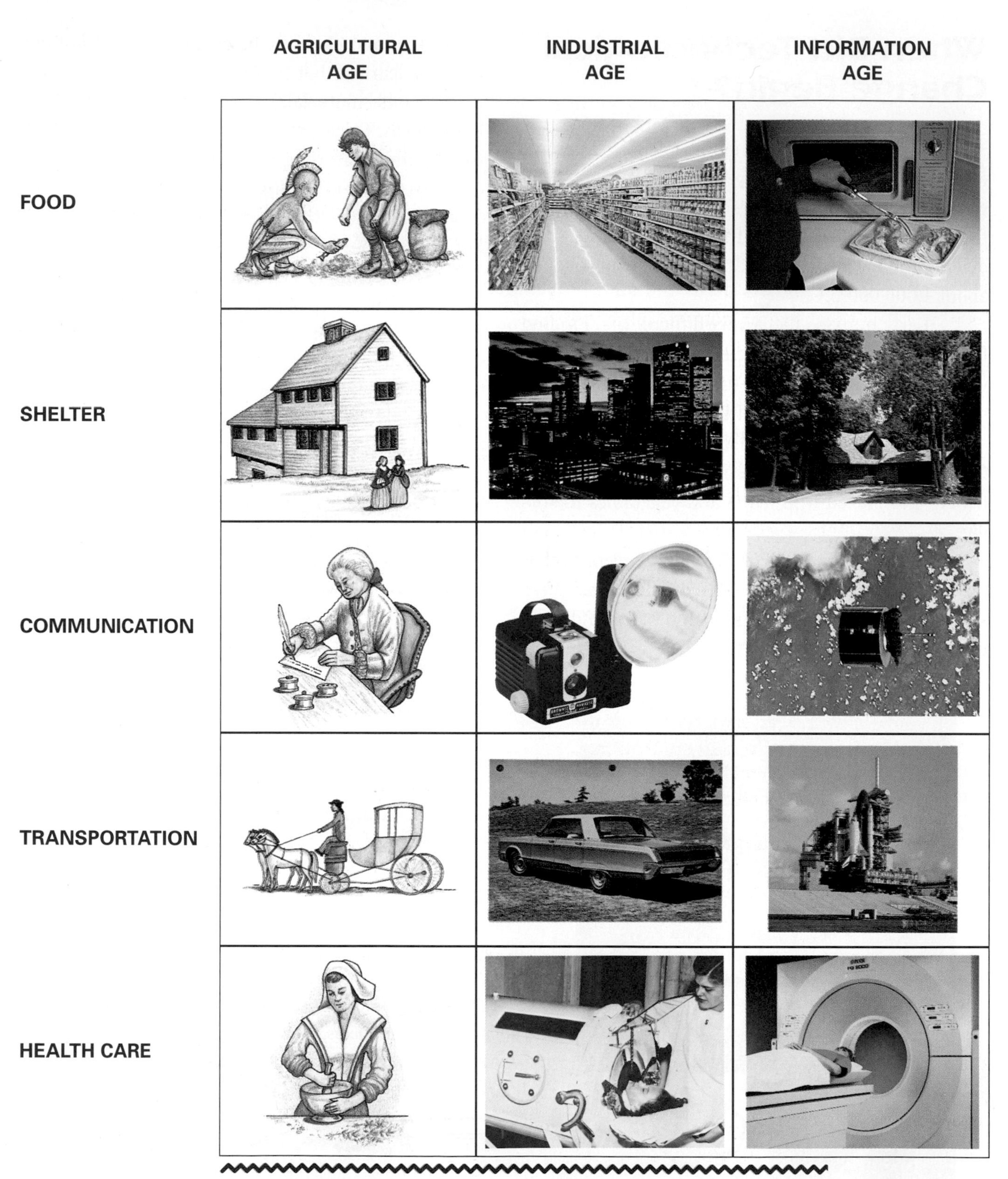

Fig. 2-2. People's basic wants and needs have not changed, but the ways in which people meet those wants and needs have changed a great deal.

Technology Satisfies People's Needs

In their attempts to satisfy their needs, our early ancestors created new technologies. The stone ax, the wheel, the canoe, and even fire-building were technologies people developed to help satisfy their needs and wants.

All people have basic needs and wants. These include food, water, shelter, communication, recreation, protection, transportation, and health care. These needs and wants have remained the same since humans first walked this planet. Although our needs and wants never really change, the products we use to satisfy them change rapidly. Fig. 2–2.

In recent years, people have used technology to customize items for use by groups with special needs. For example, some telephones today are connected to flashing lights. This allows people who cannot hear the telephone ring to know when someone is calling them.

▶▶▶ FOR DISCUSSION ◀◀◀

1. Food today comes in many different forms. Canned food and fresh food are two examples. List four other ways food can be purchased.
2. Technology has changed our approach to recreation. What types of fun things do you think kids did during the early 1800s?

The Agricultural Era

Our ancestors began satisfying their needs by developing more and more new technology. Hunters gradually became farmers. To prepare the ground for planting, the farmers designed the sickle, the hoe, and the plow. The **Agricultural Era** was about to begin.

During the Agricultural Era, people banded together and stayed in one place. They formed villages centered around farming and domesticated animals such as cattle and horses. They learned how to plant and harvest many crops. Fig. 2–3.

Large, permanent shelters replaced the nomadic tents. People in each community were assigned specific jobs. Some people were hunters, some developed farming techniques, and others cared for the animals. Craftspeople made products for sale and trade. They wove fibers into fabrics, forged metals into tools, and carved wood into furniture. Fig. 2–4.

The Agricultural Era was a period of change that lasted into the early 1800s. People were happy with the results of technological change, and they wanted more.

TECHNOLOGY TRIVIA

The medication <u>digitalis</u>, used today to help patients who have heart disease, was first used in A.D.1550 for the same purpose. It was made from the foxglove herb.

Fig. 2-3. Farming changed the way people lived and worked. Hunters became farmers as nomadic people settled in villages and began growing their own food.

Fig. 2-4. Craftspeople worked at home or in small shops. By the 1700s, these "cottage industries" produced many products for the home and the community. The woman shown here is spinning wool. Because she worked in her own home, she also had the freedom to enjoy a book while she worked.

The Industrial Era

Inventors responded to society's desire for new products with an explosion of new technologies. The **Industrial Revolution,** as it was called, began in England around 1750. The word *revolution* means "drastic change." The industrial revolution drastically changed how people lived and worked. It also marked the beginning of the **Industrial Era.**

The steam engine was one of the great forces behind the Industrial Era. People developed methods of powering machines by steam. They replaced hand-powered tools and machines with steam-powered equipment. Fig. 2–5.

Fig. 2-5. By the late 1800s, entire factories were powered by steam. In this textile factory, an overhead shaft and belt system allowed a single steam engine to run every machine in the factory.

▶▶▶ FOR DISCUSSION ◀◀◀

1. The Egyptians were among the first people to farm the land. How do you think people first discovered they could plant and grow their own food?
2. Today, only two percent of the American work force is made up of farmers. How can such a small percentage of people produce enough food for our entire population?

To meet the demand for more products, people developed factories that used **mass production** techniques. This **factory system** enabled people to produce products faster and less expensively. The **cottage industry**—craftspeople working out of homes and small shops—could not compete with the new factories. Many of the craftspeople closed their businesses and went to work in the factories for wages.

The factories provided many people with jobs. They became an important part of the European and American economies. Cities expanded quickly as people moved from rural farm areas into the cities to work in the factories.

Ideas also moved with new speed. Change began taking place faster and faster. The Industrial Era lasted into the mid-1900s.

TECHNOLOGY TRIVIA

Lazlo Biro and his brother George produced the first commercially successful ball-point pen in 1938.

IMPACT

The change to a factory-based economy affected all aspects of life. Before, most people spent their entire lives in the same area, rarely meeting strangers. Now, they moved to cities, where they met other people from different backgrounds. In the cities, there were more opportunities for education and for good jobs. Money became more important, both for the basics (people no longer grew their own food) and for status (success was measured by the things one owned). Families spent less time together; so parents had less influence on their children. For many, life became easier in some ways but more complex in others.

Major changes also occurred in transportation. Ships, trains, automobiles, and airplanes were developed or refined during the Industrial Era. They helped move products to new markets.

▶▶▶ FOR DISCUSSION ◀◀◀

1. Why do handmade products usually cost more than mass-produced products?
2. Laws preventing young children from working in factories were a direct result of the Industrial Era. What do you think brought these laws about?

■ A company called MegaIndustries is planning to build a factory near the small town of Glenville. What would be the effects upon the community? What kinds of information would Glenville residents need to gather in order to plan for expected changes?

The Information Era

The world today is the result of continuing technological change. Our basic needs are met by complicated devices. Our lives are centered around information—all kinds of information. People gather, store, and use more information than ever before. Fig 2–6.

For this reason, we live in what is known as the **Information Era.** People today gather so much information that we double our knowledge every seven years. We use this information to create new products and services to meet our wants and needs.

What are services? Many of our needs cannot be satisfied by products. To satisfy these needs, we need help from other people. Doctors, repair people, hospital workers, restaurant workers, technicians, and thousands of other people make up the **service industry.** The Information Era has caused an explosion of jobs in the service industry.

The computer is an essential tool of the Information Era. Computers do calculations almost instantly. They can scan their vast memories to provide us with instant access to information.

Fig. 2-6. Our world is centered around information that comes in many forms and on endless subjects. We have created many new technologies to help us gather, store, and communicate the information we use every day. How many examples can you find here?

Computers now perform many of the activities that people used to do. Home computers connect us through telephone lines to banks, stores, libraries, and even our doctors' offices. People can do much of their work at home and send it into their offices using modems or fax machines. Fig. 2–7. This arrangement is becoming so common that home offices are now referred to as the **electronic cottage.**

Of all the technologies that have emerged during the Information Era, medical technologies are perhaps the most amazing. Organ transplants, body replacement parts, miracle drugs, and medical machines have extended our life expectancy to 75 years and well beyond. Fig. 2–8.

Changes continue to take place in transportation, communication, production and biological technologies. Most of these changes help make our lives easier and better.

We must not forget, though, that technology has had many negative impacts on our lives. Pollution, energy shortages, and depletion of our natural resources are topics that we will discuss in detail later in this text.

▶▶▶ FOR DISCUSSION ◀◀◀

1. Has the need for higher education changed in the Information Age? Why?
2. In your opinion, what technology has had the greatest impact on your life?

■ Make a list of at least 25 types of service work. Which types have originated relatively recently? Share lists in a class discussion.

Fig. 2-7. People who worked out of their homes during the Agricultural Era made up the cottage industry. Communication devices developed in the Information Era have allowed a return to working in the home. Home offices are often referred to as the "electronic cottage."

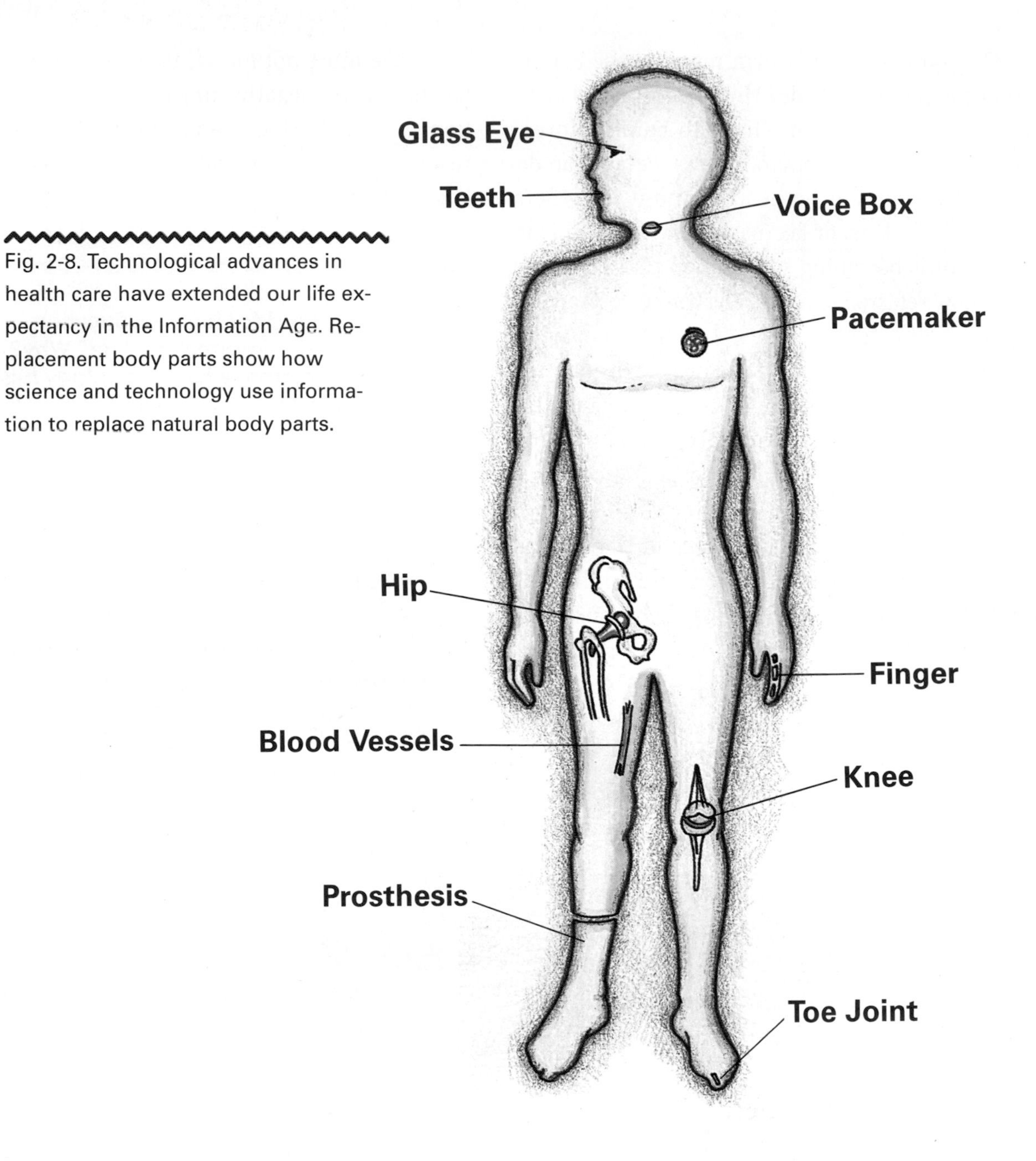

Fig. 2-8. Technological advances in health care have extended our life expectancy in the Information Age. Replacement body parts show how science and technology use information to replace natural body parts.

CHAPTER 2 Review

Chapter Highlights

- Technology is a powerful force that changes how we live and work.
- Technology produces products and services that help satisfy our physical and emotional needs.
- The Agricultural Era brought people together into communities that centered around farming and handcrafted products.
- The Industrial Era brought about machine-made products and the factory system. A new lifestyle emerged, centered around city living.
- Today, we live in the Information Era. Our lives are centered around gathering, storing, sharing, and using information to create products and services.

Test Your Knowledge

1. Give at least two examples of how today's technology can be used to benefit people with special needs.
2. The automobile and airplane were first produced in which era?
3. The plow is a tool used to make the trench in which farmers plant their seeds. List three ways people have developed to push or pull a plow.
4. Describe the differences between making products by hand and by using the factory system.
5. Give three examples of how modern technology helps people satisfy their physical needs.
6. List three products of technology that help satisfy people's need to communicate.
7. We all enjoy recreation, such as playing games, going to movies, and listening to music. How does recreation help satisfy our emotional needs?
8. Is the stone ax an example of technology? Why?
9. How has the computer changed the way people work?
10. Give three examples of how technology has increased our average life span.

Correlations

SCIENCE

1. You read in this chapter that a flashing light can signal hearing-impaired people that the phone is ringing. Find out how a TDD can help them communicate with the caller.

MATH

1. Life expectancy has increased 50 percent since 1910. If the average person of that time lived 50 years, what would be the average lifespan today?

LANGUAGE ARTS

1. Interview a parent or grandparent about a typical school day when he or she was your age. In an essay compare and contrast a typical school day today with the one that was described to you.

SOCIAL STUDIES

1. Your family car is an example of the advance made in transportation as a result of the Industrial Revolution. Find out exactly how automobiles are mass produced. If you live by a car manufacturing plant, see if you can tour the plant and see the production line first-hand.

Make a Difference

The Knowledge Master Open

Which structure is correctly paired with its system?
(a) ganglion—skeletal system
(b) kidneys—circulatory system
(c) tongue—integumentary system
(d) feathers—reproductive system
(e) pituitary gland—endocrine system

Can you answer this question? Could you answer it with the help of a team of your classmates? If so, your school might be interested in forming a Knowledge Master Open team.

Every spring, middle school/junior high students all over the world form teams to compete in the Knowledge Master Open, an academic contest run on classroom computers. Technological advances such as microcomputers and fax machines let schools compete in a national/international contest without the expense of traveling to a central site.

The students at Washington Grade and Junior High School have a scholastic bowl team which travels to other nearby schools for academic competition. When the Computer Lab Director, Mrs. Diane Hoover, received information about the Knowledge Master Open (KMO), she knew the scholastic bowl team would be interested in competing.

A panel of teachers selected eighteen students from grades 6, 7, and 8 to compete in the contest. Since the questions were about varied topics such as literature, earth science, current events, art and music, economics and law, math, history, and trivia, the teachers chose students with a variety of interests and "specialties."

Mrs. Hoover received a computer disk from KMO which contained the official competition questions. The contest disk was programmed so that it could only be run once for actual competition purposes. Only students could be involved after the computer hardware was set up.

On the day of competition, Mrs. Hoover placed the disk in the computer and entered the password. Then the students took over and the competition began! Students had two chances to answer each question. They earned five points for each question they answered correctly on the first try. They also received bonus points for speedy answers. On the second try, they earned two points for a correct answer. No points were deducted for incorrect answers.

Students were allowed to use pencil and paper during the competition, but they could not use calculators and books. Mrs. Hoover, acting as sponsor and supervisor, was not allowed to help in any way.

The competition lasted for 2½ hours. When it was over, Mrs. Hoover called Knowledge Master Central in Durango, Colorado, with the team results. The computer disk provided a summary of how well the students did in each content area of the contest.

Within a week of the competition, the Washington students received the world-wide results of the Knowledge Master Open competition. Checking the results from KMO, the team discovered that they had answered the same number of questions correctly as a school from the Philippines (183 out of 200). The Washington students, however, took a second try on 25 of the questions, whereas the students in the Philippines tried a second time on 36 questions. How exciting to compete with students thousands of miles away without having to pack a suitcase!

SECTION TWO

Technology Relies on Resources

Activity Brief
Combining the Resources of Technology

PART 1: Here's the Situation ....

Resources are all the things you depend on to get a job done. What kind of resources would you depend on to write a report on a famous inventor? You would surely depend on books for information. Your teacher would also be a valuable resource for facts on famous people. You might even watch a video on the life of the inventor. If you did, your TV and VCR would become resources. If you wrote the report by hand, your pen, paper, and a dictionary would be resources. If you used a word processor to prepare the report, your computer and printer would be resources.

Technology is also dependent on resources to get a job done. When people use technology to create a product or service, they rely on seven important resources: people, information, energy, tools and machines, materials, capital (such as money), and time. As you do this activity and as you read Chapters 3 through 9, you will learn more about the seven resources of technology.

PART 2: Your Challenge..........

In this activity, you and your teammates will form your own company and design and then manufacture a hand-held game. Through experience, you will learn how each of the seven resources of technology can be used in the creation of the game.

PART 3: Specifications and Limits..........

Your game and the way in which your company develops it will need to meet certain standards. Read the following specifications and limits carefully before you begin.

1. The finished game must be hand-held and may not exceed 25 square inches in area.
2. The game must be hand-operated with no electrical parts.
3. The game must have a well-defined way of winning or completing the challenge.
4. Your team must produce a prototype (unique, one-of-a-kind item) of the game and then manufacture one game per team member.
5. You must hand in the following:
   - all sketches and drawings
   - a log sheet of your team's daily progress
   - a sample game

Safety Notes

- As you do this activity, remember to follow all the safety guidelines your teacher has explained to you.
- Use all tools properly. Use special care with tools that are sharp.
- If you use any power tools, be sure you understand how to operate them, and always get your teacher's permission.

PART 4: Materials..........

There are many materials you might use to build your game. Here is a list of possibilities. Possible tools you might use are also given.

Materials

acetate
aluminum rivets (bumpers)
cardboard
construction paper
foam core board
glue
marbles
pine wood
poster board
rubber bands
springs
steel bearings
tin plate
washers
wood dowels
wood molding

Tools

hand tools
markers
scissors
utility knife

PART 5: Procedure...........

The game you choose to design and manufacture will be up to you. Still, there are certain steps to follow that will make your work easier.

1. Working in manufacturing teams of 3 students each, brainstorm ideas for games. Keep a list. Try themes like sports and movies.
2. Develop a sketch of an idea you think will be a marketable game. Fig.II-1 shows one possibility.
3. Present your idea to the team. Consider ideas presented by your teammates. Then, as a team, adopt an idea or combine ideas into a new design.
4. Build a prototype of the game.
5. Test the prototype. Make any necessary changes in the game design.
6. Set up the production sequence for producing the game and make a trial run. Make any changes that are necessary in your production sequence.
7. Manufacture the game.

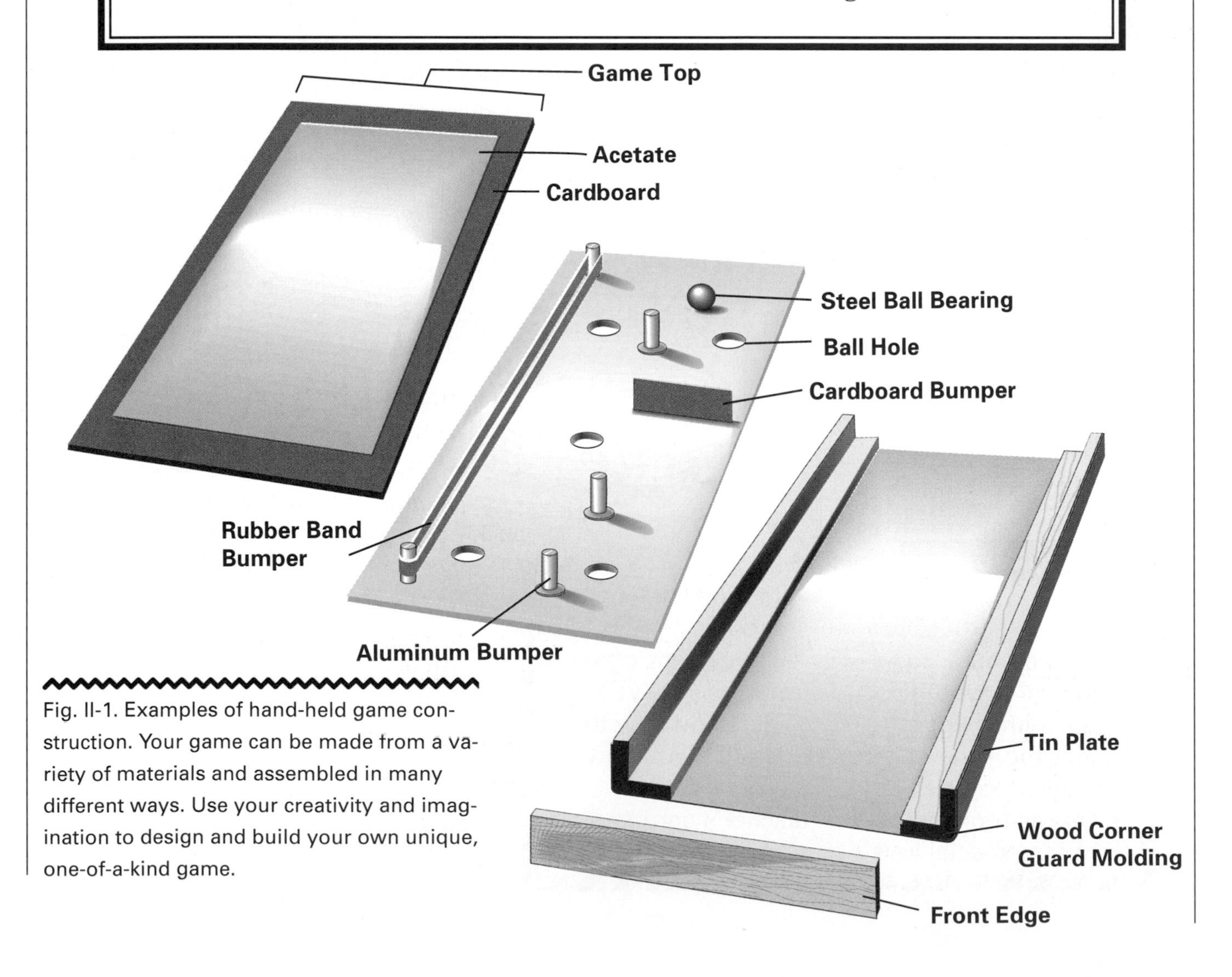

Fig. II-1. Examples of hand-held game construction. Your game can be made from a variety of materials and assembled in many different ways. Use your creativity and imagination to design and build your own unique, one-of-a-kind game.

PART 6: For Additional Help...........

For more help with this activity, look up the following terms. You'll find some of them in this book. (Check the index.) You'll find others in dictionaries, encyclopedias, and other resource materials.

brainstorm
elements of design
logo
manufacturing
material processing
product engineering
production
prototype
resources of technology
trial run

PART 7: How Well Did You Meet the Challenge?...........

After the product has been manufactured, evaluate the product and the processes you and your teammates used to produce it. How were the seven resources of technology used in the production process? Answer the following questions based on this evaluation.

1. **People**—What role did you and your teammates play in the production process?
2. **Information**—What information had to be obtained to design and manufacture your product?
3. **Materials**—Make a list of all the materials you used in manufacturing your game.
4. **Tools and Machines**—List the tools and machines you used in manufacturing your game.
5. **Energy**—You may include in your game a device to propel a marble or ball bearing. Look up potential, kinetic, and mechanical energy. How would these terms apply to using a propelling device?
6. **Capital**—If you and your teammates actually owned a company that manufactured games, how would money be spent within the company?
7. **Time**—What changes did you make or could you make in your production sequence to produce products more quickly?

Fig. II-1. Continued.

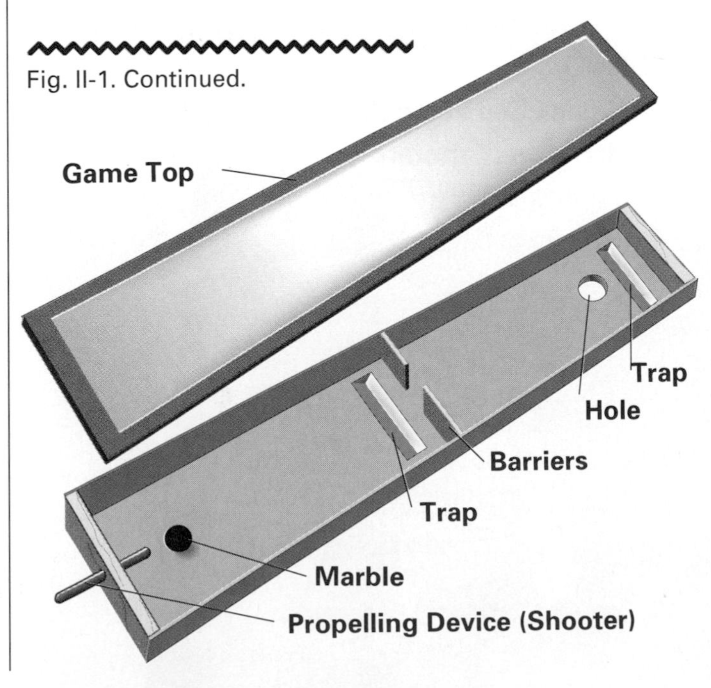

PART 8: Extending Your Experience...........

This activity helps you learn about basic resources of technology. Companies large and small around the world are dependent upon the same resources to produce products and services. You'll find more about the resources of technology in Section II, "Technology Relies on Resources."

New developments involving the seven resources of technology are continually taking place. Watch for articles about these in newspapers and magazines. Also, be alert for reports on television and radio. Share the information you learn with your class.

CHAPTER 3

Resources of Technology/People

Introduction

What if you had a great idea for a new product? How would you transform this idea into a salable item? How would you get this item onto the shelves at your local department store?

Your product, like any other product of technology, would be dependent on many resources. **Resources** are all the things needed to accomplish a goal or to get a job done.

Even simple tasks are dependent on seven resources: people, information, tools and machines, materials, energy, money, and time. In this chapter and the chapters that follow, we will study how people use these same seven resources to create technology that provides us with products and services that meet our needs and wants.

After reading this chapter, you should be able to

Define resources.

Discuss people as a resource of technology.

Explain the role of people as the consumers, creators, and managers of technology.

Words you will need

resources
consumers
gross national product (GNP)
entrepreneur
innovation
ergonomics
Occupational Safety and Health Administration (OSHA)
Environmental Protection Agency (EPA)
Underwriters Laboratory

What Are Resources?

Suppose you wanted to pop some popcorn. What resources would you need? Fig. 3–1. If you didn't know how to make popcorn, you might ask someone for instructions. *People* would be your first resource. The instructions or *information* you receive would also be a valuable resource. Now you might search the kitchen for additional resources.

Materials such as popcorn, salt, and butter would certainly be needed. Without *tools and machines* such as a popcorn popper, bowl, and measuring cup, the job could not be completed. *Energy* is also an important resource. Electrical energy is needed to power the popcorn popper. Heat energy is used to burst the tiny kernels of corn into fluffy popped corn. *Money* is the resource used to purchase other resources. Finally, all things are dependent on *time*. How long does it take to make popcorn? How long have the popcorn kernels been in your food closet? How long can your friends wait for fresh popcorn? Time is a resource we must learn to manage.

It's hard to say which resource of technology is most important. If we were to remove any one of the seven resources used to make popcorn, the result would be quite different. Perhaps the most critical resource is people. People consume, create, and govern technology.

Fig. 3-1. The products of technology are created by combining the seven resources of technology. For example, to make popcorn, you would need all the resources shown here plus one more—energy. How would energy be used to make the popcorn?

▶▶▶ FOR DISCUSSION ◀◀◀

1. How might a popcorn-popping process result if the resource of information were missing?
2. Why are people the most critical resource in technology?

People Are Consumers of Technology

Consume means "to use up." People consume the products and services of technology. Fig. 3–2. As **consumers**, we create a market for these products. If we like a product, we buy it. Products that we don't like or that don't meet our needs are left on the shelves.

How do companies know what consumers want? It is the job of market researchers to determine what new products consumers want and are willing to purchase. Consumer demand for new products keeps a constant flow of new technologies emerging.

Fig. 3-2. As consumers, we purchase the products of technology. Companies use the profits to pay salaries, buy new equipment, and finance new product development.

Through their purchases, consumers also contribute to the strength of our nation's economy. By purchasing goods and services, consumers create an increased demand for these items. To meet this demand, companies increase production and hire additional employees.

Economists study the production and consumption of goods and use this information to determine the health of our nation's economy. Fig. 3–3. A healthy economy produces a large **gross national product (GNP)**. A fall in the GNP means consumers are buying less and production is declining.

IMPACT

Consumer demand for products keeps our economy going, but many of those products create problems for our environment. For example, many people buy microwavable meals because they are fast and easy to prepare. Many of these meals come on plastic plates. Does it make sense to use a plate once and then send it to the landfill, where it may last for centuries?

▶▶▶ FOR DISCUSSION ◀◀◀

1. What role did consumers play in the production of new products during the Industrial Era?
2. How does the demand for a product and its availability affect the price of the product?

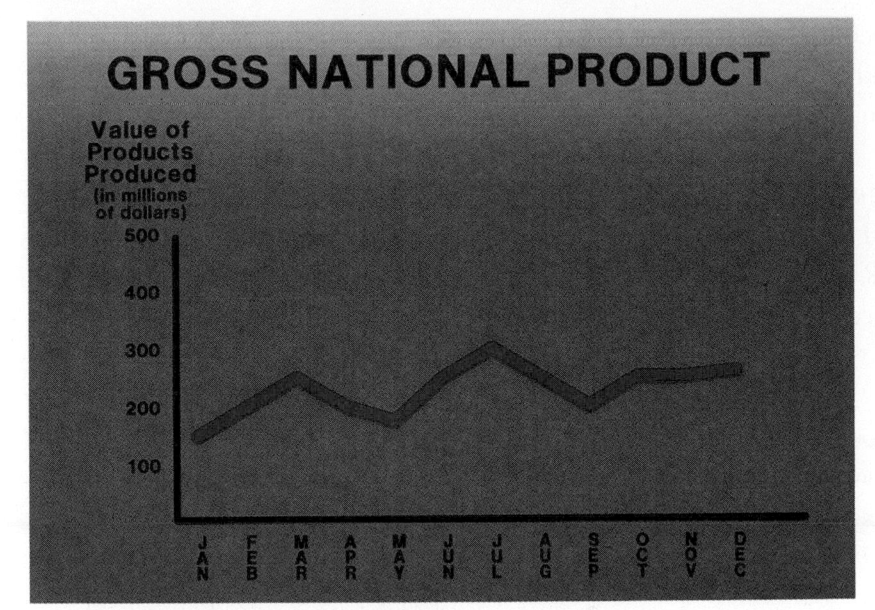

Fig. 3-3. A high GNP means that companies are producing and consumers are purchasing more products. The American GNP can be negatively affected when Americans buy products made in foreign countries rather than those made in the United States.

■ Clip an article from a newspaper or magazine that tells about a new product on the market. What consumer demand is met by the product? How well do you think the product meets the demand?

Fig. 3-4. Creative people use many techniques to bring their ideas to reality. This man is making a rendering for a new car model.

People Create Technology

People are responsible for developing, producing, and maintaining our technical world. The skills and talents of workers around the world create the products and services we use each day.

The use of technology begins when a need or want is identified. People think of products that meet that need. Fig. 3–4. People then combine the seven resources of technology to bring their ideas to reality.

During the late 1800s, Whitcomb Judson saw a need for a device to fasten boots quickly. In 1891, his creative use of resources gave the world the zipper. Fig. 3–5.

Professionals such as engineers, architects, and scientists develop technical ideas and create specific plans for making them. Technicians and skilled workers work from these plans to produce the product or provide the service.

Fig. 3-5. The simplest of ideas often have the greatest impact: Whitcomb Judson's zipper was a creative idea that solved a basic want. Judson submitted this drawing of his device to the patent office.

(No Model.)

W. L. JUDSON.
CLASP LOCKER OR UNLOCKER FOR SHOES.

No. 504,038. Patented Aug. 29, 1893.

■ People in a variety of occupations will be needed when MegaIndustries builds its factory near Glenville and the town begins to grow. Think of ways to attract people to the community. Write a newspaper ad or prepare an informational brochure to accomplish this. In your ad or brochure, describe the community and the expected changes.

Entrepreneurs

How creative are you? Do you often come up with great ideas or solutions to technical problems? You might someday make a lot of money as an entrepreneur. An **entrepreneur** is a person who forms or starts his or her own business. Your business might be based on a brand new invention or an innovation. **Innovations** are modifications to products that already exist.

Bette Claire Graham was an entrepreneur. As a secretary for a Texas bank, she made her share of typing mistakes. In a search for a quick correction process, she drew upon her knowledge as an amateur artist. Mixing white paint and turpentine, she developed the first liquid correction fluid. She sold her product to secretaries in her building. Her small company grew into a nation-wide industry. Bette Graham died in 1980. Her business was sold to the Gillette Company for $45.5 million.

Human Factors Engineering

Have you ever sat in a chair and noticed that the chair back just didn't fit the contour of your spine? It probably was not a very comfortable

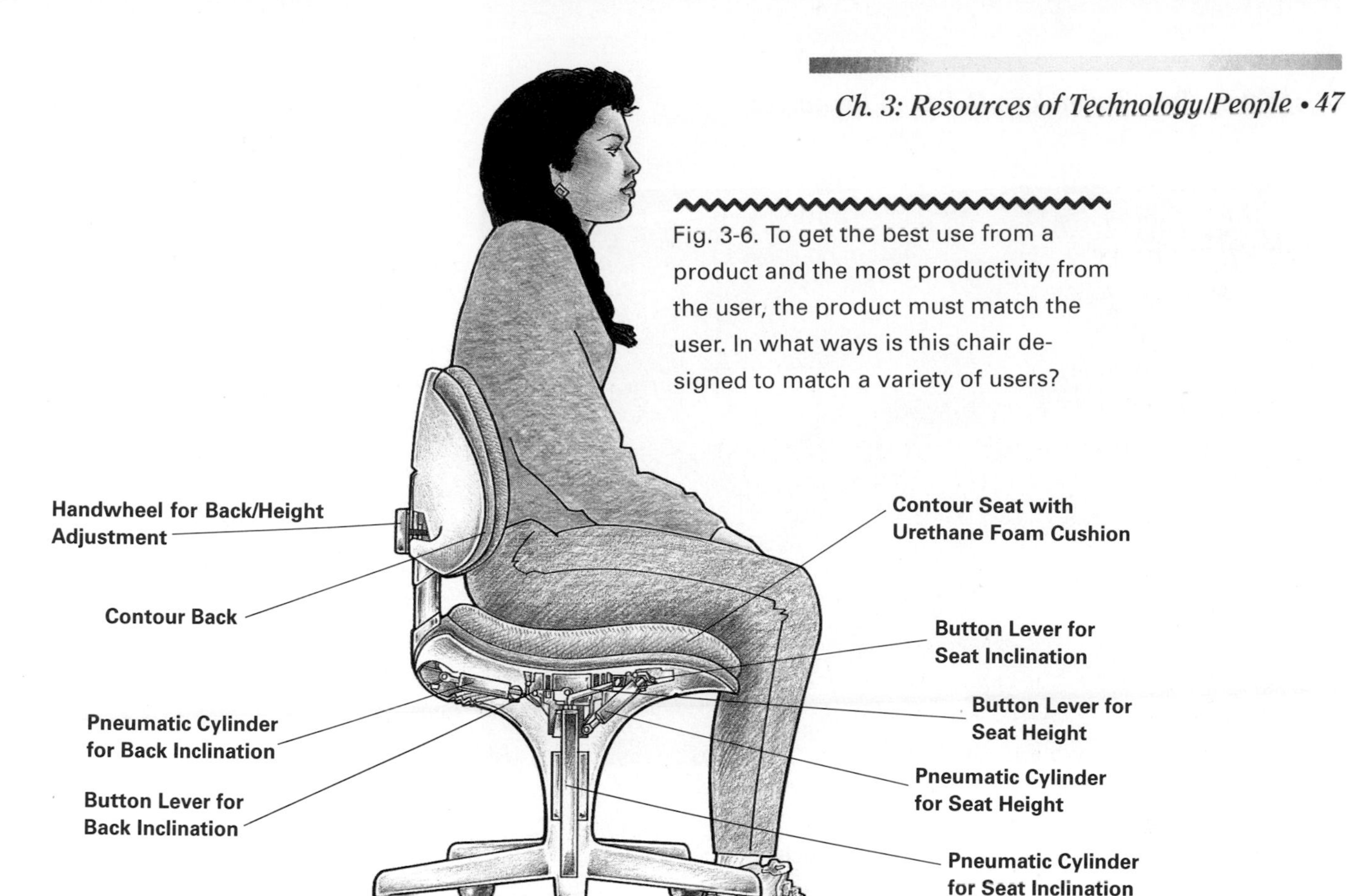

Fig. 3-6. To get the best use from a product and the most productivity from the user, the product must match the user. In what ways is this chair designed to match a variety of users?

chair—one you wouldn't want to spend a great deal of time in. Sometimes the products of technology don't seem to match the people that use them. When people create technology, they must consider "the human factor."

Human factors engineering, also called **ergonomics**, is a type of designing in which the designer studies how to match the product to the human user. Fig. 3–6. If the designer of that uncomfortable chair had studied human posture, a more comfortable chair might have resulted.

▶▶▶ FOR DISCUSSION ◀◀◀

1. Pick a product in your classroom and show how each of the seven resources of technology were incorporated into that product.
2. If you had a great idea for a product, how might you raise the money to fund your business?

■ **Select an object and think about how you would redesign it ergonomically. For example, how would you make a calculator easier to use? How about a telephone or a bicycle or a rake? Sketch your ideas and tell how your design would improve the product.**

People Govern Technology

Technology brings about change within the community and the workplace. People are responsible for governing and regulating this change. By creating rules and laws, people can maintain control

Fig. 3-7. Federal, state, and local governments create regulations that govern the creation and use of technology.

over the impacts of technology. Fig. 3–7. Regulations are designed to protect workers, consumers, and the environment from the negative effects of technology.

Worker Safety

The technological workplace can be a hazardous place. Conditions are much better now than they were in the early days of sweatshops and 18-hour work days. Fig. 3–8. This came about through regulations designed to create better working conditions. The **Occupational Safety and Health Administration (OSHA)** is the federal office that creates many of these regulations.

Fig. 3-8. In the early days of the Industrial Revolution, children worked long hours in unsafe and unhealthy conditions. Child labor laws were finally passed, making it illegal for children to work in the factories. Other regulations have improved the safety and health conditions in the workplace.

Fig. 3-9. OSHA inspectors check work sites for safety hazards. Random inspections help reduce accidents and increase a company's productivity.

OSHA regulations are designed to protect the health and welfare of workers. OSHA inspectors make surprise visits to factories and other workplaces. They make sure employees are wearing proper protective clothing, machines are safe, and the company is following the regulations set up by OSHA to protect the employees. Fig. 3–9. If violations are found, the employer can be fined.

Minimum wage laws set a limit on how low wages can be. Workers' compensation laws provide medical care and income if a worker is hurt on the job.

Consumer Safety Laws

Product safety laws are designed to protect consumers from products that might be unsafe. Products are tested to ensure that they are safe to use. Take a look at your radio or stereo. There is probably a UL label somewhere on the product. The **Underwriters Laboratory (UL)** tests products to ensure that they work properly. Fig. 3–10.

Local towns and cities use building codes to set standards on methods used in construction. Electrical codes, plumbing codes, and construction codes ensure that buildings and homes are constructed safely.

Fig. 3-10. Toys, appliances, and many other products are tested for safety and reliability by the manufacturers and by outside agencies. This symbol on a product means that it has been tested and approved by Underwriters Laboratory.

Environmental Laws

Many of the impacts of technology are unforeseen or not realized when the technology is first developed. Henry Ford never dreamed that the automobile would have such a negative effect on the environment. People have created laws and regulations to protect our environment from further damage.

The **Environmental Protection Agency (EPA)** is the office of the federal government that creates regulations to protect our air, land, and water. EPA laws regulate the dumping of industrial waste into our waters and land.

Fig. 3-11. People must protect the environment from the products of their own design. Environmental laws set regulations controlling emission of pollutants into the atmosphere.

People have also created laws to regulate the fumes coming out of smokestacks and the tailpipes of automobiles. Fig. 3–11. Other environmental laws require developers or builders to fill out Environmental Impact Statements before they begin any major construction projects. Environmental Impact Statements try to predict the impact of construction on the environment, wildlife, and the surrounding community.

TECHNOLOGY TRIVIA

Each American produces approximately 1,500 lbs. of garbage each year.

▶▶▶ FOR DISCUSSION ◀◀◀

1. If your neighbors built a second floor on their house that blocked the sunlight hitting your solar hot water collectors, should they have to take it down? Explain.

2. What would you do if the smokestack of a neighborhood factory continually puffed black smoke into the air?

CHAPTER 3 Review

Chapter Highlights

- Technology depends on seven resources: people, information, tools and machines, materials, energy, money, and time.
- Each resource of technology is necessary for the production of goods and services.
- People create, produce, and maintain the products and services of technology.
- The products people produce and consume have an impact on our nation's economy.
- People create laws to control the use of technology and its impacts on society and the environment.

Test Your Knowledge

1. List the seven resources of technology.
2. Imagine that you are the owner of a company that manufactures sunglasses. Give one example of each of the seven resources of technology you need to produce your product.
3. How does the consumption of products and services affect our nation's economic health?
4. As a consumer of products, how can you have an impact on what is sold in local stores?
5. Explain what "human factors engineering" involves.
6. What is an entrepreneur? Name at least one modern entrepreneur.
7. How are product innovations different from inventions?
8. Which federal office is responsible for creating environmental regulations?
9. Name one agency that tests products to make sure they are safe.
10. Schools, like factories, have regulations designed to protect your safety. List three safety rules that apply in your school.

Correlations

SCIENCE

1. Choose one of the following industrial materials and write a report about the health risks faced by the people who work with the material: asbestos, polyvinyl chloride, uranium, coal dust, naptha, benzene.

MATH

1. Matt has developed a protective case for Nintendo® cartridges. The production costs are $.78 each. He sells them for $1.45 each. How much profit does he make on one item? On a dozen?

LANGUAGE ARTS

1. How would you like to become an entrepreneur? Develop an idea for a business that would appeal to other students. You may invent or innovate, but consider all seven resources. Describe your new company in a news release.

SOCIAL STUDIES

1. Your book defines the economic term GNP. What is America's current gross national product? What industries in the United States have decreased the most in production? See if you can find out at least one reason why the GNP has changed in the last ten years.

CHAPTER 4

Tool and Machine Resources

Introduction

It was not by accident that the first attempt at technology resulted in a tool. Tools make work easier. Our ancient ancestors were trying to make everyday life less difficult.

Modern technology uses many different kinds of tools and machines to accomplish this same goal. In this chapter, we will look at many tools and machines and how they help change materials, energy, and information into products and services.

After reading this chapter, you should be able to

- Discuss the different families of tools and what they do.
- Identify specific tools and machines.
- Understand the importance of safe tool and machine operation.
- Understand the importance of safe conditions in a laboratory.

Words you will need

force
mechanical advantage
measuring
layout
separating
forming
combining
optical

What Are Tools and Machines?

How many tools and machines can you name? Your list would probably be quite long. Somewhere on your list, we would find hammers, saws, and wrenches. These are common tools that most people know about.

Technology is dependent on these as well as tools and machines that many people do not usually consider. For example, is a computer a tool? What about a microscope?

A good definition for tools and machines might help answer these questions. Tools and machines are devices that help make work easier.

Some tools and machines make work easier by changing the amount of force and the direction of the force you place on an object. **Force** is the push or pull that gives energy to an object. For example, a crowbar increases the force applied by your arm to pull a nail. Fig. 4–1.

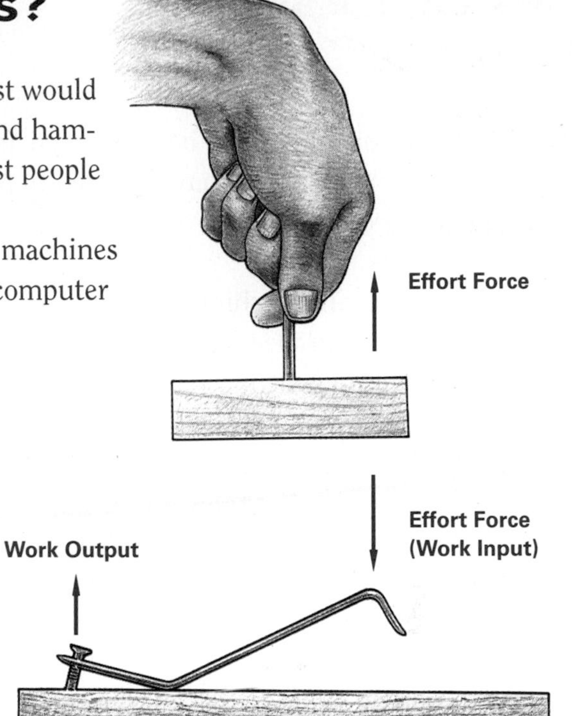

Fig. 4-1. To pull this nail, you would push down on the crowbar with less force than you would need to pull it by hand, but you would have to push through a greater distance.

Tools and machines that increase our applied force create a mechanical advantage. **Mechanical advantage** is the number of times a machine increases the force we apply. Fig. 4–2.

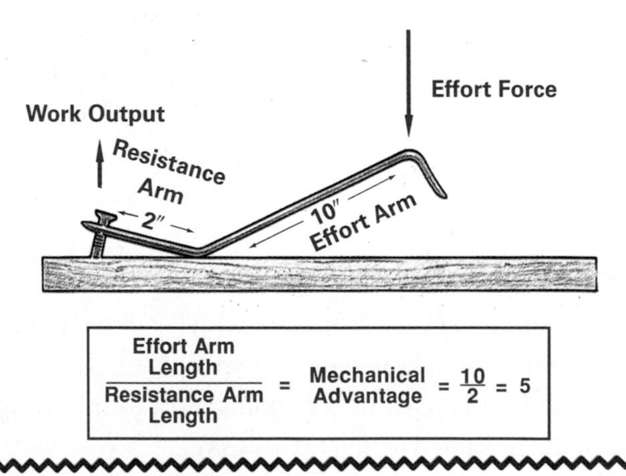

Fig. 4-2. To calculate the mechanical advantage of this crowbar, divide the length of the effort arm by the length of the resistance arm.

Not all tools and machines increase our force. Some make work easier by extending our human powers. Microscopes and telescopes allow us to see objects that are too small or too far away to be seen by the human eye alone. Computers help our brains organize large amounts of information and make instant calculations. Are computers and microscopes tools? According to our definition, the answer is yes.

▶▶▶ FOR DISCUSSION ◀◀◀

1. You have two hammers, one with a 10″ handle and the other with a 14″ handle. Which hammer gives you the greatest mechanical advantage for pulling a nail from a board? Why?

2. Make a list of the tools and machines commonly found in a dentist's office.

Uses of Tools and Machines

Tools and machines can be organized into groups. The tools in each group are designed to perform a variety of processes. The groups are:

- hand tools
- power tools and machines
- computer-controlled machines
- electronic equipment
- optical tools and machines

Different tools are designed to be used for different processes. **Measuring** tools, for example, are used to determine the size of objects. **Layout** tools are used to draw lines, angles, and circles on different materials. These lines tell us where to cut or bend the material.

Separating tools and machines cut or remove part of a material. **Forming** tools and machines change the shape of materials such as clay and metal without removing any materials. **Combining** tools and machines allow us to fasten materials together using nails, screws, or other fasteners.

Other tools and machines are designed to collect and help us examine information. A battery tester is a good example of this type of tool. It not only measures the output of the battery but uses this information to determine whether the battery needs to be replaced.

Hand Tools

Hand tools are tools for which people supply the power—muscle power. Some of the more common hand tools used in technology are found in figures 4-3 through 4-7 (pages 54–58).

HAND TOOLS FOR MEASURING

RULES

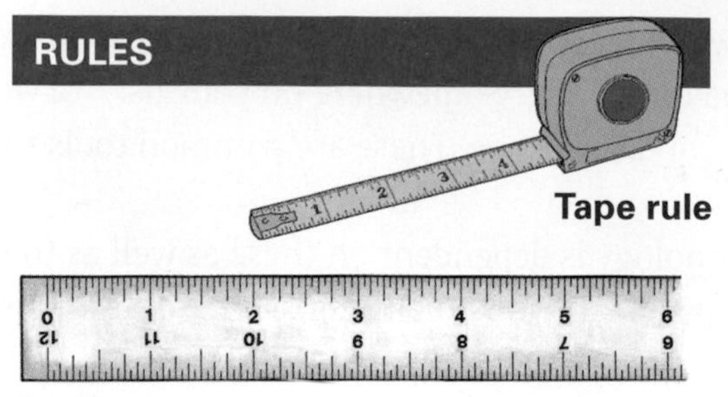

PRECISION TOOLS

Micrometer—for outside diameters and thicknesses

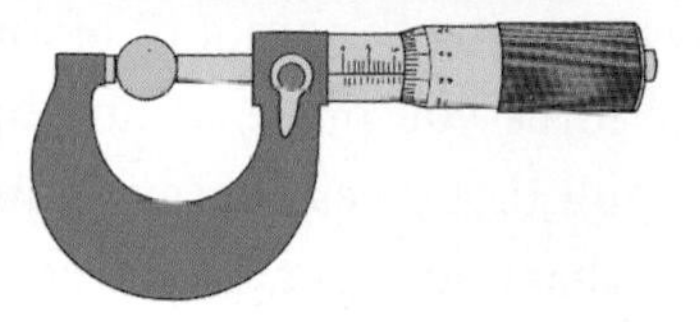

Vernier caliper—for inside or outside diameter

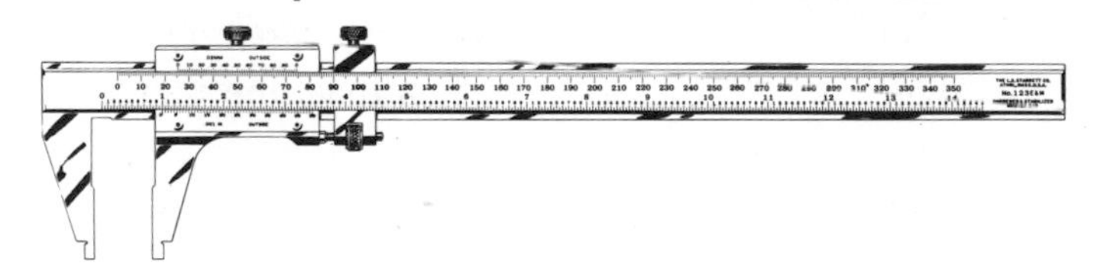

Fig. 4-3. Rules measure distances such as length, width, and thickness. For finer measurements, precision tools are needed.

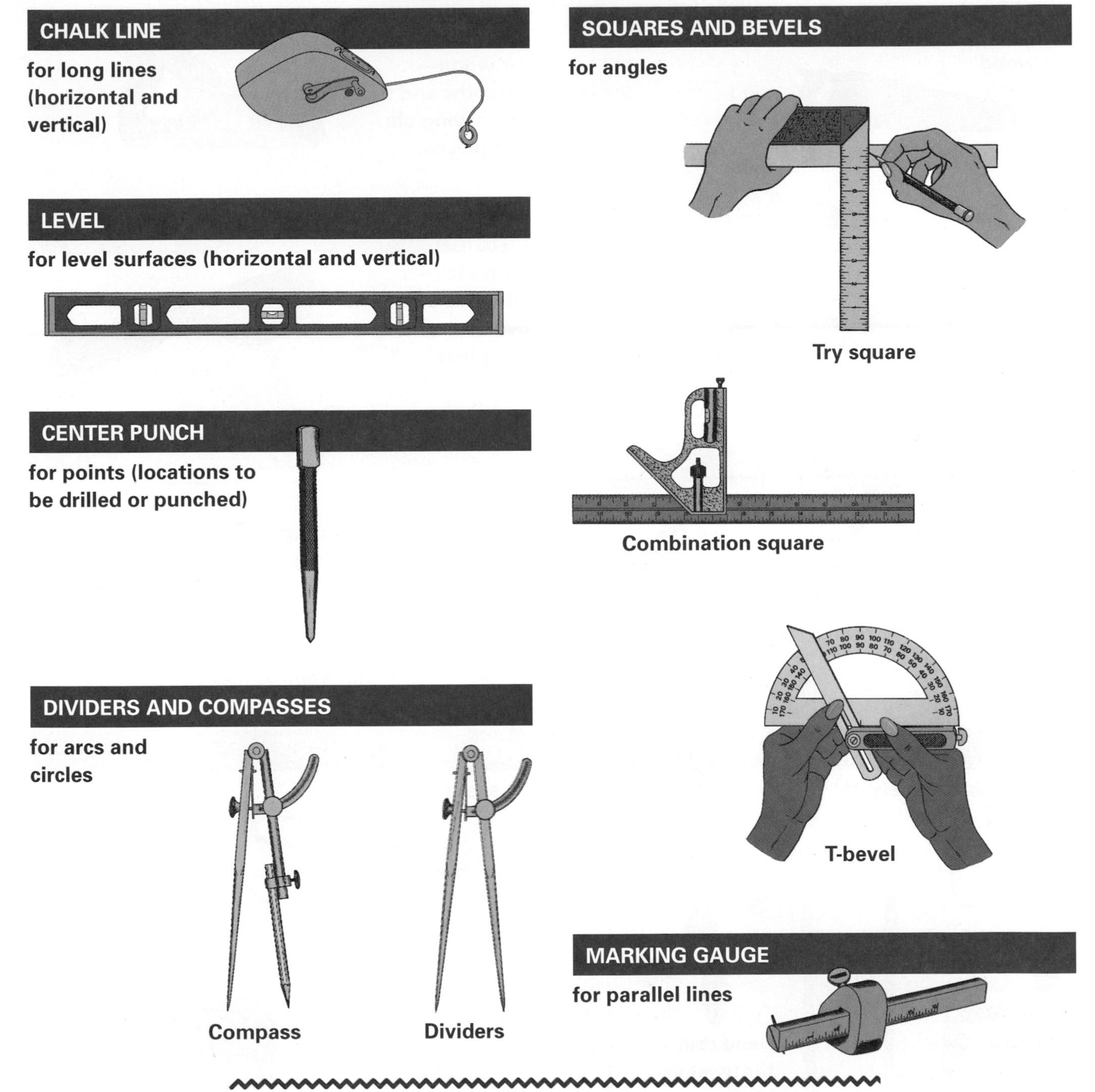

Fig. 4-4. A wide range of layout tools allows us to lay out designs accurately.

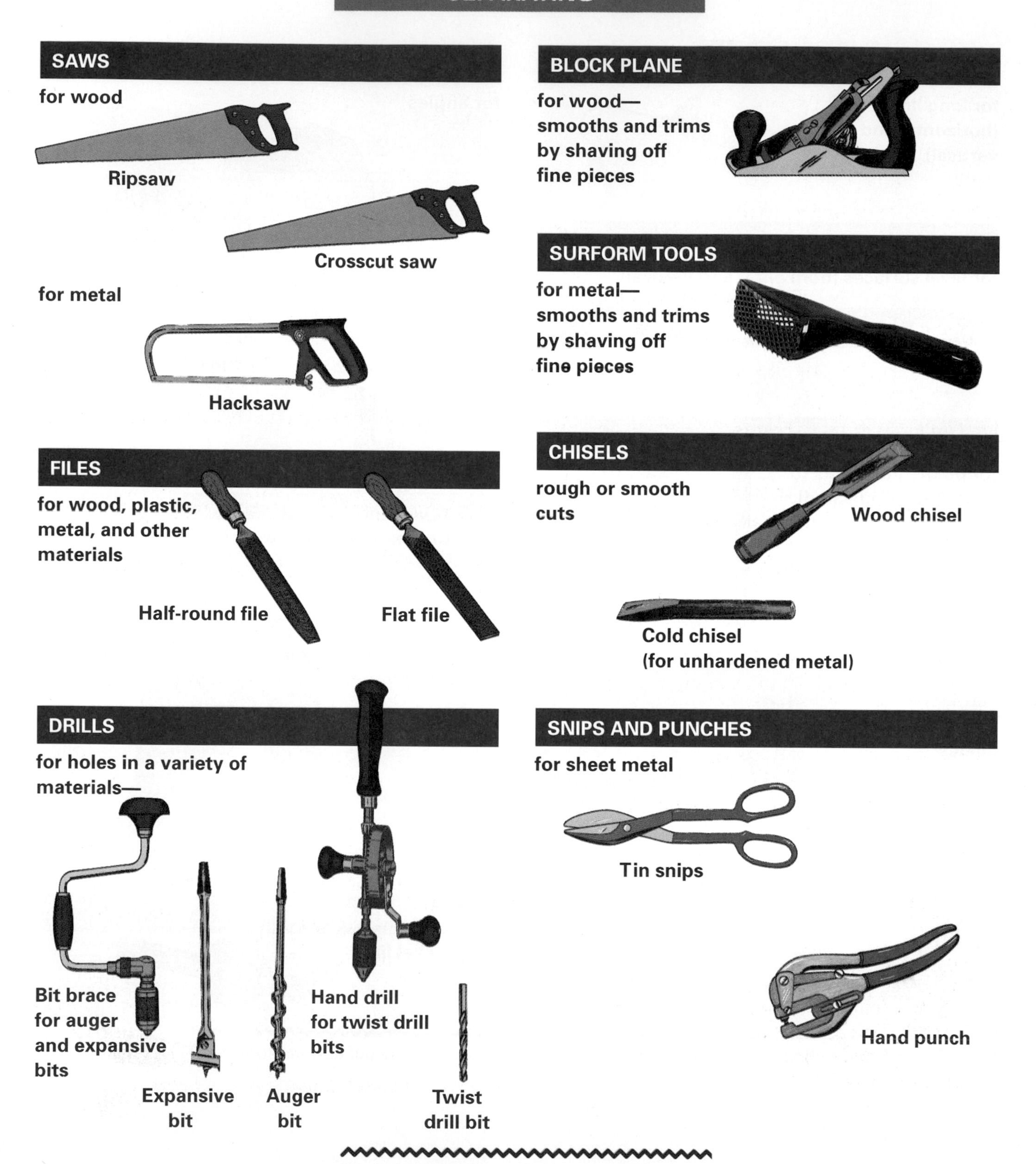
HAND TOOLS FOR
SEPARATING
SAWS
for wood
Ripsaw
Crosscut saw
for metal
Hacksaw
BLOCK PLANE
for wood—
smooths and trims
by shaving off
fine pieces
SURFORM TOOLS
for metal—
smooths and trims
by shaving off
fine pieces
FILES
for wood, plastic,
metal, and other
materials
Half-round file
Flat file
CHISELS
rough or smooth
cuts
Wood chisel
Cold chisel
(for unhardened metal)
DRILLS
for holes in a variety of
materials—
Bit brace
for auger
and expansive
bits
Expansive
bit
Auger
bit
Hand drill
for twist drill
bits
Twist
drill bit
SNIPS AND PUNCHES
for sheet metal
Tin snips
Hand punch

Fig. 4-5. Hand tools used for separating.

HAND TOOLS FOR COMBINING

WRENCHES

tighten nuts and bolts

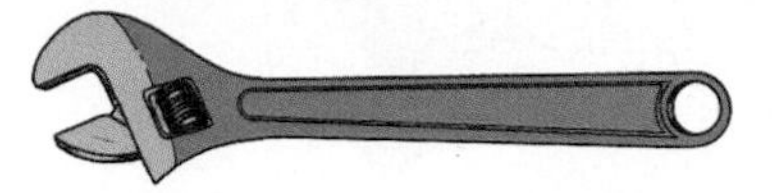

Adjustable open-end wrench

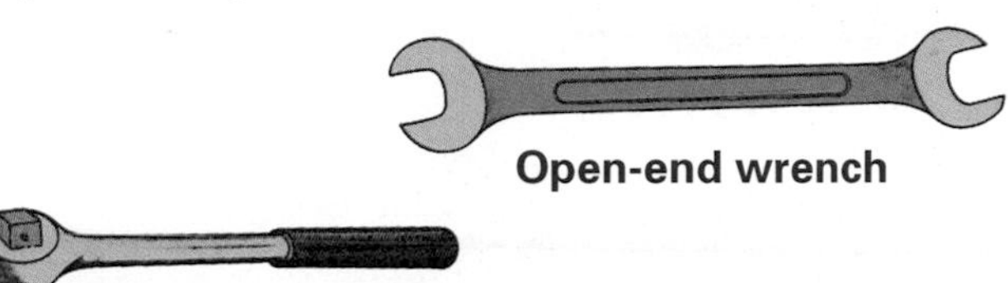

Open-end wrench

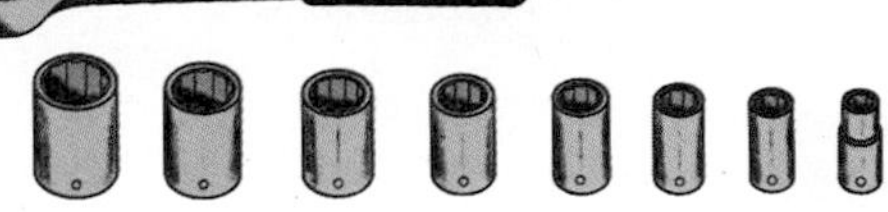

Ratchet wrench with sockets

Box-end wrench

PLIERS

hold fasteners in place during installation

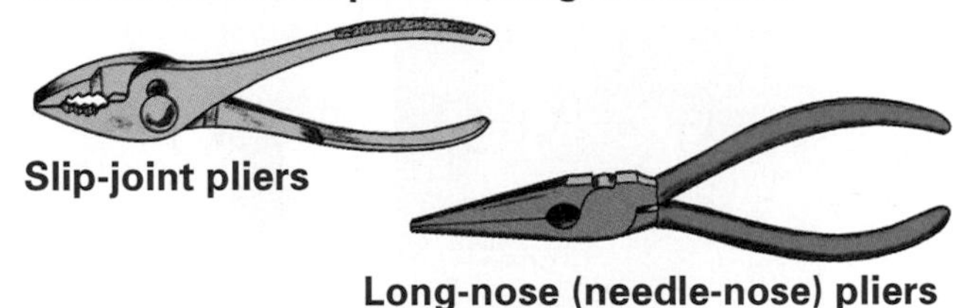

Slip-joint pliers

Long-nose (needle-nose) pliers

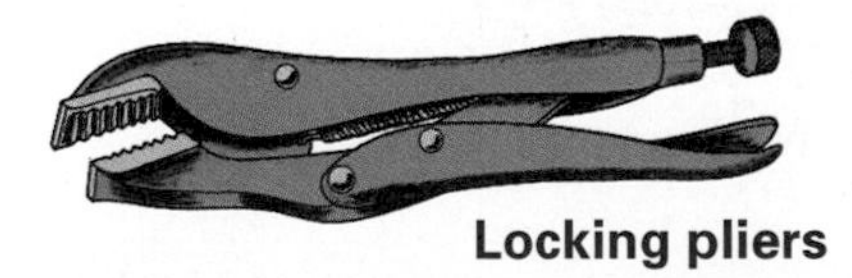

Locking pliers

HOT GLUE GUNS

melt solid glue or adhesive, which is used to combine materials

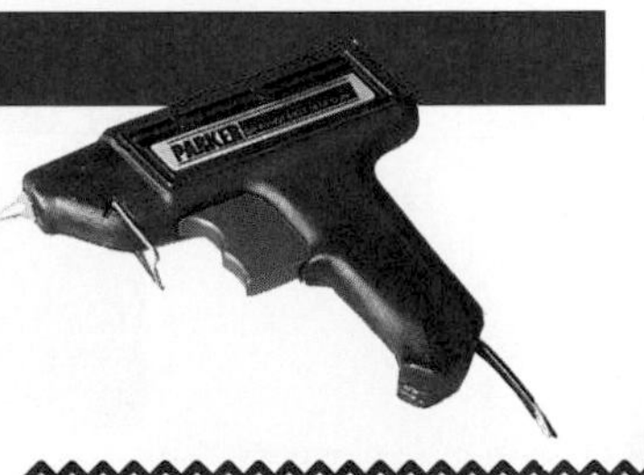

HAMMERS

apply mechanical force to fasteners

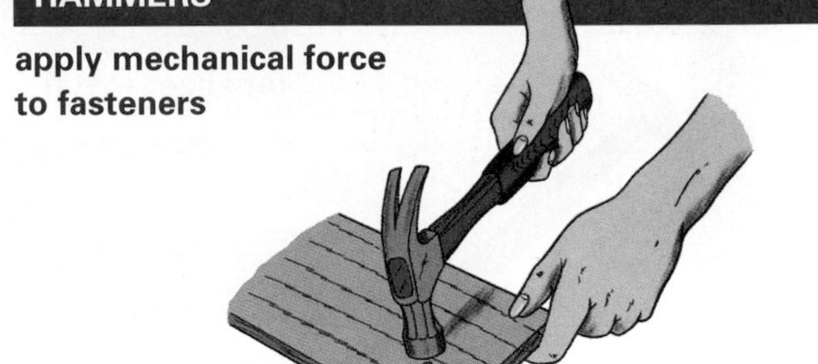

Claw hammer

SCREWDRIVERS

turn screws, which fastens materials together

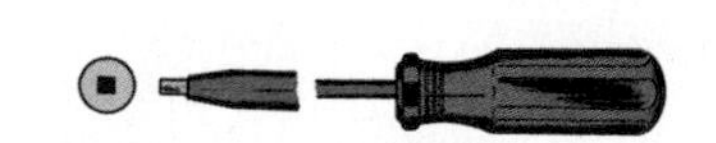

Standard screwdriver

Phillips screwdriver

Square-head screwdriver

POP RIVETERS (BLIND RIVETERS)

install rivets

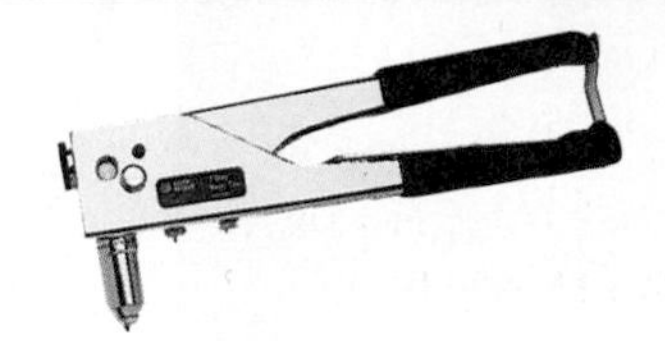

SOLDERING TOOLS

melt solder, which is used to combine materials

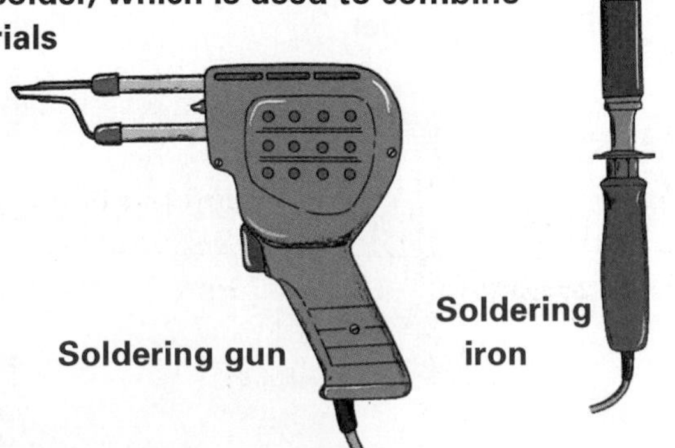

Soldering gun

Soldering iron

Fig. 4-6. Hand tools used for combining.

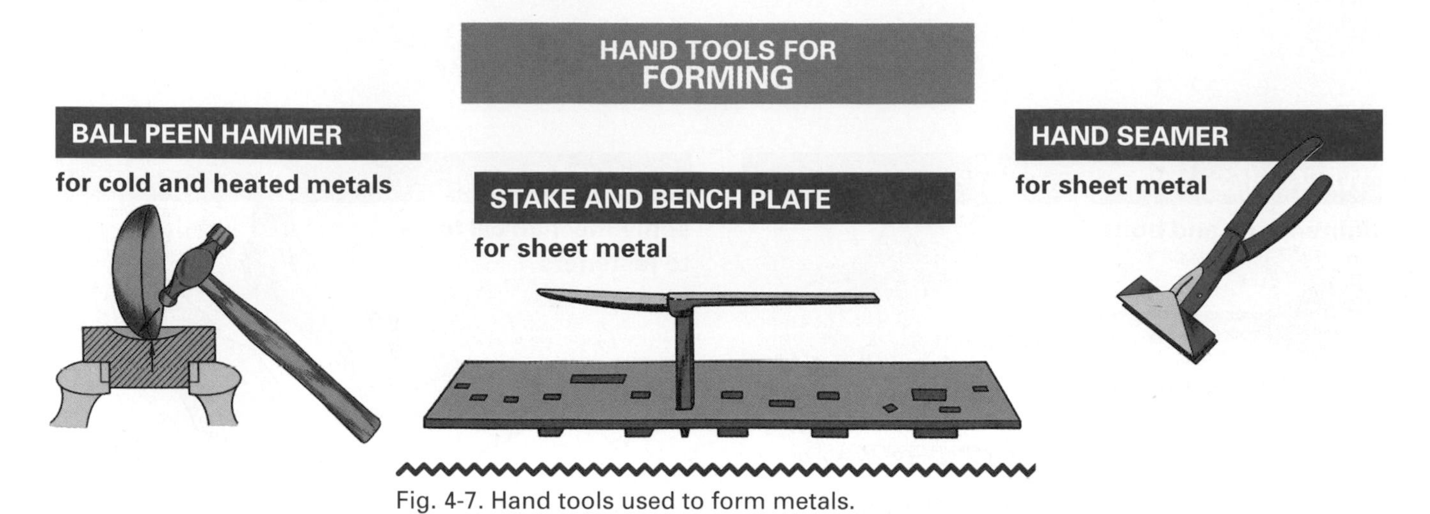

Fig. 4-7. Hand tools used to form metals.

MACHINES

HYDRAULIC

Robotic gripper

Floor jack

Backhoe

PNEUMATIC

Jack hammer

Nail gun

Dentist drill

Fig. 4-8. Have you ever felt the force inside a bottle of soda when you shake it up? This force is caused by the gases inside pressing against the bottle walls. The same force is used to power pneumatic and hydraulic machines.

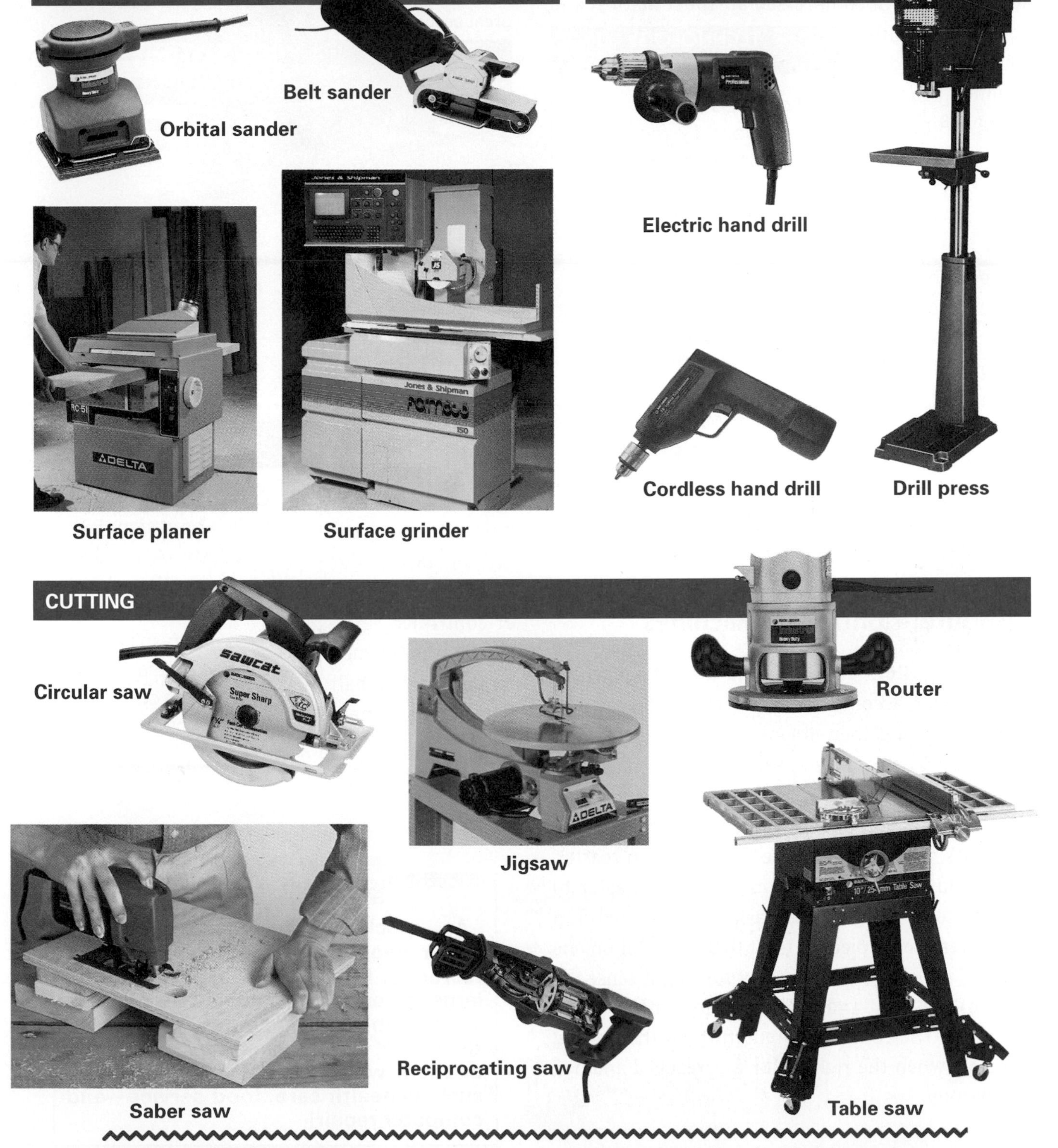

Fig. 4-9. Power tools and machines make many tasks easier by substituting electrical energy for human effort.

Power Tools and Machines

Power tools and machines use energy from other sources to accomplish work. Electrical energy and energy from compressed air (pneumatics) and compressed fluids (hydraulics) are common sources of energy for power tools. Fig. 4–8 (page 58).

TECHNOLOGY TRIVIA

The first power tools were developed in the early 1800s when the steam engine was applied to various tools. More than 1700 years earlier, however, the first known steam engine was described (about A.D. 60) by Hero, a scientist who lived in Alexandria, Egypt. Unlike later steam engines, Hero's steam engine performed no useful work.

Most power tools are handheld or portable. Machines are usually large pieces of equipment that are not moved around. Figure 4–9 (page 58) shows some of the more common power tools and machines.

Computer-Controlled Machines

Computers can be found almost anywhere work is being accomplished. In technology, computers are used to gather, organize, store, and share information. They are also used to control machines. Fig. 4–10.

To help you understand computer control of machines, let's "computerize" your kitchen toaster. You could program your computerized toaster to produce toast as light or as dark as you wanted. You could even have a piece of toast that was light on one side and dark on the other. A sensor might sense the color of the toast and feed this information to the computer. The computer would turn off the heating elements when the right color was reached and pop the toast up.

Fig. 4-10. Computers can be used to control complex machinery with great accuracy.

Using computers, operators can program machines to do a variety of tasks. For example, a computer-controlled lathe can be programmed to make a baseball bat without help from a human operator.

TECHNOLOGY TRIVIA

Computers can make milling operations 53 times faster. A part that takes an experienced machinist four hours to make on an ordinary milling machine can be made in 4½ minutes with computer control.

IMPACT

As machines become "smarter," fewer people are needed to work in factories. In the future, there may be large factories with only a few humans supervising the machines. Where will the rest of the people find work? Many of them will be in service industries, such as health care, food service—and computer repair!

Electronic Equipment

Electronic equipment is used to gather, test, and evaluate information about machines. Using sensors, electronic equipment supplies information we need to determine how well a machine is performing. Some electronic equipment can even evaluate this information and tell us where there is a problem.

Optical Tools and Machines

Optical tools and machines allow us to view things that we could not ordinarily see. Microscopes and telescopes are optical machines. Using lenses, these machines help us investigate things as small as a human cell or objects as far away as a star. Fig. 4–11.

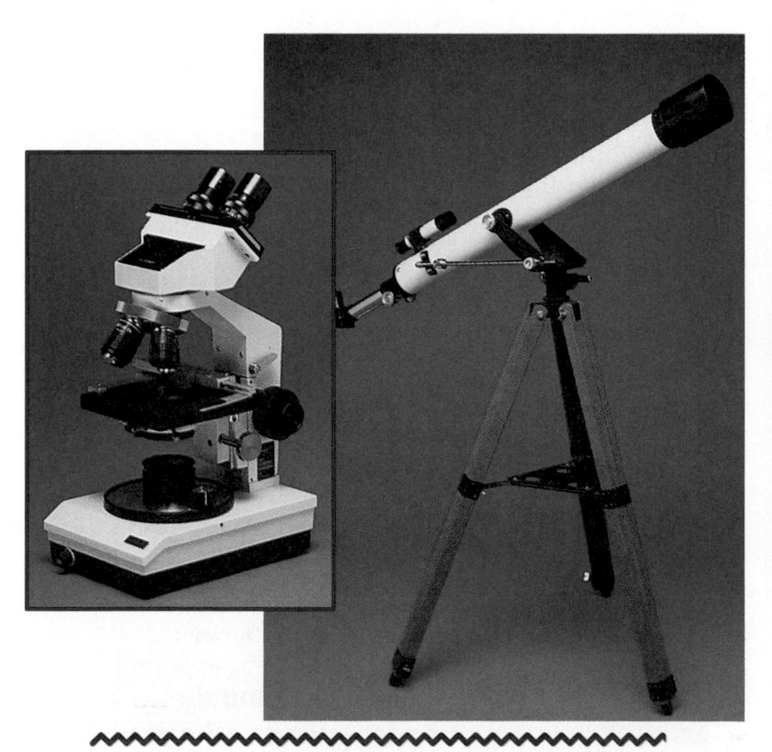

Fig. 4-11. Telescopes and microscopes are two common examples of optical tools.

Lasers are another type of optical tool. Lasers are devices that produce a concentrated beam of light that is very powerful. Fig. 4–12. Modern lasers can also be controlled by computers.

▶▶▶ FOR DISCUSSION ◀◀◀

1. How has the development of computer control and optical equipment changed supermarket checkout counters?
2. List each piece of equipment in your kitchen as a hand tool, power tool, or computer–controlled/electronic machine. Tell which ones measure, separate, or combine materials.

Fig. 4-12. Laser tools use a concentrated light beam to do tasks that range from cutting materials to performing human eye surgery.

■ **If you were to build a model of a structure, what tools in your laboratory might you use?**

Tool and Machine Safety

The safe use of tools and machines is very important. Each year, thousands of people are injured while working with tools and machines. Whether we are working for a company, at home, or in the school laboratory, there are common safety rules we must follow.

Tool and Equipment Safety Rules

1. Get proper instruction and permission before using a tool or machine.
2. Keep the work area clean and organized.
3. Keep all cutting tools sharp.
4. Always use the correct tool or machine for a particular job.
5. Always use guards and safety equipment on tools and machines.
6. Report all damaged tools and machines to the instructor.
7. Be sure machines and portable power tools are unplugged when you change their settings.
8. Handle tools and equipment with care. Abuse of tools can cause accidents.
9. Allow only one person in a machine safety zone at a time.

Personal Safety Rules

1. Keep your hair away from moving parts. Wear a hat or tie back your long hair.
2. Do not wear loose-fitting clothes while you are using tools and machines.
3. Roll up your sleeves before you operate tools and machines.
4. Always wear safety glasses, goggles, or a face shield while operating tools and machines.
5. Know where all the emergency shut-off switches are.
6. Remove all jewelry before using tools and machines.

Laboratory Safety Rules

1. Clean up spills immediately.
2. No horseplay or fooling around is permitted in the laboratory.
3. Place tools back in their proper places when you finish using them.
4. Place oily rags in a safety container.

■ Make a drawing depicting (showing without words) one of the personal or lab safety rules.

CHAPTER 4

Chapter Highlights

- Tools and machines increase our mechanical advantage and our ability to get work done.
- Tools and machines are used to change materials, information, and energy into products and services.
- The many different types of tools and machines perform a variety of operations.
- Tool and machine safety is important to your health and safety.

Test Your Knowledge

1. Why are tools the earliest forms of technology?
2. Why are a stone ax and a laser both considered tools? Explain.
3. "Tools increase our mechanical advantage." What does this statement mean?
4. What type of operation does a ruler perform?
5. What type of operation does a pencil sharpener perform?
6. What type of operation does a calculator perform?
7. What do we call the group of tools that are usually powered by humans?
8. In which group of machines does a factory robot belong?
9. List three personal safety rules that you must follow when you use machines.
10. Do safety rules apply when you are using power gardening equipment? Explain your answer.

Correlations

SCIENCE

1. Power = Force × Distance, divided by Time. Calculate the horsepower you use to go up a flight of stairs. First, multiply your weight (in pounds) by the vertical height of the stairs (in feet). Then divide by the time (in seconds) that it took to climb the stairs. Note: 1 horsepower = 550 foot pounds per second.

MATH

1. The two most widely used measuring systems are the customary and the metric systems. One customary inch equals about 2.5 metric centimeters. About how many centimeters equal one foot?

LANGUAGE ARTS

1. Write directions for using a tool you have at home. Be sure your directions include safety rules for use.

SOCIAL STUDIES

1. When Samuel Slater came to the U.S. in the late 1700s, he began the factory system as we know it today. Compare the power tools used in the factories of the early 1800s with the power tools used in today's factories.

CHAPTER 5

Material Resources

Introduction

All technologies depend on materials. Materials are the "stuff" that things are made of. Early products were limited by the material resources available. Today, products are created from thousands of combinations of materials.

During the Stone Age, people chipped the first tools from rocks. The first metal tools were made during the Bronze Age (3000 B.C.). Bronze tools were stronger than tools of stone and wood, and they could be made in more shapes and sizes.

During the Iron Age, a period beginning about 1200 B.C., people learned to remove iron from iron ore and create tools that were even stronger than bronze tools.

After reading this chapter, you should be able to

Describe and give examples of the different categories of materials.

Discuss the basic principles of material science.

Describe some properties of materials.

Words you will need

natural resources
raw materials
renewable resources
synthetic materials
hardwoods
softwoods
alloy
ferrous
ceramics
polymers
thermoplastics
thermoset plastics
composite
molecule
plasticity

Types of Materials

Today, people who create products can choose from thousands of kinds of materials. Some are found in nature, and others have been developed by humans. Fig. 5–1. Materials that are found in nature are called **natural resources**. Natural resources include land, water, air, plants, animals, and minerals. Fig. 5–2. From these natural resources, we obtain the **raw materials** we need to create products. Wood, oil, cotton, animal hides, and iron ore are some of the raw materials we process from our natural resources.

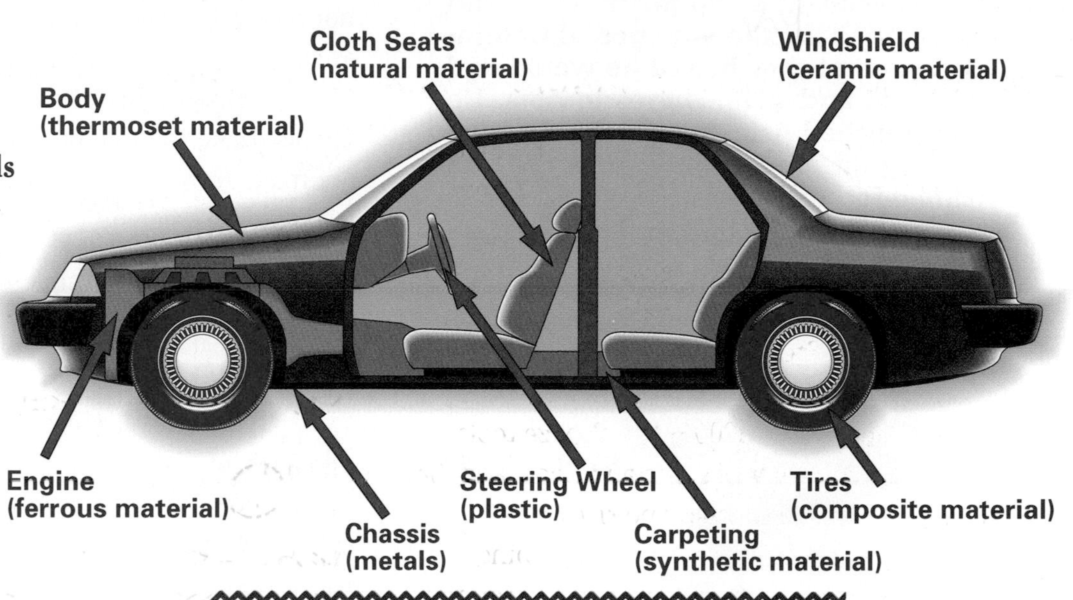

Fig. 5-1. Modern technology depends on thousands of material resources. Each material is selected based on its properties, cost, and availability.

Fig. 5-2. Nature provides many of the raw materials we use in technology. People must help protect our environment by carefully managing our nonrenewable resources.

IMPACT

Europeans might never have come to America if Spain had more silver and gold mines. In the 15th century, Spain needed silver and gold for its treasury. One reason the Spanish king and queen financed the voyages of Columbus was that they hoped he would find new gold and silver mines in Asia. He never reached Asia, but he did discover silver, gold, and a new land that he claimed for Spain.

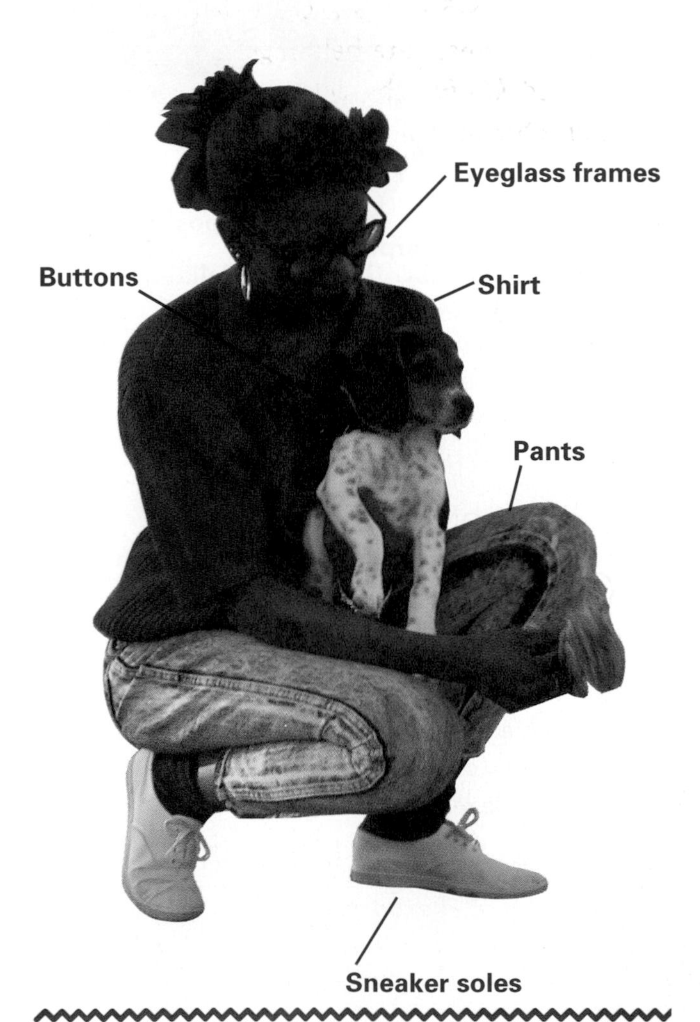

Fig. 5-3. Synthetics have replaced natural materials in many products because they are cheaper, lighter, and easier to care for.

Some raw materials are replaceable after they are removed from nature. We can grow or raise new trees, plants, and animals to replace the ones we use. These raw materials are called **renewable resources.**

Unfortunately, we cannot easily replace many raw materials. Coal, oil, and water are examples of nonrenewable raw materials. When these resources are used up or ruined, they are gone forever.

Synthetic materials are not found in nature. People make them from chemicals. (The chemicals are made from raw materials, such as oil.) Chances are, you are wearing clothing made with synthetic fibers. Fig. 5–3. Look at the labels. Perhaps your shirt is made of 50% cotton and 50% polyester. The cotton is a natural material and the polyester is a synthetic material.

Synthetics are created to improve upon the qualities of natural materials. Synthetics can be made stronger, lighter, and longer-lasting than the materials they are replacing. Synthetics can also be used instead of nonrenewable raw materials.

▶▶▶ FOR DISCUSSION ◀◀◀

1. Look around the room. List the materials you see as natural or synthetic. If the material is natural, what synthetic material might be able to replace it?

2. It seems that water should be a renewable resource, but it's not. Dead rivers result from pollution. What are some of the impacts resulting from water pollution?

Classification of Materials

Technology is dependent on thousands of different materials. Most of these materials can be arranged into five groups:

- wood
- metals
- ceramics
- plastics
- composites

Wood

Wood is a natural material that has been a valuable resource for thousands of years. Many raw materials are obtained from trees. Wood is the most obvious, of course, but some of the materials used in making paper, turpentine, varnishes, and even plastics also come from trees.

Wood can be grouped in two categories: hardwoods and softwoods. The classifications have nothing to do with how hard the wood is. Instead, they refer to the type of tree the wood comes from. **Softwoods**, such as pine and cedar, come from conifer trees (trees that have needles and cones rather than leaves). **Hardwoods** come from trees that have broad leaves, such as maple and oak trees. Fig. 5–4.

Figure 5–5 shows a cross-section of a tree trunk. Wood fibers are held together by a natural glue called *lignin*. The color and shape of these fibers give the wood its natural beauty.

Fig. 5-4. In most areas, hardwood, or deciduous, trees lose their leaves in the fall. Softwood trees keep their needles all year.

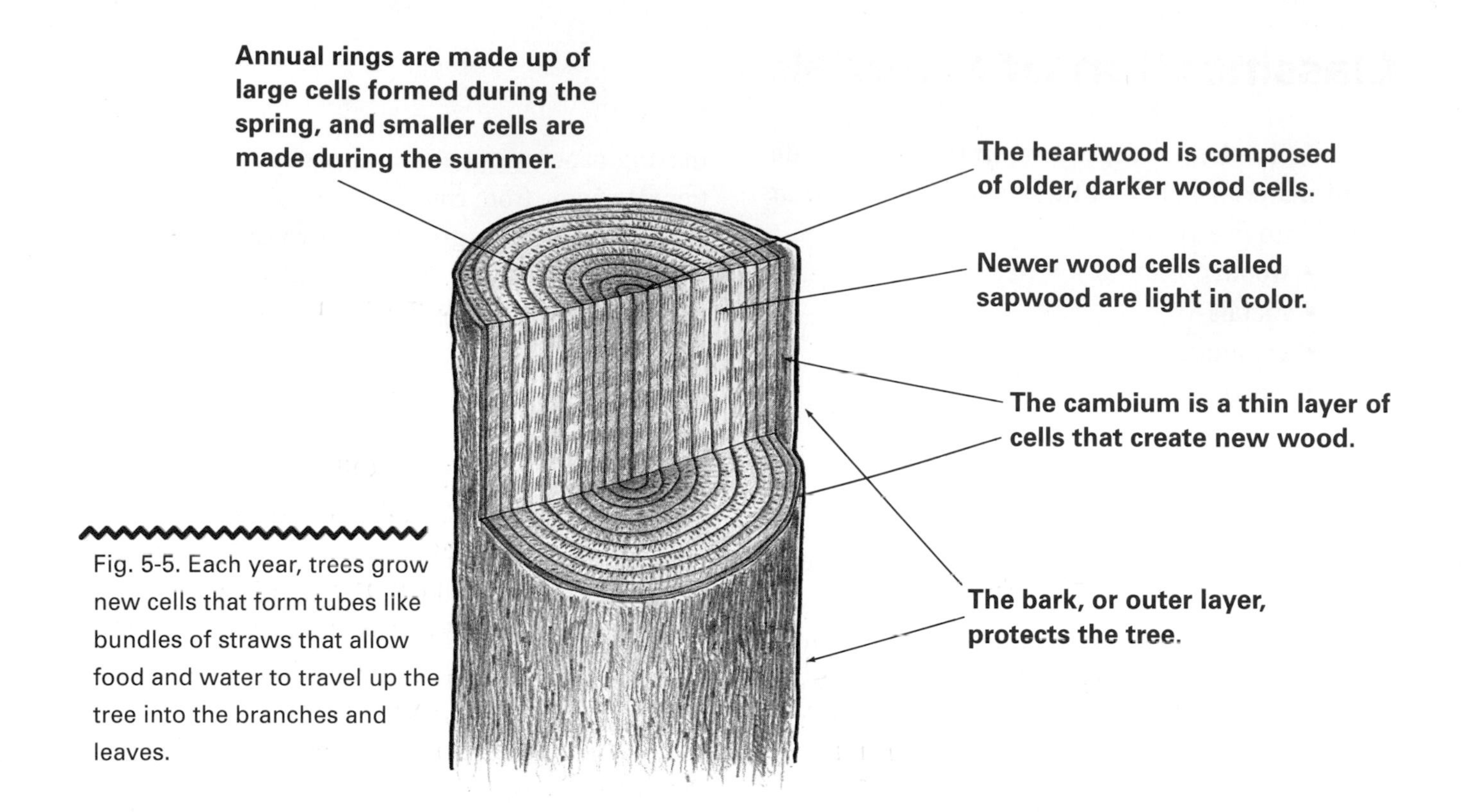
Annual rings are made up of large cells formed during the spring, and smaller cells are made during the summer.
The heartwood is composed of older, darker wood cells.
Newer wood cells called sapwood are light in color.
The cambium is a thin layer of cells that create new wood.
The bark, or outer layer, protects the tree.

Fig. 5-5. Each year, trees grow new cells that form tubes like bundles of straws that allow food and water to travel up the tree into the branches and leaves.

Metals

Scientists have discovered 107 different simple substances called *elements*. Different materials are created by combining these elements. Of the 107 elements, 70 are metals. When two or more metals are mixed, a new metal is produced. This new metal is called an **alloy.** The chart in Fig. 5–6 shows some common metals and alloys and their uses. People try to create alloys that have more useful properties than the individual metals they are created from. So far, people have created more than 70,000 alloys.

Metals and alloys can be divided into two families: ferrous and nonferrous. The word **ferrous** means "containing iron." Any metal or alloy that contains iron is a ferrous material. Steel is an example of a ferrous alloy.

Metal	Classification	Raw Material	Common Uses
Aluminum	Nonferrous Metal	Bauxite	Airplanes, Foils, Cookware
Copper	Nonferrous Metal	Chalcocite	Wires, Pipes, Coins
Iron	Ferrous Metal	Hematite	Steel Making
Brass	Nonferrous Metal Alloy	Copper & Zinc	Musical Instruments
Stainless Steel	Ferrous Metal Alloy	Iron, Chromium, Nickel	Sinks, Utensils, Knives, Forks
Nichrome	Ferrous Metal Alloy	Nickel, Iron, Chromium, Manganese	Heating Element in a Toaster

Fig. 5-6. Common metals and alloys.

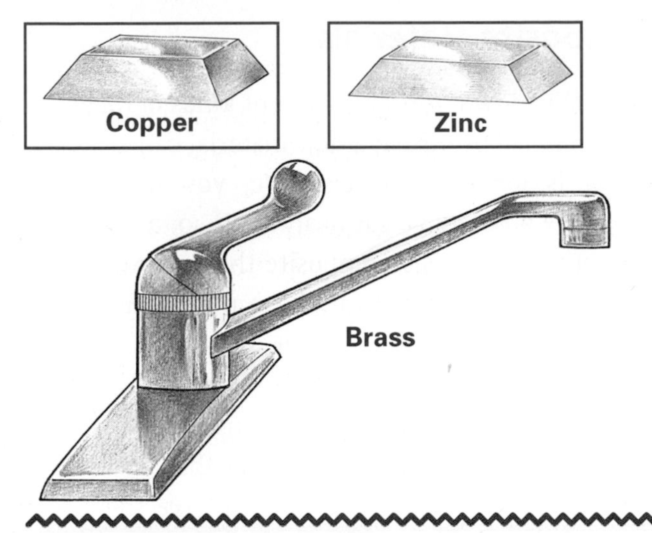

Fig. 5-7. Why is brass a better choice than copper for faucets?

Steel is made up of the elements iron, carbon, and oxygen. By changing the amount and type of element, we can change the properties of the material. For example, if we increase the amount of carbon, the steel becomes harder. When we add chromium and nickel, we create stainless steel.

Nonferrous metals and alloys contain no iron. Aluminum, copper, and tin are nonferrous metals. When we mix the elements copper and zinc, we get a nonferrous alloy called *brass*. Fig. 5–7. Brass is harder than copper and zinc and lasts longer in water than most other metals.

Ceramics

Ceramic materials are natural materials that have been used for centuries. **Ceramics** are made from minerals called *silicates*. Silicates make up one of the most abundant classes of materials found on earth. The sand, clay, and quartz you can find on beaches and river banks are all examples of silicates.

Silicates can be combined with other elements to form a variety of ceramic materials. Clay, glass, cement, plaster, abrasives, and refractory bricks are a few of the more common ceramics.

Most ceramic materials require heat to harden them. After they have been hardened, ceramics are usually strong materials that are resistant to being "eaten away" by other chemicals. Many are also resistant to the flow of electricity. Ceramics are used in a wide variety of products, from cookware and bricks to heat-absorbing tiles and memory chips.

Plastics

Plastics, sometimes called **polymers**, are synthetic materials. Most plastic materials are made from carbon obtained from petroleum and natural gas. Plastic materials have been developed to replace many natural materials. Plastics are cheaper, lighter, and stronger than most metals. Plastics last longer than wood products, especially outdoors.

All plastics can be formed easily using heat. Plastic materials can be divided into two groups based on their behavior when heated: thermoplastics and thermoset plastics. **Thermoplastics** can be repeatedly reheated and reshaped. **Thermoset plastics**, once formed, cannot be reheated and reshaped. Fig. 5–8.

TECHNOLOGY TRIVIA

In 1862, the first plastic articles—made from "Parkesine," invented by Alexander Parkes—were exhibited in London.

Fig. 5-8. This girl is drinking from a squeeze bottle made of a thermoplastic. Why is it important that the bottle be able to bend?

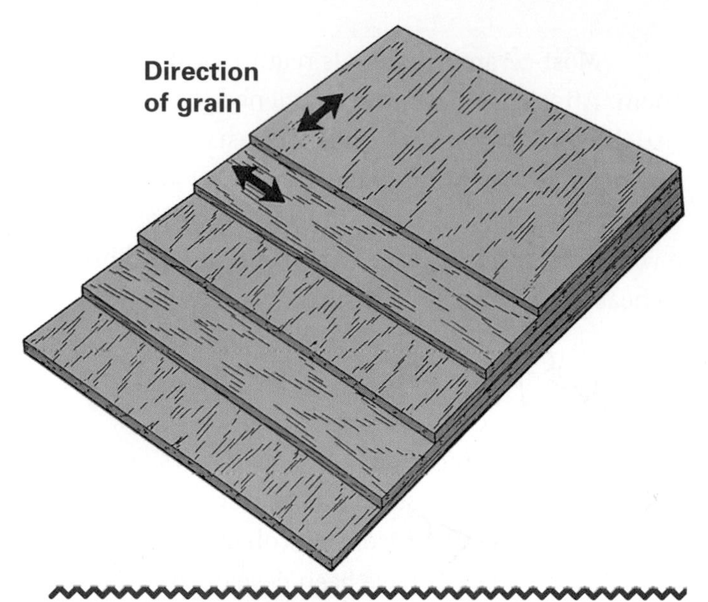

Fig. 5-9A. Each layer of wood in a piece of plywood is glued so that its grain is at a 90-degree angle to the surrounding layers.

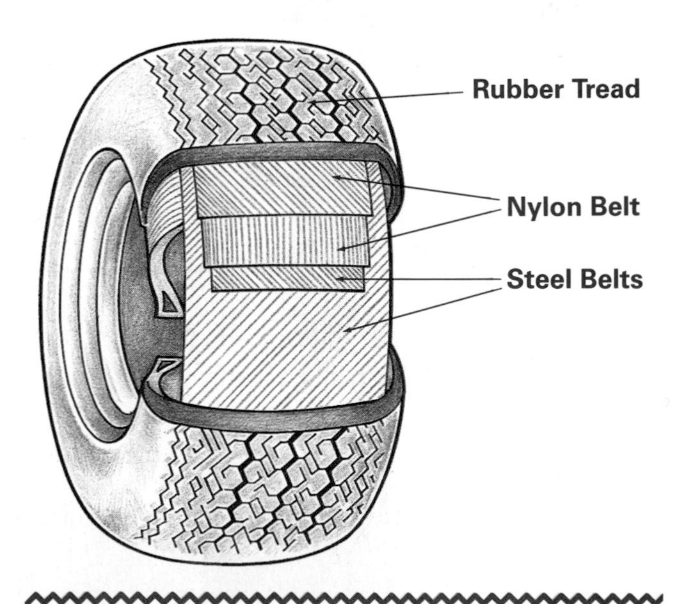

Fig. 5-9B. A tire is a composite made up of layers of rubber, nylon, and steel.

Composite Materials

When two or more different kinds of materials are combined or mixed, a new material called a **composite** is formed. For example, you might love chocolate and like to eat peanuts. If so, a chocolate bar with peanuts is a composite that's hard to beat. The chocolate and peanuts have not changed, but the composite tastes even better.

Plywood is a composite material. Layers of wood are held together with a strong glue. The composite is stronger than the materials that make it up, but the wood layers and glue still have their own qualities. Fig. 5–9 A and B.

►►► FOR DISCUSSION ◄◄◄

1. An automobile is made up of many different materials. Give examples of each of the classifications of materials discussed in this chapter.
2. If you were to build a shelter beneath the ocean, what kinds of materials would you select and why?

Material Science

People who design and create new materials are called *material scientists*. Material scientists study the chemical structure of materials and their properties.

All materials are made up of atoms. The atoms determine a material's structure. In turn, atoms combine to form molecules. A **molecule** is the smallest part of a substance that still has the properties of that substance. Fig. 5–10. If you were to take a splinter of wood and continue to split it, the smallest piece of wood there could be is called a molecule of wood. Material scientists create new materials by combining the molecules of different elements.

Each different kind of material has unique properties. The properties of a material tell how the material is expected to perform. Fig. 5–11. For

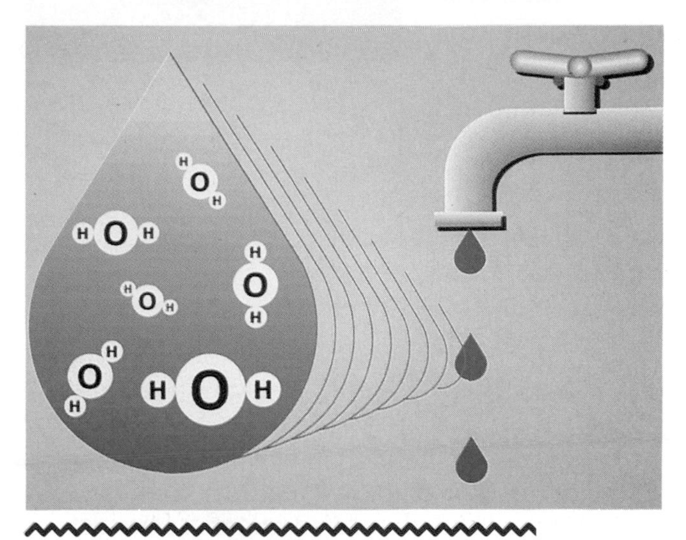

Fig. 5-10. One molecule of water contains one atom of oxygen and two atoms of hydrogen. Material scientists study how materials combine so that they can create new combinations of materials.

Fig. 5-11. The properties of materials are determined in part by how their molecules are linked. Gases expand or contract to fill different containers. Liquids take on the shape of their containers. Solids do not change their shape or size unless acted on by some force (such as a hammer blow).

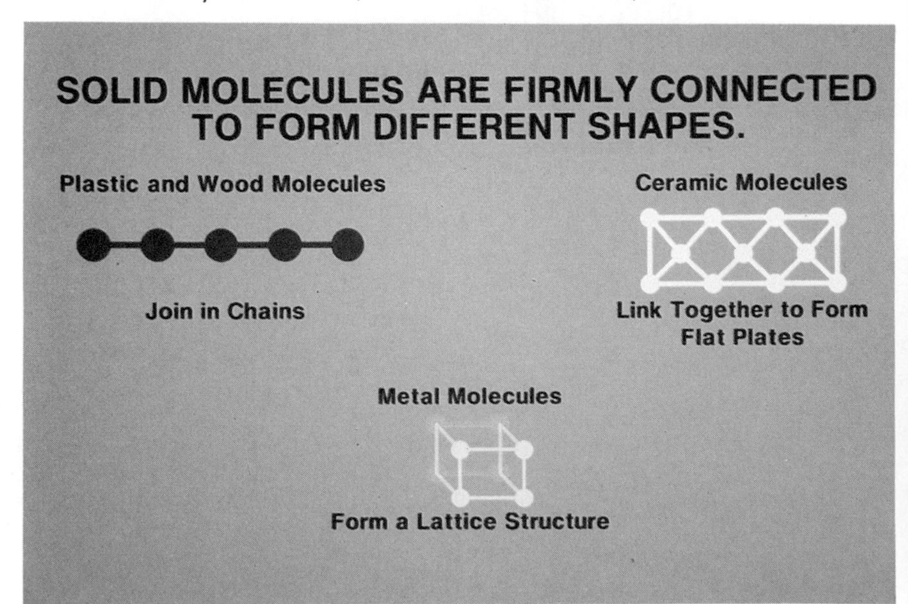

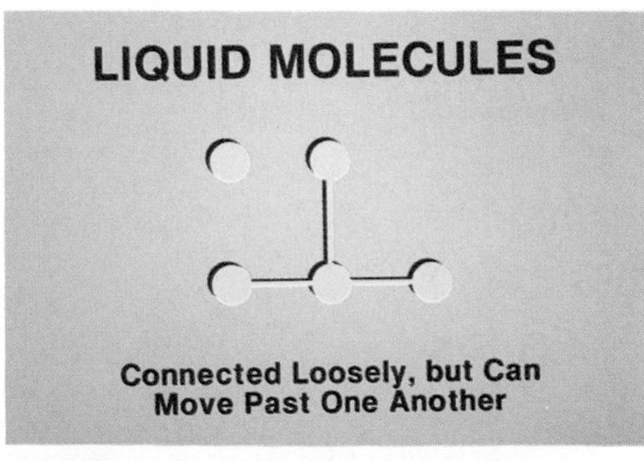

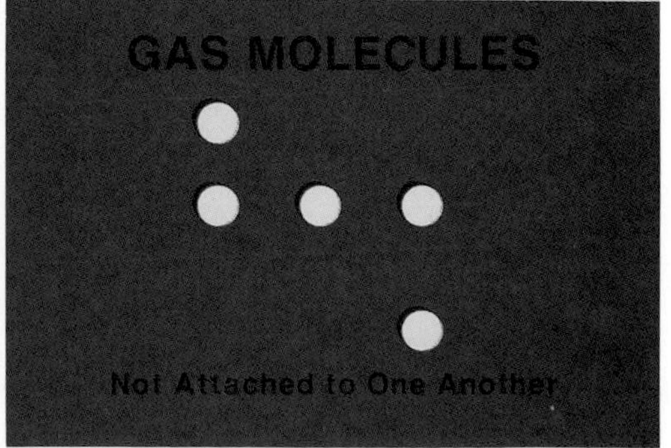

example, the microscopic molecules of clay form flat plates that easily slide over each other. This gives clay the property of **plasticity**, or the ability to be formed into shape easily and to stay in that shape.

TECHNOLOGY TRIVIA

There are at least as many molecules in a teaspoon of water as there are teaspoons of water in the entire Atlantic Ocean.

▶▶▶ **FOR DISCUSSION** ◀◀◀

1. Describe a situation in which a company might need to employ a material scientist.
2. What information can a material scientist gain by studying the chemical properties of a material?

■ **Make a display of samples of various materials available in your laboratory.**

Properties of Materials

Designers and engineers select materials very carefully. All materials have advantages and disadvantages. One way to determine whether a material is appropriate is to study its properties.

After determining how a material reacts under certain conditions, engineers may or may not use that material for a given project. For example, glass is a brittle material that has very little flexibility. Would you recommend that a manufacturer build a skateboard from a pane of window glass?

▶▶▶ **FOR DISCUSSION** ◀◀◀

1. Many natural body parts can now be replaced by artificial parts. What material properties would a metal knee joint have to possess?
2. Describe what happens to the molecules of frozen water when the ice melts and becomes a liquid.

■ **If you were to build a model of a structure, what materials might you use?**

CHAPTER 5 Review

Chapter Highlights

- Materials have always been an important resource for the development of new technologies.
- Natural materials are found in nature; synthetic materials are made by people.
- Not all natural resources are renewable.
- Most materials can be organized into the following groups: wood, metals, ceramics, plastics, and composites.
- New materials are developed by material scientists to meet new needs.
- The properties of materials determine how they will react when they are used.

Test Your Knowledge

1. Why was iron an important advancement over bronze in material science?
2. List three renewable and three nonrenewable materials.
3. List three natural resources and the raw materials with which they provide us.
4. Define *synthetic*.
5. Balsa wood is a very soft wood, but it is classified in the hardwood family. Explain this.
6. What are the two broad classifications of metals?
7. Define *alloy*.
8. How are thermoplastics and thermoset plastics different?
9. What is a molecule?
10. Explain the property of plasticity.

Correlations

SCIENCE

1. Carbon fiber composite is a very lightweight material. What are some practical uses for this material?

MATH

1. A two-ounce "Choc-o-nuts" candy bar claims to be 25 percent nuts. How many ounces of nuts are in the candy bar?

LANGUAGE ARTS

1. In a paragraph compare and contrast the use of raw materials with synthetic materials for clothing. Keep in mind cost, cleaning process, and durability.

SOCIAL STUDIES

1. Find out what natural resources Americans used to run their homes in the 1800s. How many of these natural resources do we still use today? How do we use natural resources differently than we did one hundred years ago?

CHAPTER 6

Information Resources

Introduction

Each of the four families of technology (communication, transportation, production, and biotechnology) consumes and produces large amounts of information. When we create technology, this information directs what we do and how we do it.

Information resources are needed to produce all products and services. This chapter will continue your study of the importance of information resources to technology.

After reading this chapter, you should be able to

Discuss the role of information as a resource for technology.

Define information technology.

Give examples of how information is gathered, stored, and moved.

Describe how a computer processes information.

Words you will need

data
data processing system
information technology
input
output
program
network
terminal
database
central processing unit (CPU)
binary code

Information Technology

What kinds of information would you need to start your own disk jockey entertainment business? You would first want to determine whether there is a need, or *market*, for your service. You might ask friends who are planning parties if they would consider using a deejay for music entertainment.

If, after gathering and studying this information, you decide your services are in demand, you would need to collect still more information. What kind of equipment would you need? Is it expensive to buy or could you rent it? How would you get the money to invest in your business?

After getting your fancy equipment, how would you operate it? What kind of music should you purchase: jazz, rock 'n' roll, dance music? This would depend on the interests and preferences of the people who would hire you.

Would you advertise? What is the best method of advertisement? How much should you charge for an evening of music?

As you can see, a great deal of information is required for the simplest of services. Think about the amount of information needed to manufacture automobiles or to run a large produce farm or a telephone company.

TECHNOLOGY TRIVIA

The automatic telephone switchboard was invented and patented in 1891 by an undertaker in Kansas City because he suspected that telephone operators were being paid by rival undertakers not to connect his calls.

Information starts out as **data**, or facts. In the disk jockey example, when you questioned your friends to see how many were interested in a deejay, you were gathering facts.

In technology, people gather data about a variety of topics. Market researchers gather data about what customers need and want. Fig. 6–1. Designers and engineers gather data by making working models, or prototypes, of products and testing them. Fig. 6–2.

Fig. 6-1. Market researchers interview people, conduct phone surveys, or demonstrate new product ideas to gather information about customers' likes and dislikes.

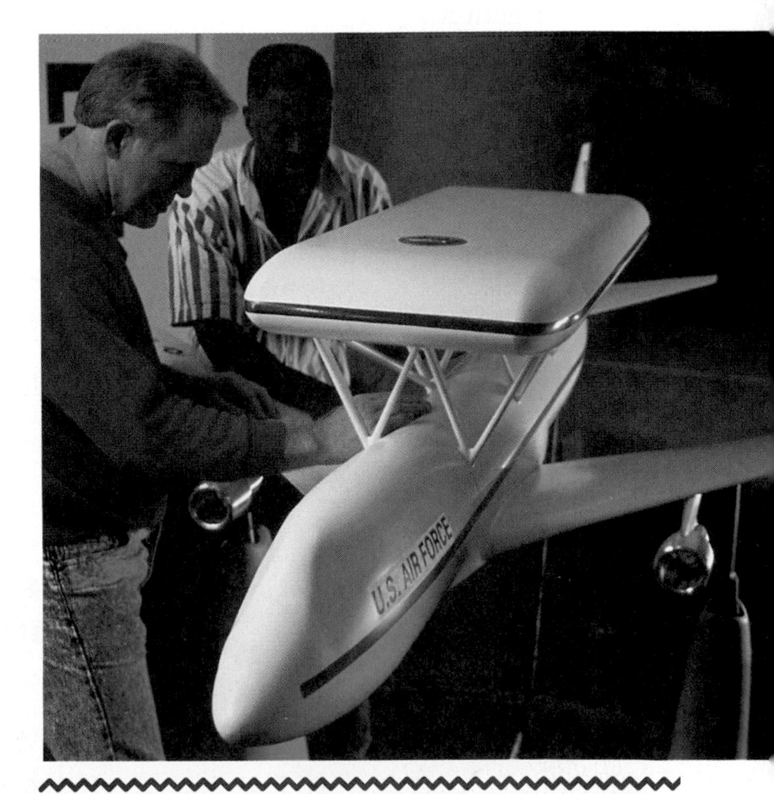

Fig. 6-2. Valuable information can be gathered by testing models or prototypes. Here, engineers are testing an aircraft design in a wind tunnel.

Material scientists gather data about the properties of different materials by conducting experiments on the materials. Fig. 6–3.

Technicians gather data about how tools and machines process different kinds of materials. The government tests products to gather data about the safety of a product and its impacts on the environment.

A **data processing system** is any system used to organize raw data into information. The data has very little meaning until it is organized into information. Not very long ago, most data was processed into information by hand in the form of documents and books. Today, the volume of data is so great that electronic methods are used to process it. Fig. 6–4.

Information technology provides a means of handling data and information. Different information systems can collect, organize, store, and send information in many forms. Data processing systems electronically process words, images, numbers, and sounds into information. Fig. 6–5.

TECHNOLOGY TRIVIA

The time delay in many international telephone conversations is caused by the vast distances the signals must travel. Even at the speed of light (186,000 miles per second) the signals take about one-fourth of a second to reach a communication satellite and bounce back to earth.

Fig. 6-3. Materials are selected for products based on information obtained by different types of testing. These people are testing the compression strength of cardboard corners.

Fig. 6-4. Data processors are people who enter data or facts into a computer. With the help of the computer, they organize the facts into information.

▶▶▶ FOR DISCUSSION ◀◀◀

1. If you were to conduct a telephone survey to gather information about America's favorite rock group, what questions would you ask?
2. Modern farming technology relies on a great deal of information. What types of information do you think would be helpful to a large-scale farm?

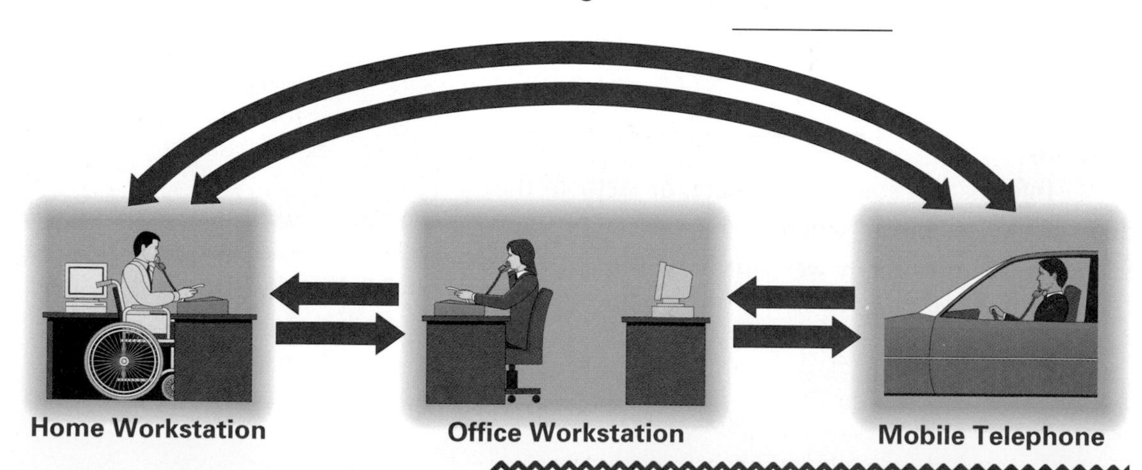

Fig. 6-5. Information technology allows us to link many smaller systems into a single, more powerful system.

Processing Information with Computers

People use computers to collect, organize, store, and move information. The process of placing data into a computer is called **input**. The computer processes the data into information according to a **program**, or set of computer instructions. The information that the computer provides is called **output**.

To get an idea of how computers can process large amounts of information, let's suppose we were going to "computerize" your school. First, we would create a computer network. A **network** links many computers to a central controlling computer and to each other. Fig. 6–6. In this case, let's suppose each member of your class would have a **terminal**, or computer station, at home that would be connected to a central computer. Your teacher would input the day's lesson onto the central computer.

You could retrieve the lesson information from your terminal at home. After reading the lesson, you could ask questions by leaving an electronic letter in your teacher's electronic mailbox. Lessons, quizzes, homework, and most other school assignments could become part of the information you and your teacher exchanged through the central computer.

The speed with which the central computer can handle information would allow all the students to work at the same time. Thousands of information transfers can occur within just a second or two.

IMPACT

Computers are helping to fight crime. With the help of computer networks, police can find out if a suspect they have arrested is also wanted in another city or state. Information about missing persons can also be distributed on a computer network. Computers have even been used to draw pictures of suspects based on the descriptions of witnesses.

▶▶▶ FOR DISCUSSION ◀◀◀

1. What type of data might an airline ticket agent input into the airline's computer terminal?
2. What type of information might an airline ticket agent be able to retrieve from the airline's computer network?

Extension
Activity

■ **Interview at least three people who are employed in different types of jobs. Find out how computers are used in each person's work. Share your findings in a class discussion.**

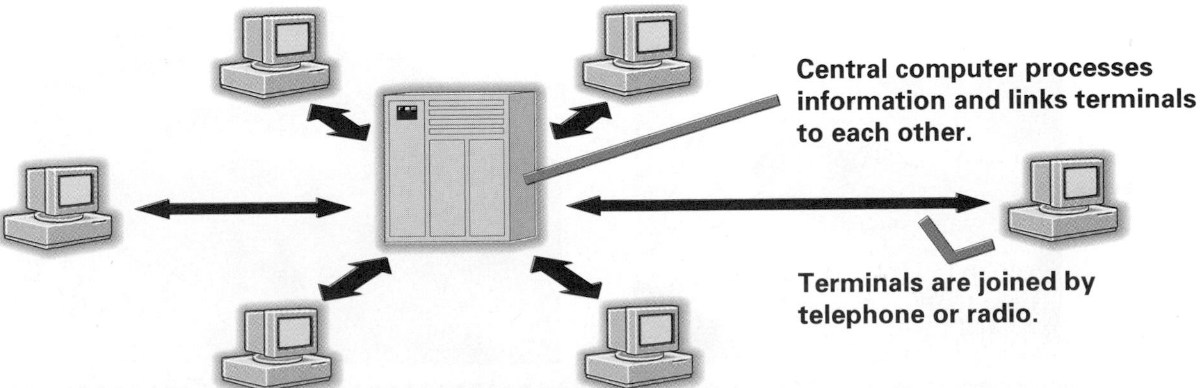
Central computer processes information and links terminals to each other.
Terminals are joined by telephone or radio.

Fig. 6-6. Computer networks link several computers to a central computer. Even terminals that are separated by thousands of miles can be joined in a network.

Collecting, Storing, and Retrieving Computer Information

Different technologies require that data be handled in different ways. For example, an automatic camera uses light sensors to measure the amount of light in a room. Fig. 6–7. Using this data, its computer calculates the proper exposure. A video game uses a joystick to enter your responses, as data, into the game's computer. In order to handle data and information in different forms, information technologies have created many devices.

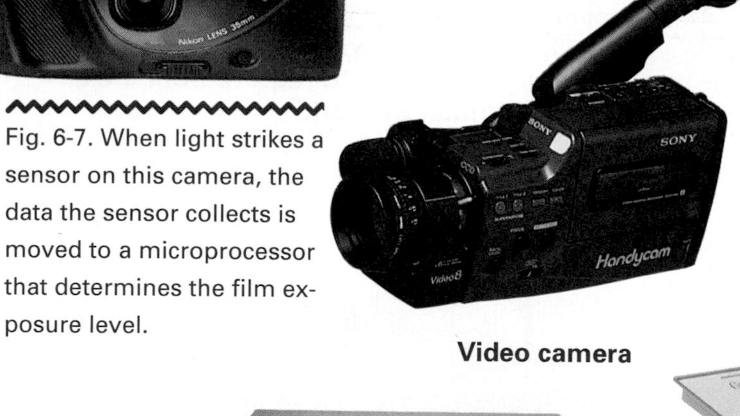

Fig. 6-7. When light strikes a sensor on this camera, the data the sensor collects is moved to a microprocessor that determines the film exposure level.

Video camera

Information Collection Devices

Computers collect data through input devices. The keyboard is the most common input device. However, many other types of input devices have been designed to meet specific input needs. Fig. 6–8.

Keyboard

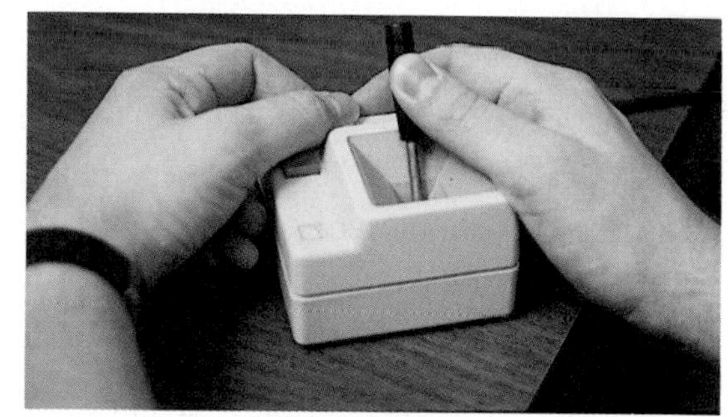

Joystick

Digitizer

Disk drive

Mouse

Bar code scanner

Fig. 6-8. Because data comes in many different forms, inputting data requires many different kinds of input devices.

For example, a bar code scanner can enter data into a computer. The scanner reads coded information from a series of black lines on a white background. These lines are called a *bar code*. Bar code scanners make data input faster in some situations.

Information Storage Devices

Computer data is usually stored or recorded on magnetic floppy disks or hard disk drives. These disks store information until you instruct the computer to retrieve the information. Figure 6–9 shows other information storage methods.

Computer Output Devices

After data is processed, you can retrieve it from the computer in a variety of forms. For example, if you asked a librarian for books written about camping, he or she might conduct a search through the library computer's database. A **database** is a collection of information organized around a topic.

After the computer located all the camping titles in its memory, it would display them on the monitor screen. If you wanted a copy for yourself, the librarian could send the output to a printer.

▶▶▶ FOR DISCUSSION ◀◀◀

1. What do you think are some of the hazards of electronic information processing?
2. A public computer in a shopping mall has a "touch screen" monitor. People touch different areas of the screen to call up information. Is this monitor an input device or an output device? Explain your answer.

Compact disk

Hard disk (inside the drive)

Cassette tape

Floppy disks

Fig. 6-9. Computers can use many different kinds of storage devices.

Inside the Computer

How does a computer work? It's a mystery to most people. All computers, no matter how big or small, work in the same way and contain four basic parts. Fig. 6–10.

Parts of a Computer

First, you have to give the computer some instructions. You do this by using an input device. The "brain" of the computer is the **central processing unit (CPU)**. This unit carries out the instructions. The computer's memory unit stores these instructions for the computer's use. The output device displays the results of the computer's work.

Binary Code

Data is passed back and forth through the parts of the computer as bursts of electricity. The bursts make up a code that represents the words and numbers entered into the computer.

The code is called **binary code** because it is based on the *binary* number system. In binary, only two numbers are used: 0 and 1. These numbers are represented in binary code by combinations of short bursts of electricity and "spaces" of no electricity. This system works much like Morse code. The computer sees all information as a string of bursts of electricity. Fig. 6–11. The input and output devices change words and numbers into binary code and then back again into words and numbers.

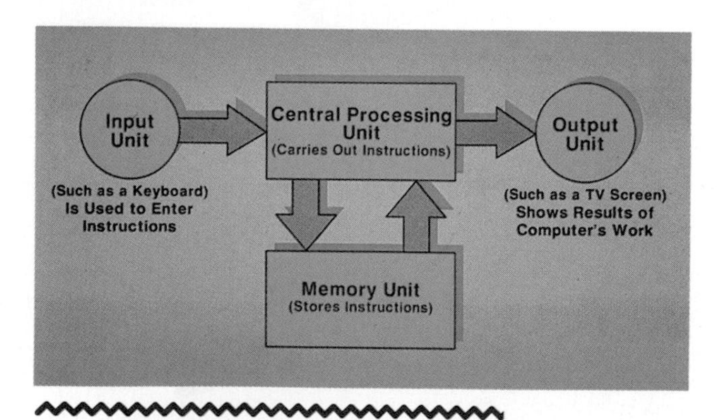

Fig. 6-10. Computers, calculators, and microprocessors all work the same way.

CHANGING WORDS AND NUMBERS TO COMPUTER LANGUAGE

Letter or Number Entered into Computer	Binary Code	Electrical Code
"B"	1000010	1 0 0 0 0 1 0 on off off off off on off
"9"	1001	1 0 0 1 on off off on

Fig. 6-11. Computer binary code is made up of 1s and 0s. 1s represent a burst of electrical energy; 0s are "spaces" of no electricity. Electronic parts inside the computer act as switches to turn the bursts of energy on and off.

▶▶▶ FOR DISCUSSION ◀◀◀

1. A calculator has the same basic parts as a computer. Explain this.
2. How could you use the lights in your room to signal a binary code?

■ Using a word processor or database, make a list of all tools and materials available in your laboratory. Give those that could be used for model-building a special code. Print a list of possible model-building tools and materials.

CHAPTER 6

Chapter Highlights

- All technologies depend on information resources.
- Information technology is any electronic means of gathering, storing, organizing, and moving information.
- The computer is the main tool used in information technology.
- Data and information can be collected, stored, and retrieved in many ways.
- A computer changes data into bursts of coded electrical energy.

Test Your Knowledge

1. How is data different from information?
2. What do we call a system used to process facts into information?
3. Define *information technology*.
4. What type of data might an engineer need to gather when developing a product?
5. Describe a computer network.
6. List three places in which computer networks might be used.
7. Why are bar code scanners used at supermarket checkout counters?
8. List three computer output devices.
9. What does a computer program do?
10. What four parts do all computers have in common?

Correlations

SCIENCE

1. Find out how a bar code scanner works.

MATH

1. The binary number system is based on powers of 2 and uses only 0's and 1's. See if you can count to ten in this system. The chart below will help.

Powers:	2^3	2^2	2^1	2^0
Values:	8	4	2	1
1=				1
2=			1	0
3=			1	1
4=		1	0	0

LANGUAGE ARTS

1. Many grocery stores are now using computerized cash registers. The bar code on packages is scanned for price. What do you think are the benefits to the store and to the consumer? Write your ideas in a brief essay.

SOCIAL STUDIES

1. When were computers first developed? What technology had to be developed before it was possible to build electronic computers?
2. If you have computers in your home or school, how are they used?

CHAPTER 7

Energy Resources

Introduction

When people create and use technology, they consume energy—lots of energy. Energy is the force that makes all things move and work. Our people-made world is an energy-hungry machine.

We consume energy to transport people and products from place to place. We rely on energy to fuel the engines in our cars, trucks, trains, planes, and ships.

We consume energy to make electricity. Electricity feeds our homes, businesses, schools, and hospitals with power. Machines, appliances, and electric lights gobble up electricity at an enormous rate.

We consume energy when we change materials into products. Just think of the heat energy required to make steel, melt glass, and create food products. This chapter will help you become more familiar with our energy needs and how we fulfill them.

After reading this chapter, you should be able to

Define energy.
List the major forms of energy.
Discuss the major sources of energy.
Explain energy conversion.
Discuss energy-related problems.

Words you will need

energy
mechanical energy
chemical energy
atomic energy
kinetic energy
potential energy
law of conservation of energy
fossil fuels
photovoltaic cells
generator
hydroelectric plants
nuclear fission
nonrenewable resources

Work and Energy

What is energy? **Energy** is the ability to do work and create movement. People are able to do work because of the energy they receive from food. When your body moves, it uses energy. When your body plays or works hard, it uses greater amounts of energy.

Technology uses energy for the same reasons. Fig. 7–1. Automobiles get energy by burning gasoline. Energy allows a car to move. To make the car move faster, the engine has to work harder. The harder the engine works, the more energy it consumes.

▶▶▶ FOR DISCUSSION ◀◀◀

1. Which of the four families of technology (communication, transportation, production, or biotechnology) do you think consumes the most energy? Explain your answer.
2. Make a list of all the devices you use each day that require some form of energy.

Products

Transportation

Energy

Fig. 7-1. Of all the energy we consume in this country: 37% is used to create products, 36% to transport people and goods, 16% to heat and power our homes, and 11% to run businesses.

Forms of Energy

Energy comes in a variety of forms. Work can be accomplished using any one or a combination of forms of energy. The most common forms of energy are:

- mechanical energy
- electrical energy
- light energy
- heat energy
- chemical energy
- sound energy
- atomic energy

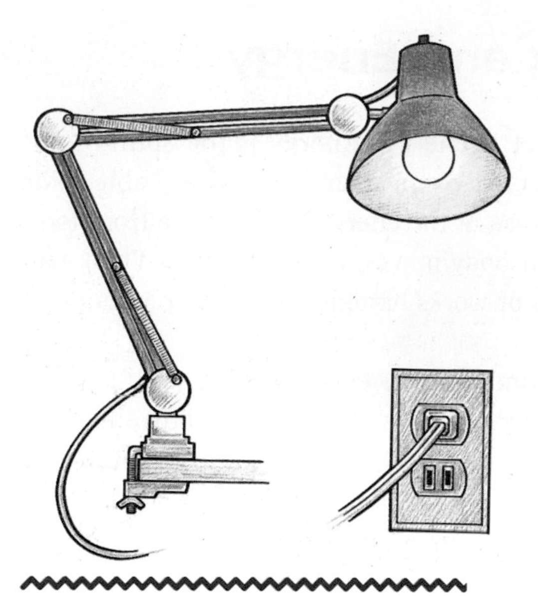

Fig. 7-3. Electrical energy is the flow of electrons through a wire.

Relationships Among Forms of Energy

Electric motors consume electrical energy for power. Fig. 7–2. These motors provide mechanical energy for many types of machines. **Mechanical energy** is the energy found in moving things. Electrical energy can also produce light energy, although the sun provides most of the light energy on earth. Fig. 7–3.

We have all experienced heat energy. Rub your hands together quickly and feel the heat build up from the friction.

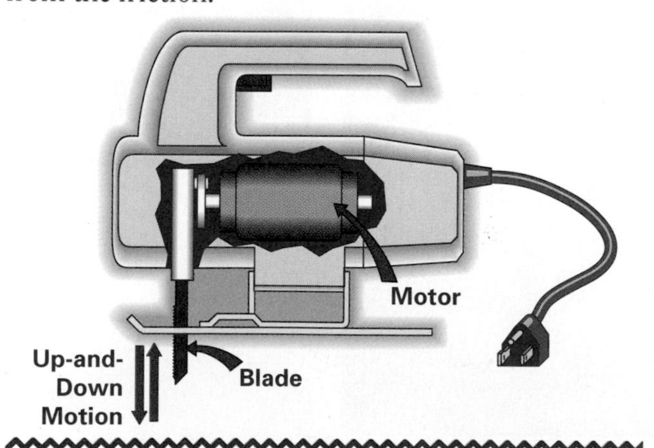

Fig. 7-2. Mechanical energy is the energy of motion. The motor in this saber saw uses electrical energy to create mechanical energy, which moves the blade up and down.

Heat from friction is also what causes a match to light. When the match head is rubbed against the striking surface, the heat releases the **chemical energy** stored in the match head. The match ignites, giving off heat and light. The sound made by striking the match is really sound energy vibrating the molecules of the surrounding air. Fig. 7–4.

Atomic energy is the energy stored in the nucleus of an atom. Nuclear reactors release a great deal of energy by splitting the nuclei of atoms. Fig. 7–5.

Energy Conversions

Changes in energy forms are called *energy conversions*. The most common energy conversion involves potential and kinetic energy. **Potential energy** is energy at rest, or stored energy. **Kinetic energy** is energy put into motion. Potential and kinetic energy changes take place all around us. Fig. 7–6.

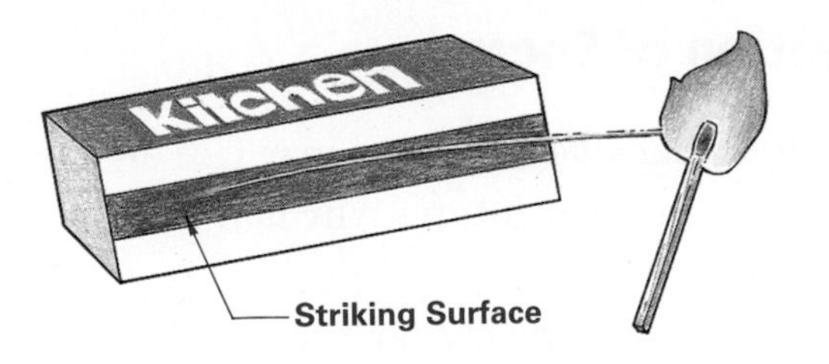

Fig. 7-4. Heat energy is often used to release the energy stored in chemicals. Sound energy vibrates the molecules in the air, creating the noise we hear.

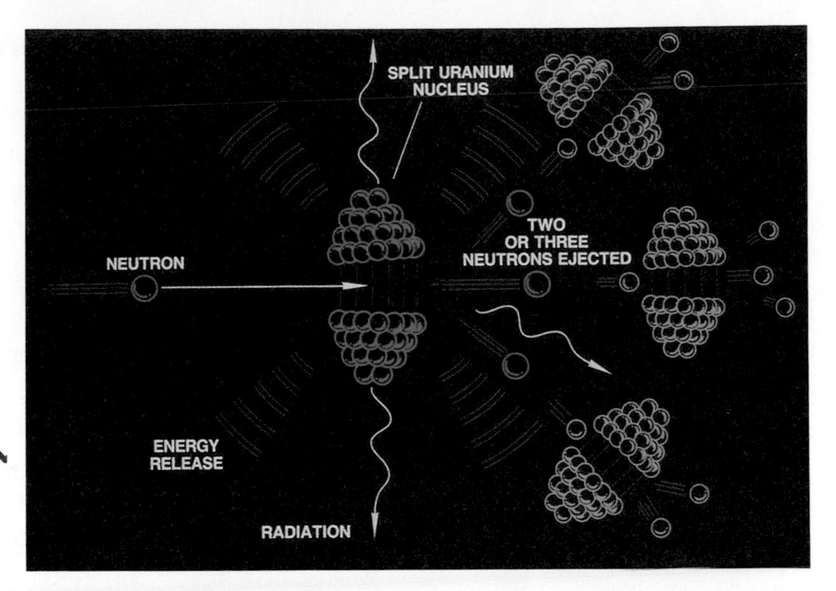

Fig. 7-5. A great deal of energy is released when an atom splits. Uranium-235 is often used in nuclear reactors.

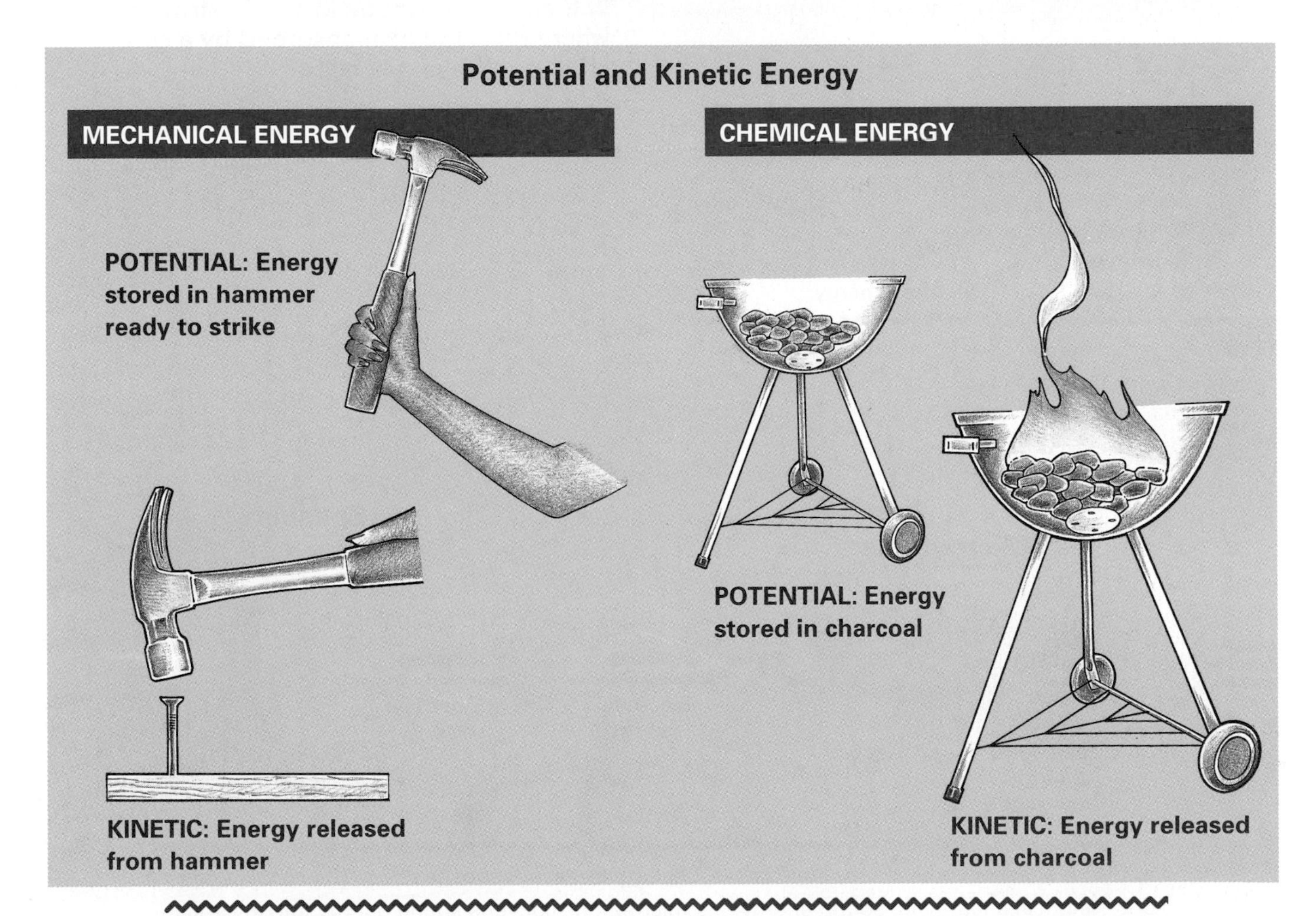

Fig. 7-6. All stored energy is potential energy, regardless of the form (mechanical, electrical, chemical, etc.). Kinetic energy is any form of energy in motion.

Other Energy Conversions

Energy changes other than potential and kinetic changes also take place. When a hammer hits a nail, for example, mechanical energy is changed into sound energy.

Have you ever felt a nail after it has been pulled from a board by a hammer? It's hot! Mechanical energy has been changed to heat energy.

In some cases, many energy changes occur to do what seems to be a simple task. For example, how many energy changes take place when you use a flashlight? Figure 7–7 shows these changes. Other examples of energy conversions are shown in Figure 7–8.

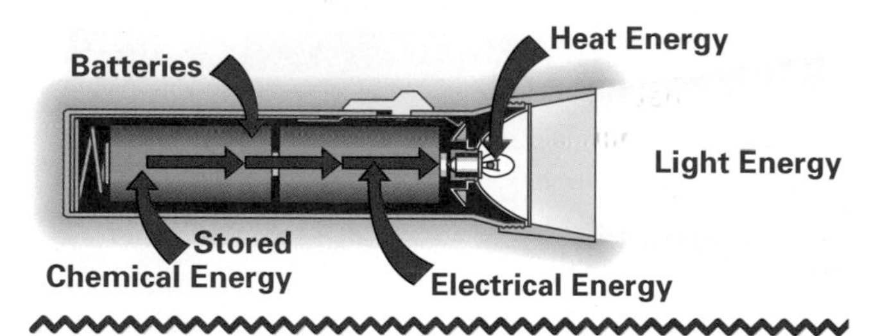

Fig. 7-7. Several energy conversions take place every time you switch on a flashlight.

Conservation of Energy

We often hear that the world may face an energy shortage in the near future. Why don't we create new energy?

We don't create new energy because it is not possible. Energy can be changed from one form to another, but it cannot be created or lost. This is called the **law of conservation of energy**.

►►► FOR DISCUSSION ◄◄◄

1. **List the different forms of energy changes that take place during a bicycle ride. Start with the breakfast you eat before the ride.**
2. **If energy cannot be lost or destroyed, where does the heat absorbed by a concrete driveway go at night?**

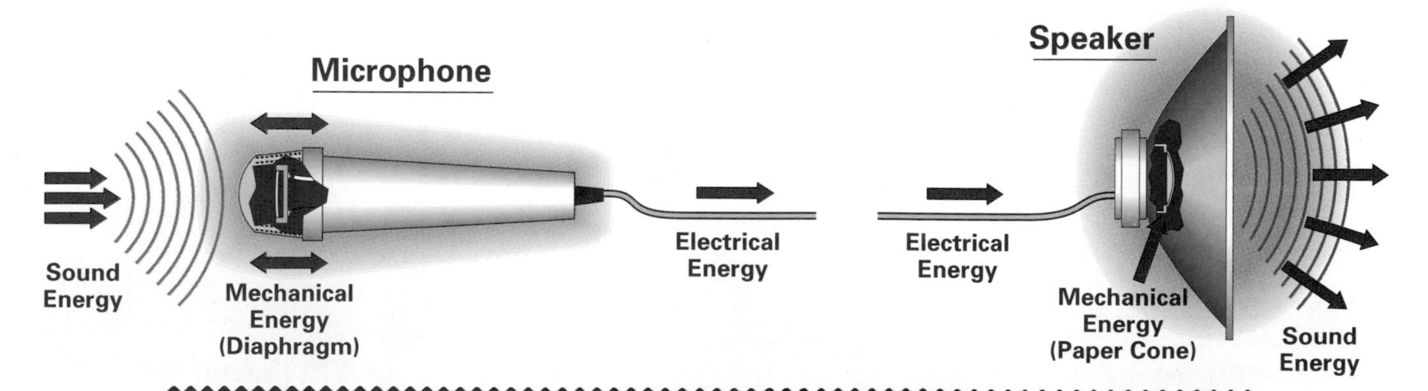

Fig. 7-8. A speaker and a microphone convert energy forms in opposite ways. The speaker converts electrical energy into sound energy; the microphone converts sound energy into electrical energy.

Where Does Energy Come From?

Where do we get the chemical energy we need to power an automobile? How do we change mechanical energy into the electrical energy we use at home? We are dependent on energy resources supplied by nature. Using technology, we have created methods for converting these resources into more useful energy forms.

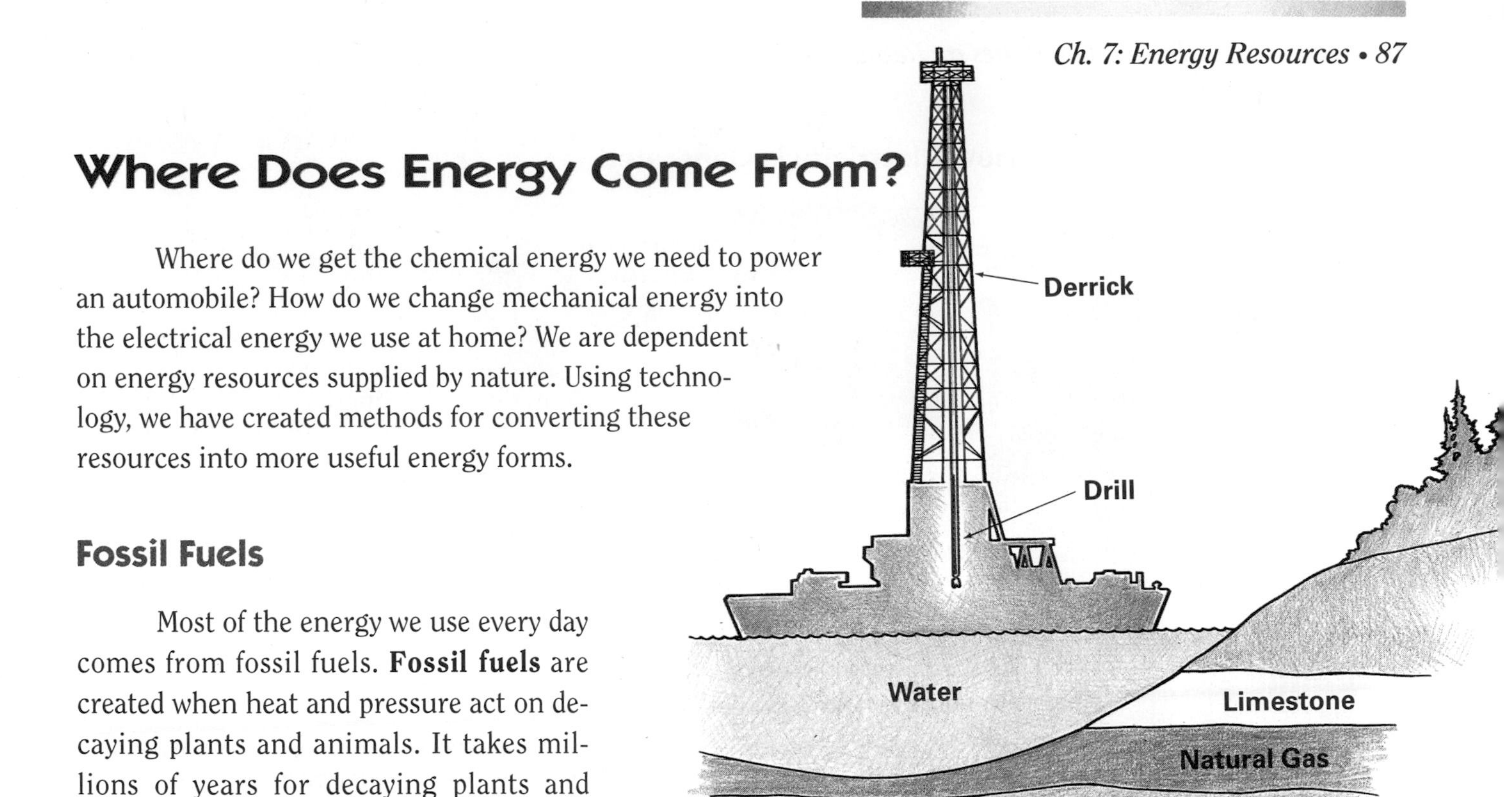

Fig. 7-9A. Oil and natural gas are created from decaying plants and animals. Millions of years of pressure and heat change the plant and animal remains into oil and natural gas.

Fossil Fuels

Most of the energy we use every day comes from fossil fuels. **Fossil fuels** are created when heat and pressure act on decaying plants and animals. It takes millions of years for decaying plants and animals to become fossil fuels. Fig. 7–9 A and B.

The three main fossil fuels are oil, coal, and natural gas. Fossil fuels provide us with gasoline, oil, and kerosene. When fossil fuels are combined with oxygen, they burn extremely well. This makes them very valuable as a source of energy. Figure 7–10 shows how fossil fuels are used to generate electrical energy.

Fig. 7-9B. Coal is created when decaying plant and animal remains are compressed under sand, stone, or clay for millions of years. There are four types of coal: peat, lignite, bituminous, and anthracite.

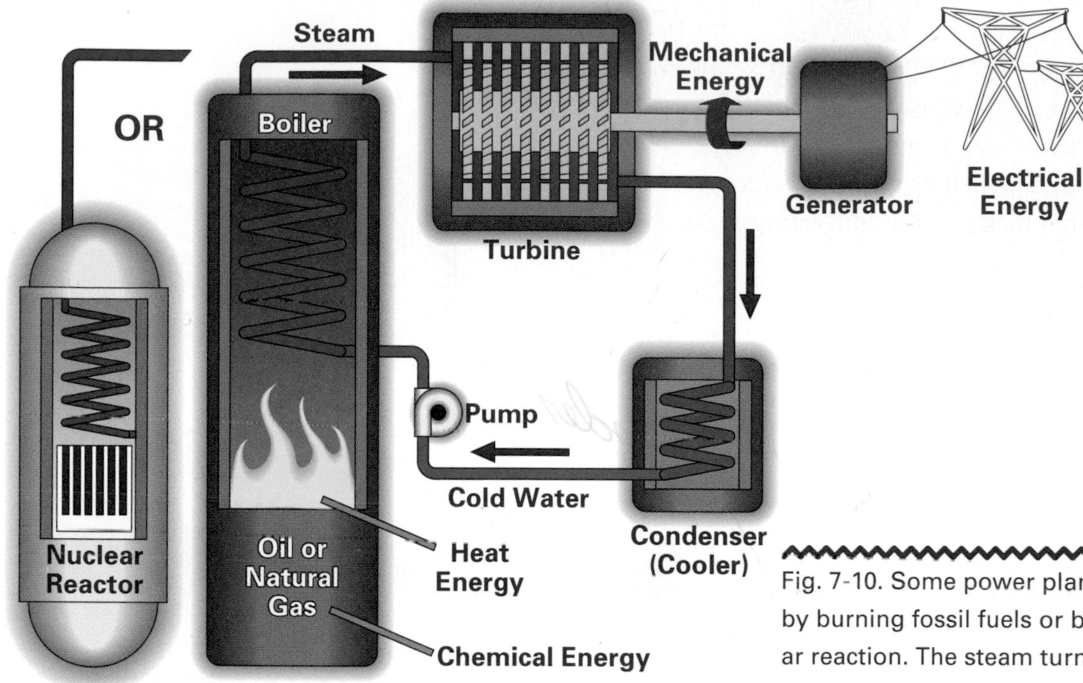

Fig. 7-10. Some power plants produce steam by burning fossil fuels or by starting a nuclear reaction. The steam turns a turbine that is connected to a generator. The generator changes mechanical energy into electrical energy.

Solar Energy

The light energy from the sun can be used to create heat energy and electrical energy. **Photovoltaic cells** (solar cells) change light energy directly into electrical energy. Fig. 7–11. Your calculator may use solar cells to provide electrical energy. The sun also provides a great deal of heat energy that can be used for heating air and water. Fig. 7–12.

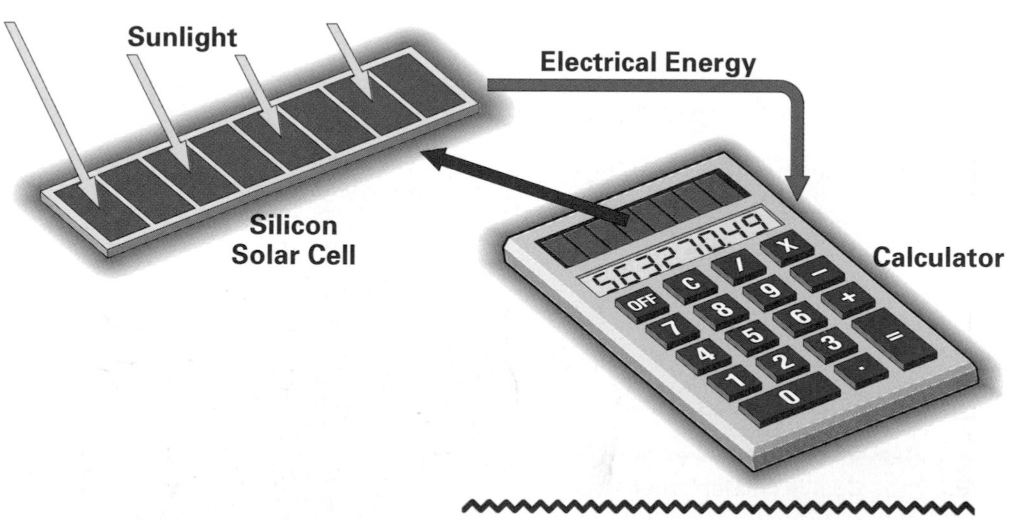

Fig. 7-11. A solar cell changes light energy directly into electrical energy.

Fig. 7-12. Solar collectors on the roof of this building change light energy into heat energy, supplying most of the hot water needs of the owner.

Wind Power

Wind is created by the uneven heating of the earth by the sun. Wind has been used for centuries to create mechanical energy.

Today, wind power is used to turn generators. **Generators** change mechanical energy into electrical energy. Fig. 7–13.

Fig. 7-13. Several types and shapes of blades can "catch" the wind to power generators. Some blades change position with the direction of the wind.

Water Power

The mechanical energy in falling or flowing water has also been used for centuries to do work. Today, **hydroelectric plants** provide much of our country with electricity. Tidal generators and wave generators also use moving water to create electricity. Fig. 7–14.

> **TECHNOLOGY TRIVIA**
>
> Most early factories were built near water. Waterwheels, pulleys, and belts transferred the energy of the moving water to the factory's machines. During the 1700s, steam engines began to replace waterwheels.

Nuclear Energy

Nuclear reactors produce electrical energy by harnessing the energy in an atom. **Nuclear fission** takes place in a nuclear reactor. Fission is the splitting of the nucleus of an atom into smaller nuclei. During fission, a tremendous amount of heat is released. This heat is used to create electricity. Fig. 7–10.

■ Would wind, water, or solar energy be most appropriate for use in your community? Build a small model of a device for the energy source that you selected.

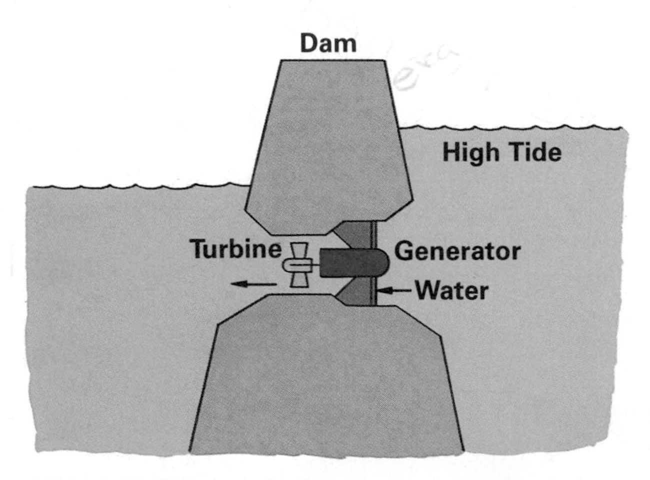

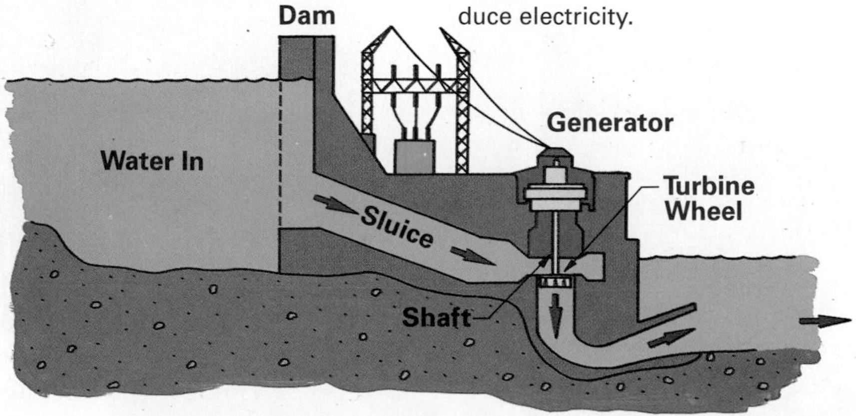

Fig. 7-14. The mechanical energy provided by moving water is used to turn a turbine. The spinning turbine is connected to a generator to produce electricity.

Alternative Energy Resources

The energy resources you just learned about seem to satisfy all our energy needs. Why should we look for alternative resources? There are two important reasons.

Many pollution problems are associated with burning fossil fuels and the use of nuclear fission. Automobiles, power plants, and manufacturers release smoke and chemicals into our air. Smog, acid rain, and polluted natural resources are the result.

IMPACT

The burning of fossil fuels releases carbon dioxide and other gases into the air. As these gases build up, they prevent heat from escaping into space. Temperatures increase. Eventually, this "greenhouse effect" could cause drastic changes in the earth's climate.

Nuclear energy produces dangerous nuclear waste. This radioactive waste will be with us for thousands of years. Finding a place to store it is a big problem. The possibility of a nuclear accident also exists.

The second reason for finding alternative energy resources is our shrinking supply of energy sources. Our supply of oil, coal, natural gas, and nuclear fuels is limited. These sources are known as **nonrenewable resources**.

Nonrenewable resources take millions of years to replace. We cannot continue to use these resources without thinking about future generations and their energy needs. Renewable resources such as wind, water, sunlight, plants, and animal waste must become alternatives to fossil fuels. Figures 7–15 and 7–16 show some alternative energy resources and how they might be used.

Fig. 7-15. Plants and animal waste make up the energy source known as *biomass*. After processing, these materials can be burned to provide energy.

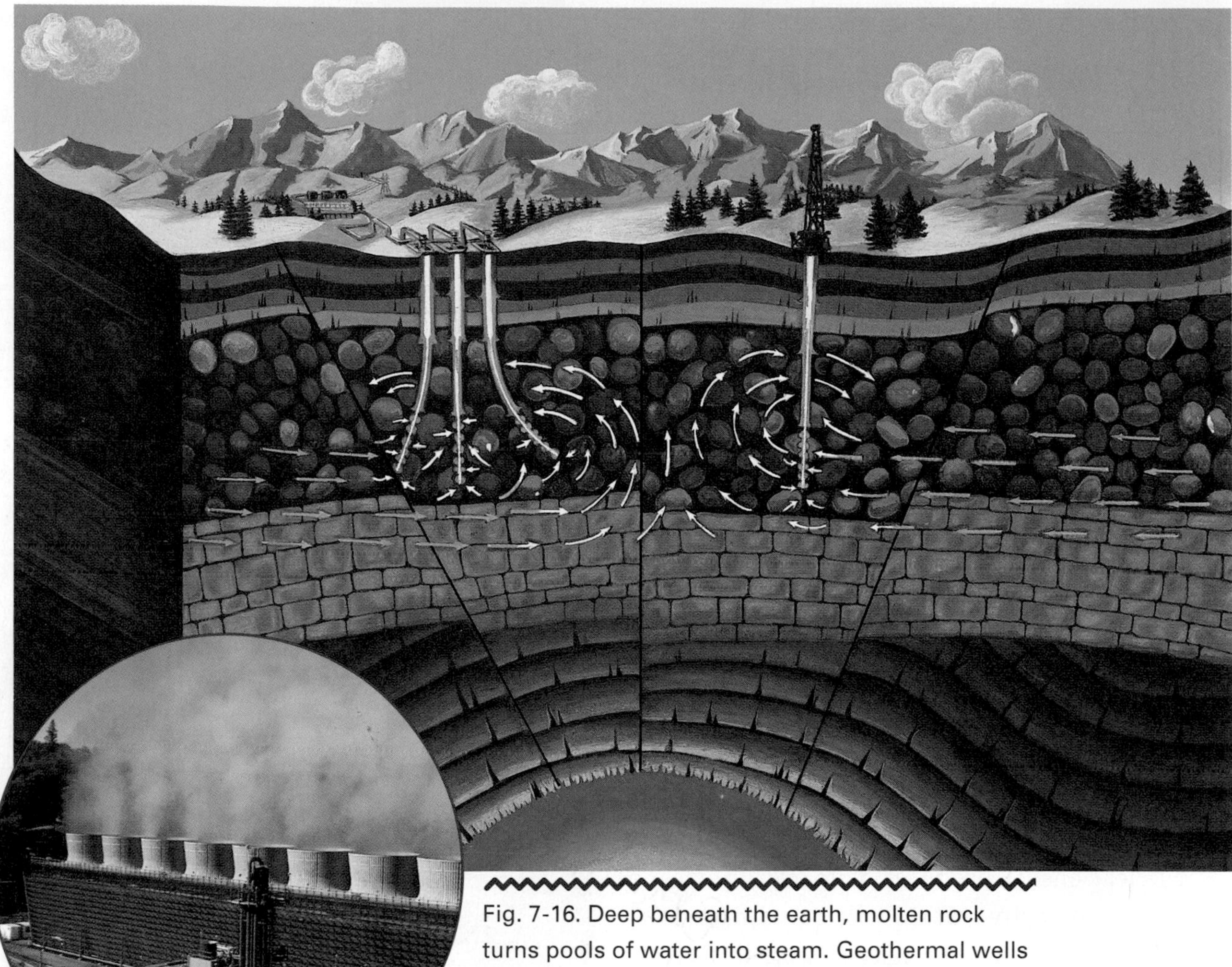

Fig. 7-16. Deep beneath the earth, molten rock turns pools of water into steam. Geothermal wells use this natural heat energy to provide steam to turn turbine generators.

▶▶▶ FOR DISCUSSION ◀◀◀

1. Solar energy seems like a great source of free heat and electricity. Can you think of any problems that might occur?
2. What would be your reaction if you found out a nuclear power plant was going to be constructed in your community?

■ Collect newspaper and magazine articles about alternative energy resources. With other members of your class, organize these into categories and post them on a bulletin board.

CHAPTER 7 Review

Chapter Highlights

- Technology requires large quantities of energy to make electricity, transport people and products, and process materials.
- Energy is the ability to do work and cause movement.
- Energy can be changed from one form to another, but it cannot be created or destroyed.
- Energy sources include fossil fuels, solar power, water power, wind power, and nuclear power.
- Alternative energy sources are needed to replace sources that are becoming scarce or that pollute our environment.

Test Your Knowledge

1. Why is energy such an important resource of technology?
2. Define *energy*.
3. Give two examples of potential energy.
4. Explain how the examples of potential energy you listed for question 3 could become kinetic energy.
5. List and describe four common forms of energy.
6. List and describe the energy conversions that take place when you turn on your Walkman.
7. Why do we consider fossil fuels to be nonrenewable?
8. Describe how hydroelectric plants generate electricity.
9. What are some of the negative environmental impacts resulting from energy production?
10. Describe two alternative energy sources.

Correlations

SCIENCE

1. In many chemical reactions, heat is given off. Is this always true? Put a thermometer into a 16-oz. glass with a quarter cup of vinegar. Record the vinegar's temperature. Now add a teaspoon of baking soda. What happens to the temperature?

MATH

1. Matt's car averages 23 miles per gallon of gas. He fills the 18-gallon tank and starts a 453-mile trip to his grandmother's house. Will he get there without stopping to refuel?

LANGUAGE ARTS

1. Imagine a nuclear power plant is to be built in an area near your home. What would be the concerns raised in your community? Hold a class discussion about the advantages and disadvantages.

SOCIAL STUDIES

1. Find out the various forms of energy used to heat American homes around 1900–1940. What kinds of problems were encountered with these? Which ones turned out to be most efficient?
2. Do you think any of those early methods of heating homes were better than today's? Why or why not?

CHAPTER 8

Capital and Time

Introduction

Capital and time are the hidden resources of technology. Their role in technology may not be as visible as people, materials, energy, machines, and information, but they are just as important.

When we say the word *technology*, what comes to mind? Most people think of fancy machines and robots. Some people see products made of new synthetic materials. Other people think about satellites and instant communication. Your parents may think about the rising costs of energy needed to run your home.

The purpose of this chapter is to help you understand the importance of capital and the role of time in technology. Have you ever considered the amount of money transferred from place to place in the production of a product? Have you ever heard the phrase "time is money"? Is there a connection between time and money in technology?

After reading this chapter, you should be able to

Define capital.
Discuss how money is spent in technology.
List some sources of capital.
Discuss the role of time in technology.

Words You Will Need

finance
capital
interest
stock
dividend
cash flow
profit

Capital Resources

What is capital? Is it just money? How do we calculate how much a business is worth? How do we raise the money to start a company? These are all questions about finance. **Finance** is the management of money. People who run businesses must manage their money wisely.

Money is only one form of capital. **Capital** includes all of the buildings, properties, equipment, and goods a business owns. Fig. 8–1.

In Chapter 6, we discussed starting your own disc jockey entertainment service. After being in business for one year, what capital resources might you have? The money you made would certainly be considered capital. Your equipment and record collection, and even the van you use to travel from job to job would be capital.

▸▸▸ FOR DISCUSSION ◂◂◂

1. Your school is like a business. List all the things in your school that could be considered capital.
2. Make a list of the capital you would need for an after-school business.

Money

Supplies

Equipment

Land

Vehicles

Buildings

Fig. 8-1. A company's wealth, or capital resources, may include cash, buildings, equipment, land, goods, vehicles, and any other items the company owns.

■ A donor has given the city of Glenville $500,000 for community improvements. Have your class act as the city council. For what projects might the money be spent?

Sources of Capital

The producers of technology need capital, usually in the form of money, to purchase other resources. Fig. 8–2. Materials, energy, equipment, and salaries cost a great deal of money. Where does this money come from?

Let's imagine that your deejay business is doing so well that you wish to expand it. Your plan is to hire a second deejay, provide the deejay with equipment, and make even more money. How could you finance this expansion?

You could use your personal savings as start-up money. You might also consider a bank loan. When you borrow money from a bank, you have to pay interest. **Interest** is a fee charged by the bank for the use of its money. Fig. 8–3.

Many large corporations sell **stock** to raise capital. When you buy stock, you own a small piece, or share, of the company. Stockholders receive a **dividend**, or money for each share of stock they own. Fig. 8–4.

Fig. 8-2. Capital resources are used to purchase other resources a company needs to produce its products or services.

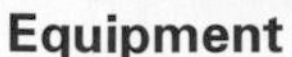

Equipment

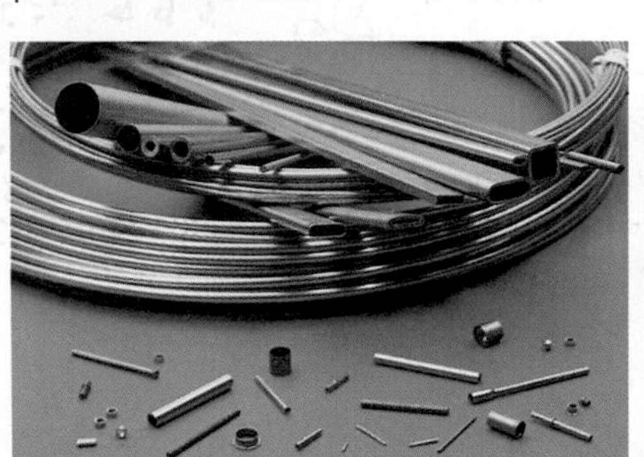

Materials

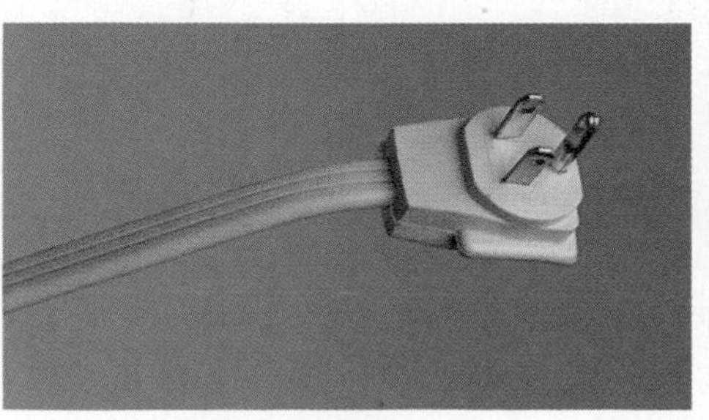

Energy

Salaries

Fig. 8-3. Before a bank loans a company money, it researches the company. Companies that have good credit histories have a good chance of getting a loan.

▶▶▶ FOR DISCUSSION ◀◀◀

1. What might be some of the risks and benefits of using personal capital to start a business?
2. If a bank charged you 5 cents interest on every dollar you borrowed, how much interest would you pay on a $2,000.00 loan?

Profit and Expenses

What kind of cash flow does your deejay entertainment business have? **Cash flow** is a comparison of a business's income and expenses. To maintain a positive cash flow, your company needs to make more money than you spend on expenses.

Your **profit** is determined by subtracting your expenses from your income. Profit is the money left over after expenses. Fig. 8–5 shows what your cash flow might look like for a single month.

Fig. 8-4. Stock certificates are issued to stockholders to show partial ownership in the company.

▶▶▶ FOR DISCUSSION ◀◀◀

1. What unforeseen expenses might occur over which a company has no control?
2. What might you do with the profits you made during your first year in business?

D.J. Entertainment Company
Monthly Income & Profit Statement

Expenses		*Income*	
Advertising	*$ 10.00*	*Party 9/12*	*$100.00*
Loan Payment	*60.00*	*School Dance 9/20*	*150.00*
New Records	*25.00*		
Repair of Equipment	*10.00*		
Transportation	*3.00*		
Total	*$108.00*	*Total*	*$250.00*

Income	*$250.00*
Expenses	*$108.00*
Profit	*$142.00*

Fig. 8-5. An income and profit statement shows your cash flow each month. Profit is calculated by subtracting expenses from income. This company made a profit of $142.00 for the month shown.

Technology and the Resource of Time

Time is a measure of how long it takes for something to happen. It is a factor that influences the production and consumption of products. Therefore, people who create technology must consider how to use time wisely.

Fig. 8-6. Companies conduct time-motion studies to see how long it takes to complete specific tasks.

Time in the Workplace

"Time is money" is a favorite phrase used by many people. In technology, it is an accurate statement. The faster products can be produced, the cheaper they can be sold. Fig. 8–6. Henry Ford used this concept when he developed his assembly line. Saving time is also one reason for the introduction of robots into the workplace. Robots work faster, cheaper, and with fewer movements than humans.

Time is money for the worker also. Many employees are paid based on how many hours they work. Employers may use punch clocks to keep an accurate record of how many hours each employee works. Fig. 8–7.

Many processes or changes done on materials require strict management of time. For example, certain plastics must *cure* (sit) for a period of time before they can be used. In biotechnology, cells may have to *incubate* (be kept at a certain temperature) for days before they start to reproduce. Fig. 8–8.

Fig. 8-7. Some employees are paid by the hour. A time clock is used to keep track of how many hours they work.

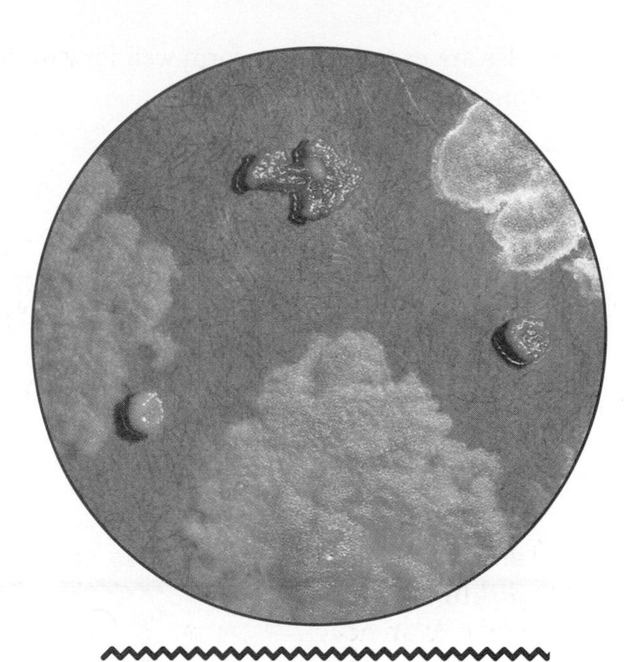

Fig. 8-8. Microorganisms grown for biotechnology may require strict timing to ensure good cell growth.

Time and the Consumer

People who use the products of technology are also dependent on the measurement of time. When you buy a roll of film, do you check the expiration date? The expiration date is placed there by the manufacturer. The manufacturer guarantees the film's performance through this date. Fig. 8–9.

Products of technology can be seasonal. This means they are used only at certain times during the year. Consumers can save money by purchasing items off-season, when there is less need or demand for the product.

Consumers should also consider a product's durability and life expectancy. How long should you expect your bicycle to work before you have to replace it? Items such as appliances and

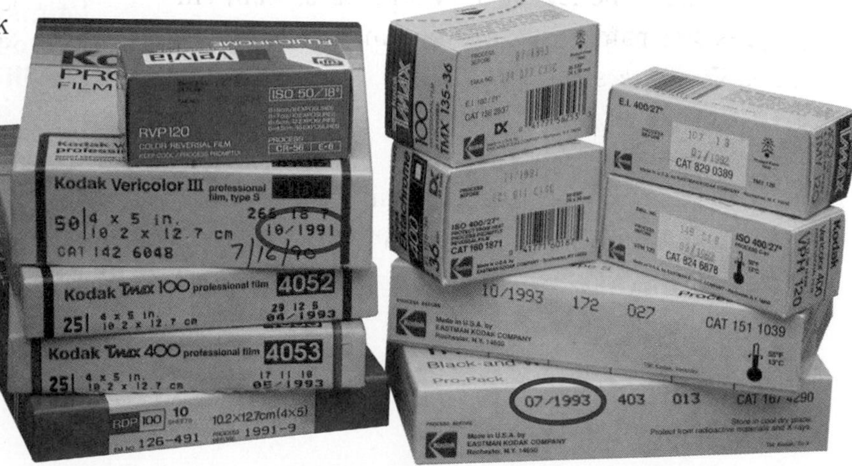

Fig. 8-9. Many products have a "shelf life." A product's shelf life is the length of time it can be stored before the manufacturer can no longer predict the results of using it.

automobiles are expected to perform well for several years before they need to be replaced.

■ **Visit a store that offers for sale many different types of items. Find at least two types of products besides food products and film that are marked with expiration dates.**

Technology Influences Time

New technologies often decrease the amount of time needed to accomplish a task. Communication technologies allow companies to transfer information instantly. Computers can do complicated calculations in a flash.

Advances in transportation technology move people and products from place to place faster than ever before. The rate at which products can be moved from the factory to the store affects the price of the product. The rate at which food and medical products can be moved affects their quality.

▶▶▶ FOR DISCUSSION ◀◀◀

1. Give three examples of how a restaurant owner might have to manage time.
2. How could you streamline your school day so it would end half an hour earlier?

IMPACT

Medical technology has given us more time. We have learned how to prevent or cure many diseases that once killed millions. As a result, most of us will live into our 70s or 80s.

CHAPTER 8 Review

Chapter Highlights

- Capital consists of money, buildings, property, equipment, and all the other goods a business may own.
- Companies can raise capital by making bank loans, selling stock, or using their own funds.
- Profit is the money left over after all the company's expenses have been paid.
- Time influences the way products are made, how much they cost, and how people use them.

Test Your Knowledge

1. List the types of capital that might be owned by a large farm.
2. List three sources of capital a company might use to finance expansion.
3. What is interest?
4. Why might stockholders not receive any dividends from a company in which they hold stock?
5. What do you call the money left after your company deducts expenses?
6. If your company had poor cash flow, should you be concerned? Explain.
7. Give two examples of why, in technology, "time is money."
8. List three products that have expiration dates.
9. Explain one method used in technology to speed up production.
10. List three seasonal items. When would be the best time to purchase each item?

Correlations

SCIENCE

1. Milk has a "sell by" date stamped on the carton. Open a container of milk on this "sell by" date and pour some milk into seven small containers. Put an airtight cover on each and refrigerate. Check one container each day for a week. Did any of the milk spoil? When?

MATH

1. Suppose you borrow $3500 from a bank at 6% interest. How much money will you owe the bank at the end of one year? Use the formula Interest = Principal × Rate × Time.

LANGUAGE ARTS

1. Think about the phrase "time is money." In a paragraph, give your own explanation of this phrase.

SOCIAL STUDIES

1. Where did Henry Ford get the idea for assembly line production? Find out how he set it up in his car production plant. How fast could Ford's workers manufacture a car? What did this do to the price of cars?

CHAPTER 9

Using Systems to Combine Resources

Introduction

The word *system* is used often in technology. What is a system? Can you name some everyday systems?

You may have seen the word *system* used to describe machines. You may even have some small systems at home. Computer systems and stereo systems are found in many homes today. Have you ever seen an advertisement in a newspaper for an intruder alarm system or low-cost heating system?

You may watch a TV news broadcast covering a debate over the proposed construction of a solid waste disposal system in your community. Each day, you may travel along a complex highway system on your way to school.

Systems—thousands of them—are all around us. Some are gigantic, like our nation's highway system. Some systems are as small as a home computer system. The purpose of this chapter is to help you understand how systems are used in technology to combine resources.

After reading this chapter, you should be able to

Define and give examples of systems in technology.

Explain the similarities among systems.

Diagram simple systems.

Separate large systems into smaller subsystems.

Discuss the impacts of systems.

Words you will need

system
input
process section
output
system diagram
feedback
subsystems

What Is a System?

A **system** is a group of parts working together to achieve a goal. Let's look at a common system: a ten-speed bicycle. When you use a ten-speed, what result do you hope to get? What goal do you want to achieve? You probably want to move fast or climb a hill with the least amount of effort provided by you, the rider. Fig. 9–1.

Goal = To travel up the hill in the fastest time possible with the least amount of energy.

Fig. 9-1. Systems are designed to help people achieve a goal by producing a desired outcome.

A ten-speed bicycle is a system designed to increase the amount of speed and force your legs provide to the wheels. This system enables you to travel faster and climb hills easier.

Large or small, systems help satisfy our needs and wants by doing work for us. Systems can help us to achieve desired outcomes. Figure 9–2 shows some other common systems and their outcomes.

TECHNOLOGY TRIVIA

Road construction was one of the great triumphs of the Roman Empire. By the time the empire fell in the 5th century A.D., its road system included more than 50,000 miles of roads in Europe and the Middle East. Some of these can still be seen today.

Goal = To find the solution to a multiplication problem: 5 x 2.

Goal = To provide music.

Fig. 9-2. The needs and wants of people determine the goals they have for the systems they use.

▶▶▶ **FOR DISCUSSION** ◀◀◀

1. **You have just learned what a technological system is. We are also surrounded by natural systems, such as our solar system. List at least three other natural systems.**
2. **What are the desired results of a large, well-designed transportation system?**

Fig. 9-4. People input or give commands to systems in many different ways every day.

How Do Systems Work?

All systems, large and small, have many things in common. For a system to provide a desired result, it must be able to accept commands from the people using it. This is called **input**. A ten-speed bicycle uses gear-shift levers and pedals to input commands from the rider. Fig. 9–3. Input commands tell the system what to do. Figure 9–4 shows some ways in which people input commands into a system.

Fig. 9-3. The rider gives the bicycle a command (input) by pulling or pushing on the gear-shift levers. The sprocket reacts by moving the chain to a different gear combination.

The part of the system that does the actual work and achieves the desired result is called the **process section**. The gears on a ten-speed bicycle make up the process section of that system. By selecting different combinations of gears, the rider can increase the speed of the bicycle and force applied to the wheels. Fig. 9–5.

The process section of a system achieves its goal by combining the seven resources of technology. Figure 9–6 shows how a ten-speed bicycle combines the seven resources of technology to provide the desired result.

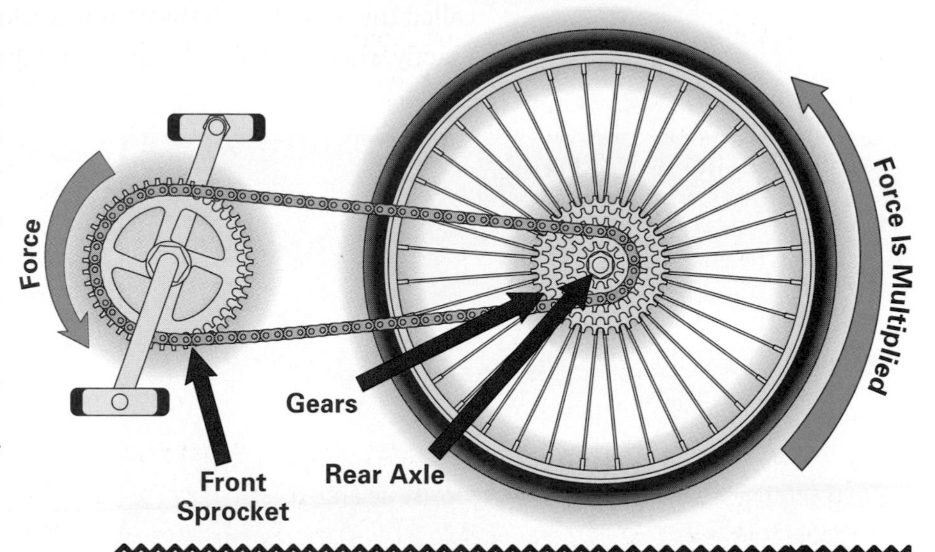

Fig. 9-5. In a bicycle, the force applied to the larger front sprocket is multiplied at the rear axle. In a ten-speed bicycle, the size of the front sprocket and rear gears can be changed to vary the force applied to the wheel.

Fig. 9-6. The process section of any system combines the seven resources of technology to achieve a desired outcome.

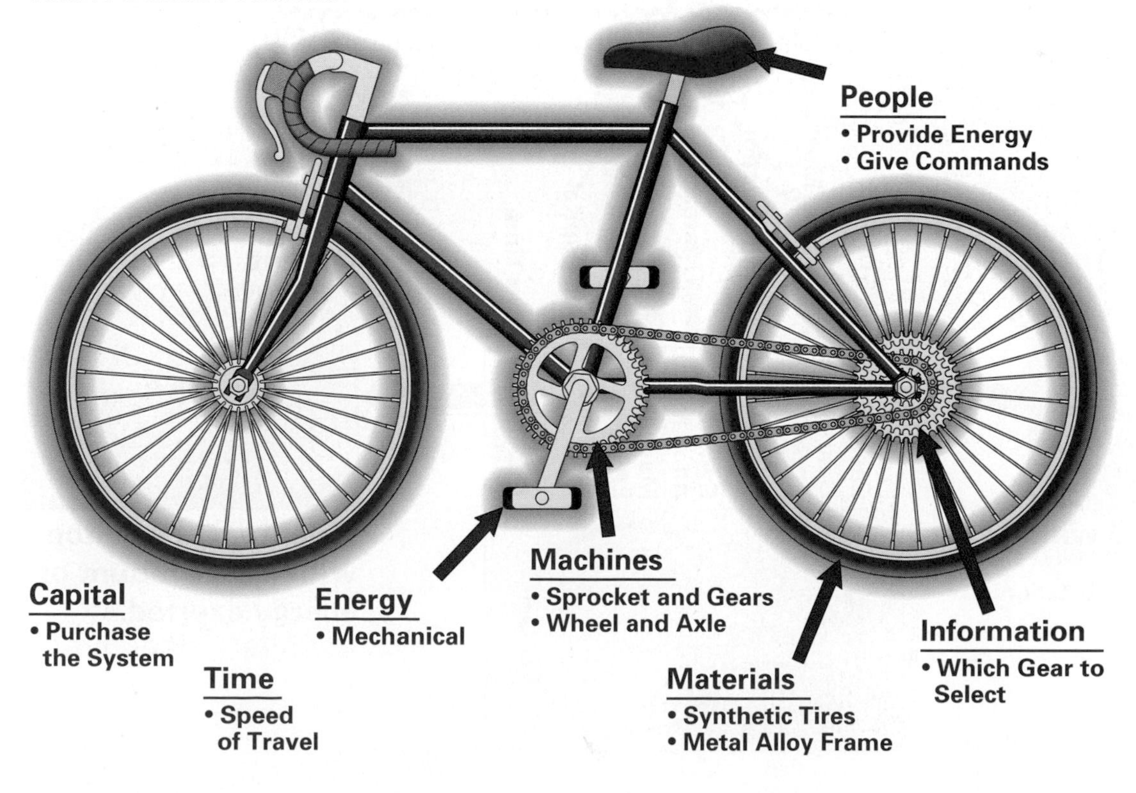

The actual result produced by the process section of a system is called the **output**. If the system is working correctly, the output should be exactly what we expect. What should the output of a ten-speed bicycle be?

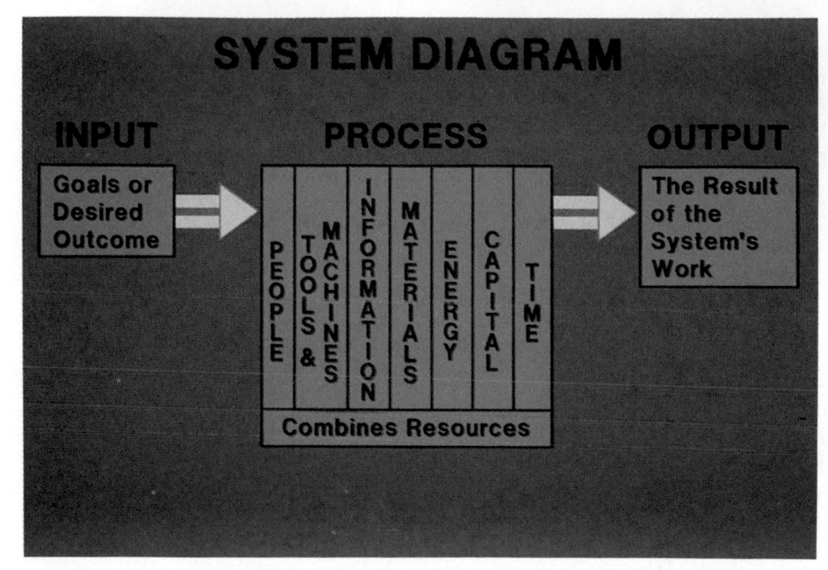

Diagramming Systems

System diagrams, or charts, make it easier to understand how a complex system works. Fig. 9–7. Figure 9–8 shows a block diagram of a ten-speed bicycle. The diagram shows the relationship between the input, process, and output sections of a system.

Fig. 9-7. A system diagram is a chart that shows how the three basic sections of a system work together to achieve a desired result.

Fig. 9-8. Example of a system diagram for a ten-speed bicycle.

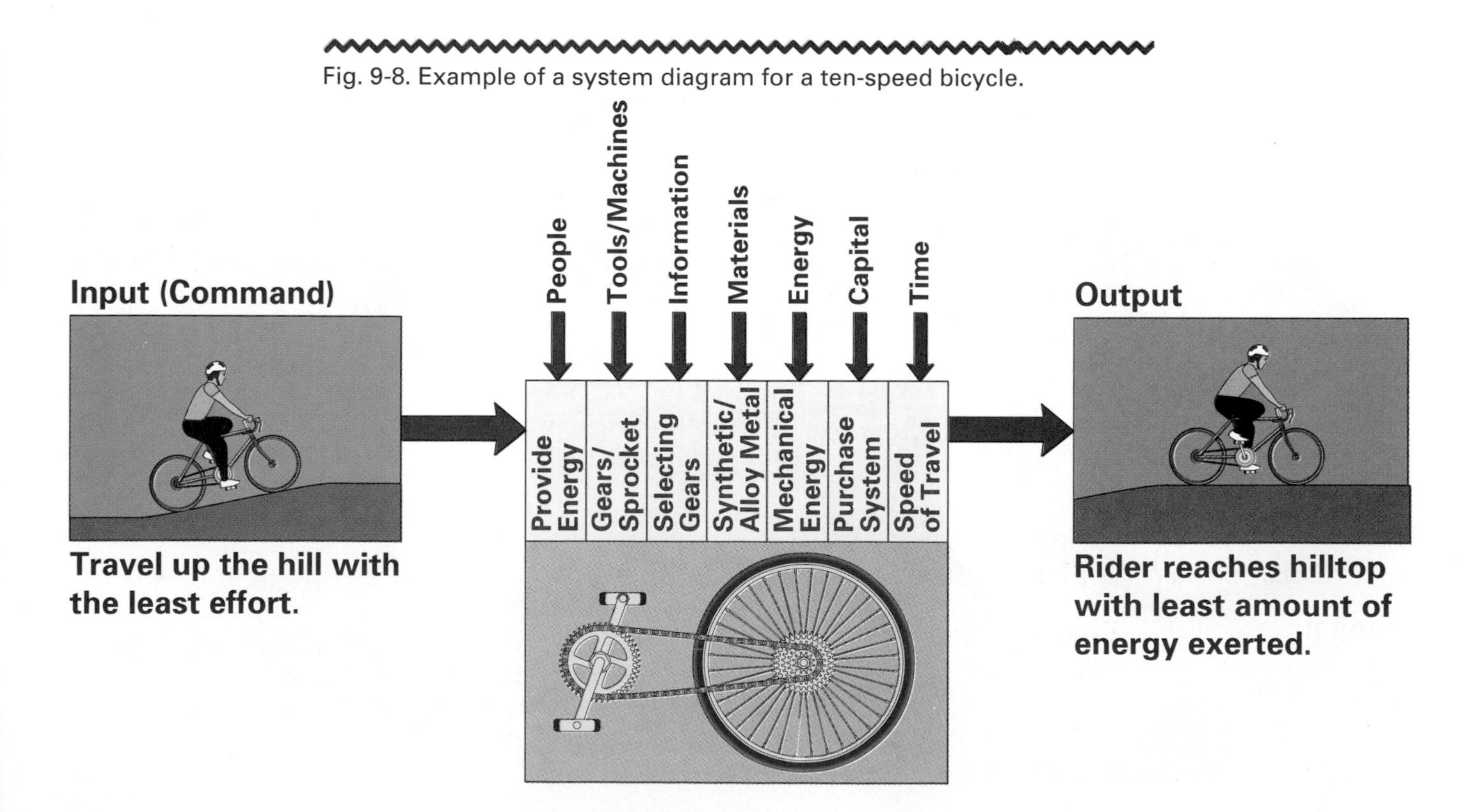

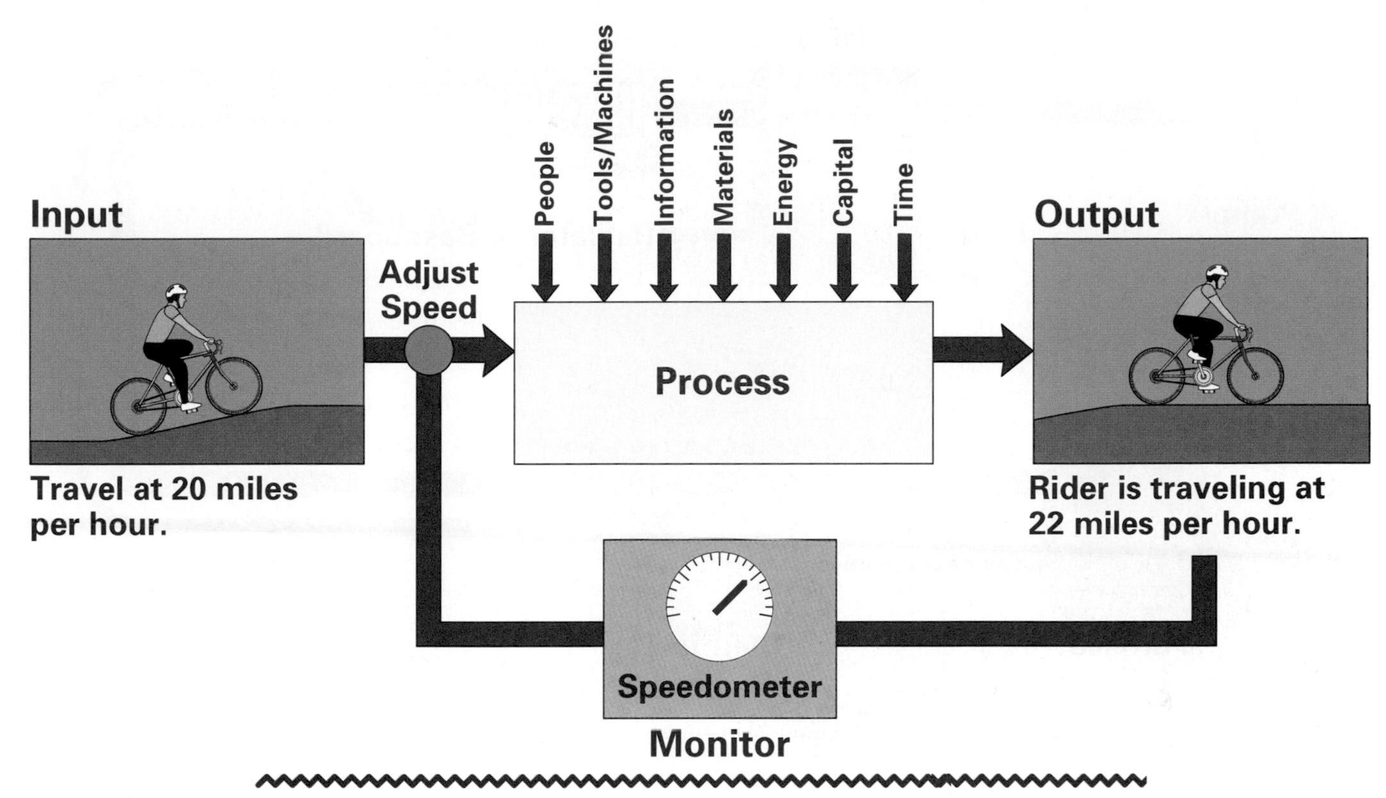

Fig. 9-9. The feedback loop in a system monitors what the system is doing.

Feedback

Suppose your goal is to travel 20 miles per hour on your bicycle. How do you know when you have accomplished your goal? You might install a speedometer on your bike. The speedometer would show you your traveling speed. In technology, we call this **feedback**.

Feedback is information about the output of a system. Many systems use feedback to monitor how a system is working. When we know how a system is working, we can adjust the system to help reach our goal.

If the speedometer tells us we are traveling at 22 miles per hour, we can adjust the system by pedaling slower or changing gears. Figure 9–9 shows how we add the feedback loop into our block diagram.

Many systems must be monitored and adjusted by people. Other systems are automatic—they monitor and adjust themselves. Fig. 9–10.

►►► FOR DISCUSSION ◄◄◄

1. Make a list of the technical systems, large and small, found in your school.
2. Draw system diagrams for two of the above systems.

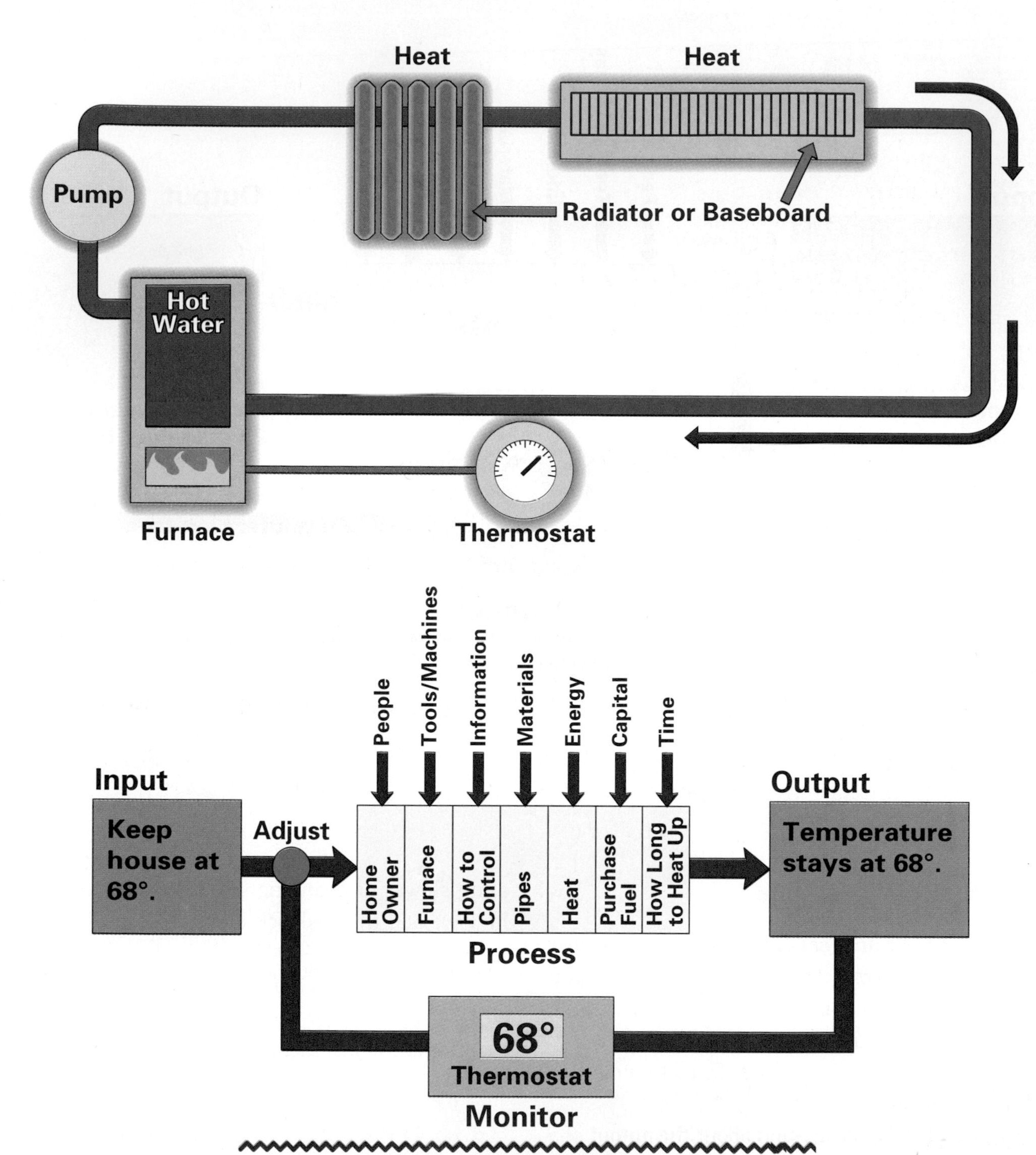

Fig. 9-10. The heating system that keeps your house warm is an automatic system. The thermostat monitors room temperature and turns the furnace on and off as needed.

Systems and Subsystems

Systems may be made up of smaller systems called **subsystems.** How many subsystems are in a ten-speed bicycle? The handlebars and fork make up the guidance subsystem. The pedals and gears make up the power subsystem, and the tires make up the suspension subsystem. Can you think of any other subsystems?

Both large and small systems can be broken down into smaller subsystems. Breaking systems down into subsystems makes it easier to study them. Fig. 9–11.

Systems and Subsystems

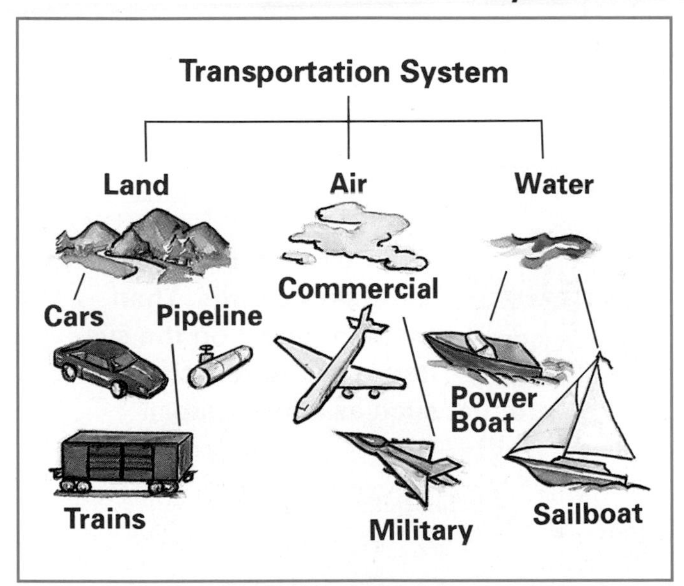

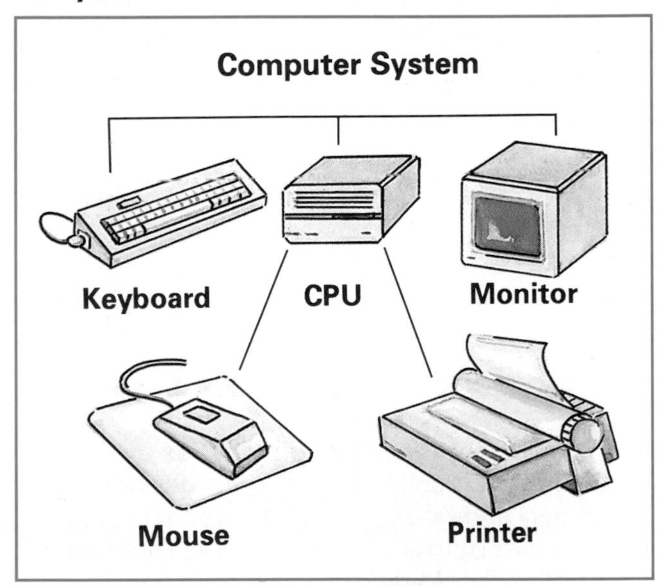

Fig. 9-11. Most systems can be broken down into smaller subsystems.

▶▶▶ FOR DISCUSSION ◀◀◀

1. Your home is a very complex technical system. List at least four subsystems found in your home.
2. Systems become more powerful when we expand them by adding additional subsystems. What subsystems can we add to a computer system to make it more powerful?

System Output

In Chapter 1, we learned that the four families of technology (communication, transportation, production, and biotechnology) provide us with the products and services we use each day. The four families of technology can be considered systems. The products and services we use are part of the output of these systems.

Systems usually have more than one output. For example, the main output of a paper mill is paper. A secondary output of the mill may be wood chips and sawdust. Some mills turn these chips into pellets that can be burned to produce electricity.

The outputs of a system are not always desirable. The same system that produces paper and pellets may also produce air pollution and terrible smells. When people design systems, they must be aware of the undesirable outputs of the systems.

▶▶▶ FOR DISCUSSION ◀◀◀

1. A new coal-fired power plant has been built in your community. List two desirable and two undesirable outputs of that system.
2. List two outputs for each of the four families of technology.

■ Make a diagram showing the inputs, processes, and outputs involved in planning and building a model community.

IMPACT

Outputs of technological systems may have good or bad impacts. Sometimes, the same product has both. For example, hospitals use many disposable (throw-away) products to prevent the spread of germs from one patient to another. Once they have been used, these products become dangerous waste. You've probably heard about needles and other hospital waste washing up on beaches. Preventing the bad impacts of technology without giving up the good ones is one of our toughest challenges.

■ Do research on recycling procedures. Then write a report on the system used to recycle a product such as a newspaper.

Chapter Highlights

- Systems are designed to produce desired outputs to help people meet their wants and needs.
- All systems are similar in that they all have input, process, and output sections.
- The process section of a system combines the seven resources of technology.
- A system diagram charts how a system works.
- Subsystems are smaller systems within a larger system.
- Systems can have multiple outputs.

Test Your Knowledge

1. What is a system?
2. When you use a calculator, what is your desired result?
3. What three parts do all systems have in common?
4. What part of a system combines the seven resources of technology?
5. What does the feedback loop do in a system?
6. Draw a system diagram to explain how a washing machine works.
7. What is a subsystem?
8. List the subsystems found in a home stereo system.
9. What are the four systems that provide all the products and services we use each day?
10. List one positive and one negative output of an automobile.

Correlations

SCIENCE

1. Consult a biology textbook to identify three of the systems in the human body. List the parts for one of those systems.

MATH

1. If you travel 20 mph (miles per hour) on your ten-speed bike, how long will it take to go 5 miles? 10 miles? You increase your speed to 22 mph. Will it take you more or less time to go the same distance?

LANGUAGE ARTS

1. Brainstorm and list as many systems as you can. Compare your list with your classmates' lists and make a final copy of all the systems named to display in the classroom.

SOCIAL STUDIES

1. Long before the construction of its vast highway system, the United States established an extensive railroad transportation system. On a map of the United States, diagram the states connected by the first railroad transportation system.
2. List the inputs, processes, and outputs of a transportation system.

Make a Difference

The President's Environmental Youth Awards Program

From conducting recycling projects to planting trees—from creating environmental awareness to helping save endangered animals—young people have been using their talents to improve their world.

By participating in the President's Environmental Youth Awards Program, students in grades kindergarten through 12 have created projects to promote community involvement with important environmental issues.

Since 1986, when the U.S. Environmental Protection Agency began sponsoring an annual national environmental competition, thousands of successful projects have been launched by students from throughout the nation.

At River Trails Junior High School, a teacher and her students launched a "grassroots" environmental action group, Project P.E.O.P.L.E. (People Educating Other People for a Long-lasting Environment). The students prepared a booklet outlining 80 ways that each person can reduce pollution and improve the environment. They encouraged individuals, schools, and businesses to join the organization and to pledge to do as many things on the list as possible. Project P.E.O.P.L.E. soon grew to more than 600 members in nine states, England, and Japan.

In Texas, a high school student started a three-city recycling program—and during a one-week pilot project, residents and students recycled seven tons of newspapers and 300 pounds of aluminum cans. Proceeds from the project were then donated to the three cities to begin a permanent recycling program.

Through a letter-writing campaign, students at a Utah elementary school started a "Leaf It to Us" Crusade for Trees. They raised funds, received two small grants, planted 182 trees on public lands—and encouraged students across the state to do the same. The students also worked with the state legislature to pass a bill to create many more grants for planting trees.

While he was a junior high student, Eric Champlin chose barn owls for a science-fair project. Hoping to increase their population, he built and placed nesting boxes in suitable barns throughout Ohio—and included tapes for attracting owls.

In still another project, students at Crosby-Ironton High School cleaned up an abandoned mine pit to make it suitable for swimming, fishing, and boating.

A group of Indianhead Council Boy Scouts traveled to Costa Rica on still another project—to help save the endangered leatherback sea turtle. The Scouts gathered information, discouraged turtle-egg poachers, and planted trees to prevent soil erosion. They helped cut the loss of turtle eggs from 85 percent to 5 percent.

In a "Save the Rain Forest" project, a student-teacher group from Dodgeville High School enlisted 9,000 schools in their cause and raised more than $150,000 for rain forests throughout the world.

Students throughout the country have also worked on a wide variety of other projects—from improving water quality in lakes, to turning a vacant lot into a "garden" classroom, to starting a recycling business—and all with one goal: to improve and protect the environment.

Amazonian Rainforest

Using Resources to Solve Technical Problems

Activity Brief
Using Technology to Overcome Physical Handicaps

PART 1: Here's the Situation ....

No one can do everything. Disabilities exist in different forms and in varying degrees of seriousness in all people. Serious disabilities, such as not being able to walk, are often referred to as handicaps. Today, advances in technology are freeing more and more people from limitations resulting from physical handicaps. While it's true that "no one can do everything," it's also true that practically everyone can do something and most can do a lot of things. Devices can be developed to use what a person is capable of doing to accomplish things that he or she is otherwise unable or finds difficult to do.

In order to develop a device to aid a handicapped person, several technical problems must be solved. As you do this activity and as you read Chapters 10 and 11, you'll learn for yourself how the resources of technology can be used to solve all sorts of technical problems.

PART 2: Your Challenge...........

Your challenge in this activity will be to invent a device that will aid a person with a physical handicap. Figs. III-1 through III-3 show some examples. Your invention can be:

- intended to aid any type of physical handicap. (Examples—limitations in body motion or senses)
- a modification of an existing device. (Example—wheelchair modification)
- a totally new concept. (Example—a new device unlike any other invention)
- a working prototype of a real device. (Example—an invention that really works, such as a special fork or spoon)
- a simulated model. (Example—a cardboard model of a special house design equipped with special devices)

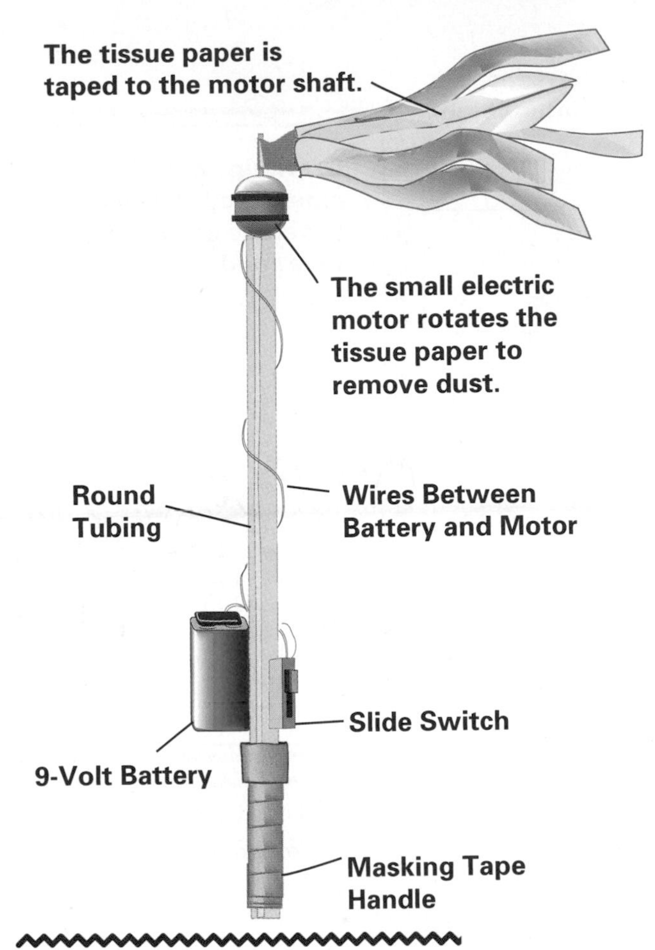
The tissue paper is taped to the motor shaft.
The small electric motor rotates the tissue paper to remove dust.
Round Tubing
Wires Between Battery and Motor
Slide Switch
9-Volt Battery
Masking Tape Handle

Fig. III-1. House dusting device for people confined to a wheelchair.

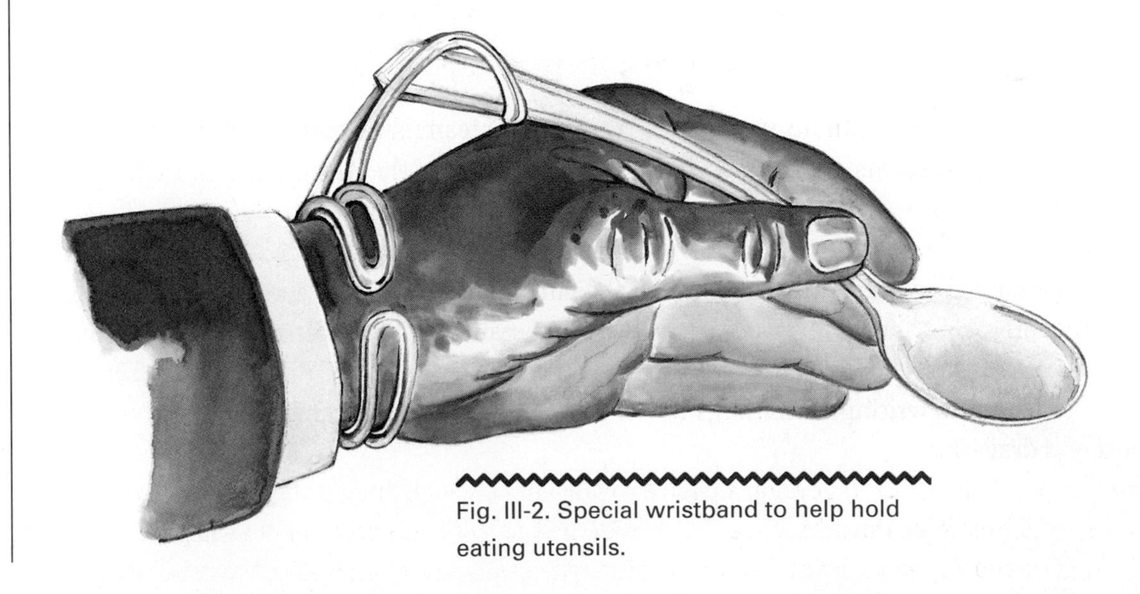

Fig. III-2. Special wristband to help hold eating utensils.

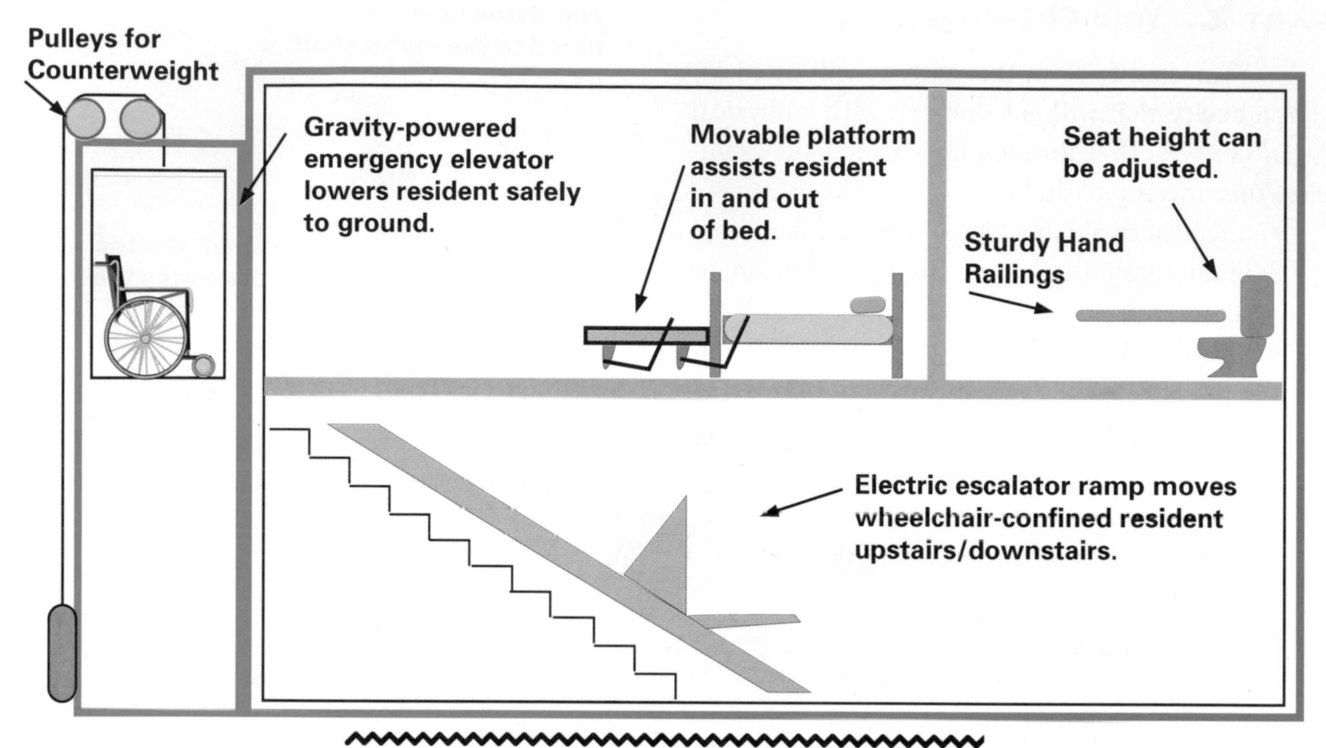

Fig. III-3. Model of a house design for a wheelchair-bound resident.

PART 3: Specifications and Limits

Your invention will need to meet certain standards. Read the following specifications and limits carefully before you begin.

1. Your invention must:
   - be a working device or a non-working model
   - help a person with a physical handicap
   - be safe
   - be documented (supported in writing) with a technical report and drawings
2. Your design team is responsible for:
   - development of the invention
   - writing a technical report
   - preparing a set of drawings
   - giving a presentation to the class describing the operation of the invention and how it helps people

PART 4: Materials..........

Any material safe to use and inexpensive can be used to build your device or model. To the left is a list of possibilities. Also given is a list of tools you might use.

Materials

cardboard
paper
string
wood
wood dowels
wire
plastic
glue
fasteners
metal
rubber bands
motors
large syringes and tubing

Tools

utility knife
scissors
power tools
screwdrivers
wrenches
hot glue gun

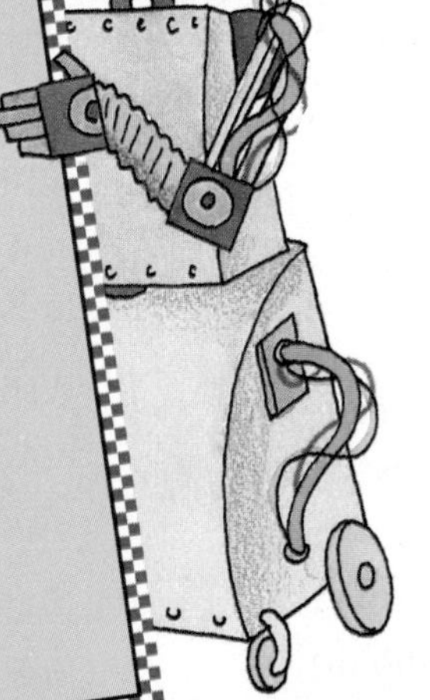

Safety Notes

- Consider always the safety of the people who would use the device you invent.
- As you do this activity, remember to follow all the safety guidelines your teacher has explained to you.
- Use all tools properly. Use special care with tools that are sharp.
- Before using any power tools, be sure you understand how to operate them and always get your teacher's permission.

PART 5: Procedure..........

Your invention and how you build it will be up to you. Still, there are certain steps to follow that will make your work easier.

1. Working in groups of 2 or 3 students each, brainstorm possible ideas for devices to aid the handicapped. Then evaluate your ideas and exclude those that are too complicated, expensive, or unsafe.
2. Draw sketches of promising ideas to be sure all members of the team understand what the inventions would be.
3. Do research on devices for the handicapped to find out what has already been done.
4. Interview someone with a physical handicap to share ideas about possible products. Take written notes or, with permission from the person you are interviewing, make an audio or video recording of the interview. Share your interview with the class.
5. What do you hope to achieve? Review possible designs and select what you think is your best idea for an invention.
6. Prepare a bill of materials. Gather needed materials and tools.
7. Cut and fasten materials together to form a rough design. Continue testing and making improvements until you achieve your final design.
8. Create technical drawings of your invention. Describe the whole design and important parts or features (details) using orthographic projection.
9. Write a technical report describing how your design team developed the invention.
10. Prepare and deliver a short presentation to the class describing your invention and the design problems you solved during its development. Demonstrate how your device can be used.

PART 6: For Additional Help...........

For more help with this activity, look up the following terms. You'll find some of them in this book. (Check the index.) You'll find others in dictionaries, encyclopedias, and other resource materials.

- bill of materials
- brainstorm
- cost-effective
- orthographic projection
- problem-solving method or approach
- prototype

PART 7: How Well Did You Meet the Challenge?...........

When you've completed your invention, evaluate it. Have you achieved what you intended? Work with your teacher and classmates to evaluate all the inventions. Consider the following qualities for each and then discuss responses to the questions.

- usefulness
- safety
- innovativeness (newness, originality)
- cost-effectiveness

1. What features do you particularly like in some of the other inventions? Why?
2. Which inventions are best or most practical for aiding handicapped persons?
3. Which inventions could help persons with more than one handicap?
4. How could your own invention be improved?
5. How did you use each of the basic resources of technology in the development of your invention? (people, information, materials, tools and machines, energy, capital, time)
6. If you were to design and build another device to aid the handicapped, what would you do differently?

PART 8: Extending Your Experience...........

This activity helps you realize some of the many ways in which technology can be used to make people's lives easier. Also, through experience, you have gained understanding in how basic resources of technology are used to solve technical problems.

Think about the following questions and discuss them in class. You'll find more about using resources to solve technical problems in Chapter 10, "The Design Process/Problem Solving Process," and Chapter 11, "Designing Graphic Solutions."

1. What handicaps were not addressed by your class? How can technology be used to help overcome these handicaps?
2. Are there handicaps that are too complex for today's technology to be effective in correcting? What future technological breakthroughs would be useful in overcoming complex handicaps? (Use your imagination. In technology, things that are only ideas today may become realities tomorrow.)

CHAPTER 10

The Design Process: A Problem-Solving System

Introduction

The Dictionary of Occupational Titles, published by the United States Department of Labor, lists more than 20,000 jobs. "Problem solver" is not one of them. Yet, if you spoke with people employed in many of those jobs, you would probably learn that much of what they are responsible for is designing solutions or solving problems.

There are, however, many occupations that have the words *design* or *designer* in their titles. Fig. 10–1. Are a designer and a problem solver one and the same? As you continue reading, substitute one term for the other—does either term fit? In this way, you can make up your own mind.

After reading this chapter, you should be able to

Briefly explain each step in the problem-solving process.

Give examples of market research and consumer research.

Describe the use of models and prototypes in the problem-solving process.

Evaluate the use of feedback in product improvement.

Words you will need

market research
consumer research
design criteria
brainstorming
mock-up
prototype
field-test

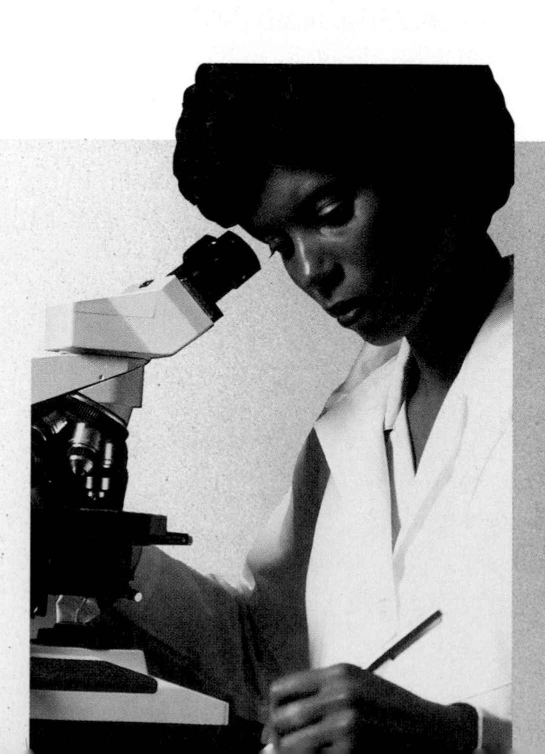

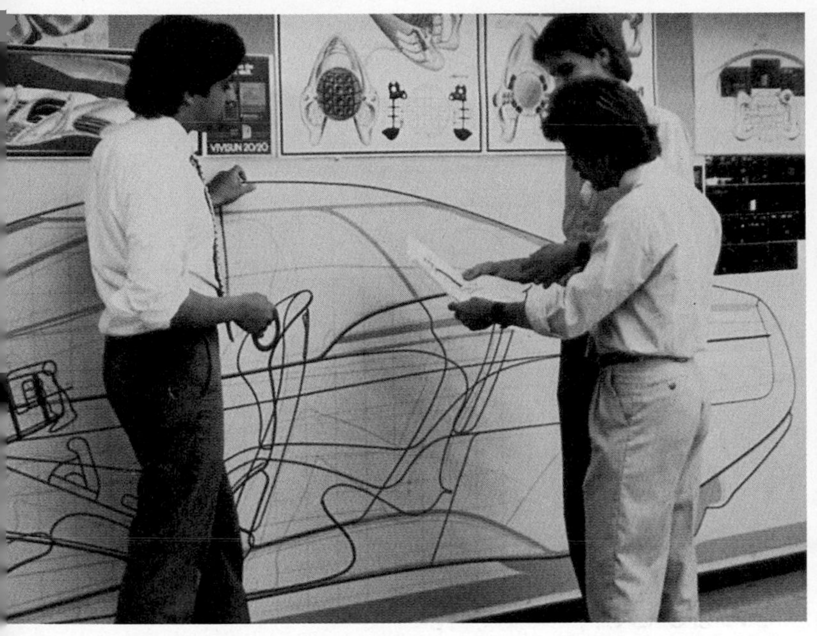

Fig. 10-1. Although *problem solver* is not a job title, many jobs place major emphasis on solving problems. Often, the solution to a problem involves designing a product.

A Basic Approach to Solving Problems

How do we go about designing a solution to a problem? Is there a process or system that we can use? Yes, there is. Although there are many different kinds of problems, most of them can be solved using a standard approach.

Will following this approach always guarantee the "best solution"? Hardly. Solutions to problems, like the world in general, are rarely perfect. In fact, it is not always possible to get a group of people to agree on exactly what the problem is.

In any event, the most that can be expected is for you to do your best. The material that follows will help you do just that. The basic problem-solving approach is shown in Figure 10–2 A and B.

Should you always follow the steps in this order? Yes, but some steps are of greater importance than others, and some can be combined with others. Also, you may need to repeat a few of these steps more than once, especially the last two. Let's take a closer look.

Recognize the Problem

You might not have realized it, but you solve many problems on a daily basis. Did you have difficulty finding your shoes this morning? How will you decide where to sit for lunch? You see? All of these are problems that we have to solve every day.

Once you have identified a problem, ask yourself if it is stated clearly and accurately. Is it too general or overly restrictive? Can everyone involved understand it?

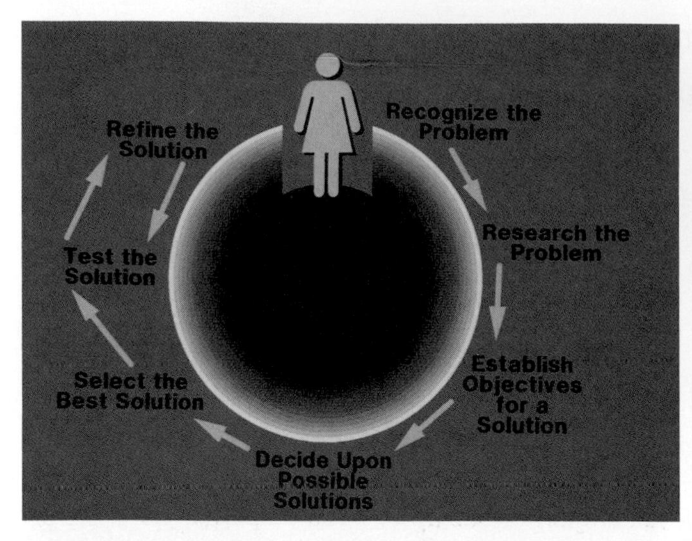

Fig. 10-2A. The problem-solving process provides a logical outline to follow, whether you're designing a product or solving another type of problem.

INPUT
Recognize & Research Problem
Establish Objectives for Solution
Adjustment
PROCESS
Decide Upon Possible Solutions
Select Best Solution
OUTPUT
Best Solution
TESTING

Fig. 10-2B. Like all other processes, the problem-solving process can be thought of in terms of inputs, processes, and outputs. The "testing" loop allows the solution to be refined.

If a problem is poorly stated, much of the effort you spend trying to solve it may be wasted. Consider for a moment this problem statement: "Design an inexpensive toy." Could you design an acceptable toy from that description? What else would you want to know? In fact, many necessary details are missing here. You will nearly always need to know some specifics about the nature of the problem. Remember, it is more difficult to solve a problem that you do not understand. Fig. 10–3.

TECHNOLOGY TRIVIA

Solving a problem can take a while. It took more than 20,000 engineers a billion hours of design time to develop the Space Shuttle.

Fig. 10-3. Just figuring out exactly what the problem is can take some serious thought. Taking time at the start to define the problem can save time later on.

Research the Problem

Gathering information is often the task given least attention in a design problem. People want to "get on with it," and not spend time in preparation. There are good reasons, however, for spending time researching your problem and gathering information. Mistakes made at the beginning of a product's design may flaw the final product. In addition, it is easier to correct errors during the "paper" stage than during manufacturing or installation.

How do you research a problem? Here are some general rules. First, try to focus on what will be most helpful. It makes little sense to gather all kinds of information on a topic, since not all of it will apply to the task at hand. Second, limit your research. Don't bother to research something with which you are already familiar. Finally, keep in mind that the material you gather should be practical. Avoid complex data that no one can understand.

Your approach to research will vary according to your topic.One approach is to talk to people who can help you. This includes anyone who might act as a resource: people in the same situation as you, experts, or people who sell or distribute your type of product. The list depends a great deal on the design problem itself.

You can also do library research. The library contains a wealth of information on almost every topic. You will find information in books, newspapers, magazines, trade periodicals, and clip file folders. Fig. 10–4. Many libraries also offer computerized database searches. Be sure to check with the reference librarian—this might prove to be your most rewarding inquiry.

Don't forget to check out your competition. This type of input is called **market research**. Are similar items already in use? You don't want to duplicate something that already exists. On the other hand, you might chance upon something that will guide or inspire you. Market research is a good way to discover what is popular in the marketplace. You can then decide whether to join the trend or provide something different.

Survey users of products similar to your proposed product. This is often called **consumer research**. What do people want? What do they like the least about the materials now being marketed?

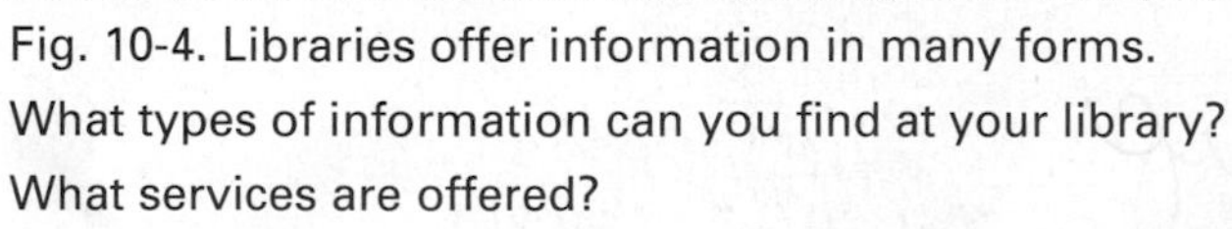

Fig. 10-4. Libraries offer information in many forms. What types of information can you find at your library? What services are offered?

■ Suppose you are a designer of athletic shoes. Survey your friends and family to learn what they like or don't like about the brands now on the market. Using your survey results, decide how you would improve the design of athletic shoes.

Establish Objectives for Solution(s)

Now it is time to decide on the goals that the solution must fulfill. These goals are often called **design criteria**. You can ask yourself questions to help establish your goals. Of course, your questions will depend on your specific problem. Some of the questions you might ask are:

- Is color a factor?
- Who will be using it?
- What materials should be used (or avoided)?
- Are there any environmental concerns?
- How will this item be used?
- What is the suggested price range?
- Why is this product needed?
- What are the legal considerations?

How many questions or goals do you need? You need as many as necessary to provide a framework to design your product. You don't want to stifle your creativity or overwhelm yourself with too many requirements, but you don't want to omit significant guidelines, either.

Decide on Possible Solutions

All ideas or solutions do not come from a single source. If they did, everyone would rush to the same place whenever they had a problem. How do you generate possible solutions to a problem? There are several ways to find possible solutions. Read through the following suggestions, and see what appeals to you:

- Brainstorming—One or more people just pour out ideas. In the **brainstorming** approach, you gather all of the ideas you can, without trying to make them perfect. Fig. 10–5.
- Listing—Compile a list of all the characteristics of the problem at hand. Later, you can use your list to suggest ideas that might work.
- Discussion—Talk the situation over with someone who is not directly involved. Ask questions. Then try reversing roles with the person. It never hurts to hear someone out.
- Sleep on it—figuratively, that is. Taking a rest and coming back to the problem allows you to look at the situation with a fresh eye.

Fig. 10-5. Brainstorming allows people to share all of their ideas about a problem. Since the goal of brainstorming is to gather all sorts of ideas, no one needs to feel that an idea is silly or not worth mentioning.

Evaluate Suggested Solutions

There are many ways to evaluate solutions. The "best" way depends on the solution you're evaluating. In general, however, you should compare each solution with your original goals. Fig. 10–6.

The characteristics you should consider are remarkably similar to those you considered when you set your goals. The characteristics you might need to look at include:

- Appropriateness of size, weight, color, etc.
- Ease of use
- Probable cost
- Reliability, durability
- Potential hazards
- Appeal to users
- Effect on environment
- Manufacturing concerns

Select the "Best" Solution

This is not as easy as it sounds. Sometimes there doesn't seem to be a "best." You might even have to reject all the possible solutions and start all over again.

To decide on the best solution, you may need to make **mock-ups**. Fig. 10–7. Mock-ups are models (typically full-sized) that look like the real thing but do not necessarily work or contain the actual materials or mechanisms.

Test the Solution

It is now time to make the item and discover how well the process worked. The first working model of a device is called a **prototype**.

Fig. 10-6. Evaluating the suggested solutions allows you to make comparisons. Some assessments can be measured objectively by using instruments. Others rely more on judgment.

Fig. 10-7. These students have made a mockup of a mailbox they have designed. What can this mockup tell them about the product?

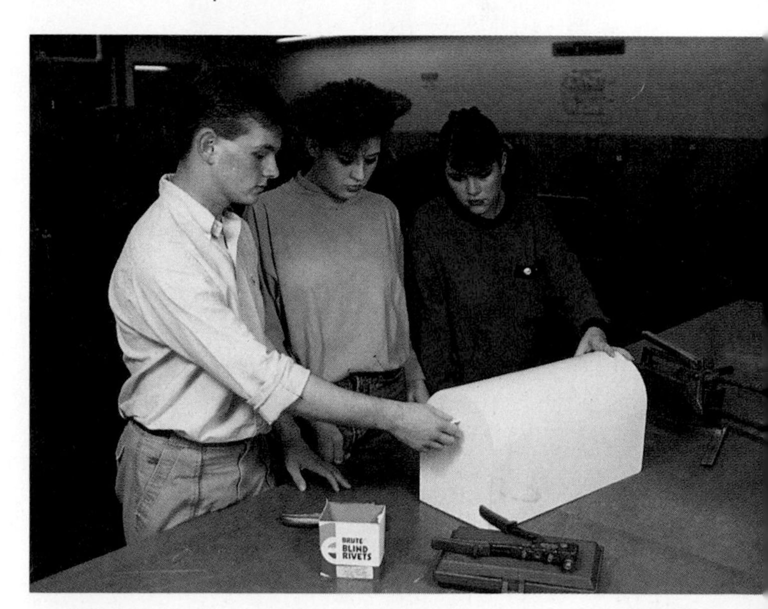

Next you must **field-test** your product, or use it under realistic circumstances. If the design problem is an automobile, for example, you will need to determine how well it works and whether it is safe to use. Fig. 10–8.

IMPACT

Did you know your community could have an impact on the nation's marketplace? Companies often test-market new products in selected cities. If the product does well in those cities, it will be mass produced and sold nationwide. Have you ever been asked to sample a new food or beverage in a store and give your opinions about it?

Fig. 10-8. Making a working prototype helps you improve the product or device before manufacturing begins in earnest.

Refine the Solution

It is rare for a proposed solution to remain as is at the prototype stage. As more people test and use it, they may suggest improvements. You should try to encourage such feedback. This process of testing and modification can continue for many cycles. Fig. 10–9.

►►► FOR DISCUSSION ◄◄◄

1. Why should people bother to follow a model, or process, to solve problems?
2. Do all jobs involve some type of problem solving? Explain your answer.

■ **A large river flows through your model community. Using the problem-solving process, decide on the best way to get people across the river on a daily basis.**

Fig. 10-9. To develop a solution to its greatest potential, it is not unusual to go through several cycles of testing and modification.

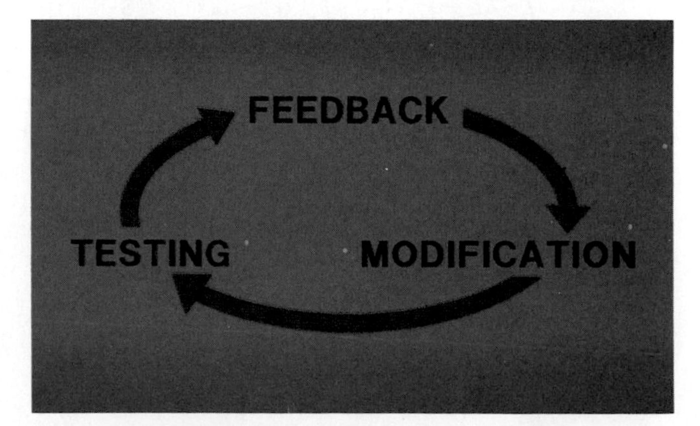

Chapter Highlights

- People have been solving problems for thousands of years.
- The problem-solving process gives us a helpful method for designing a solution.
- A clear statement of the problem is a good start toward its solution.
- Establishing goals provides a means for proposing and evaluating solutions.
- People often modify or revise their solutions as a result of feedback.

Test Your Knowledge

1. Explain why there may not be one perfect solution to a problem.
2. What are the basic steps of the problem-solving process?
3. Name three design criteria you might consider if you were designing a new book bag for students.
4. Why is it important for a problem to be stated clearly?
5. Name three general rules to follow when you research a problem.
6. What is market research?
7. Why should you establish goals for suggested solutions?
8. What are four ways to come up with ideas of possible solutions to a problem?
9. Why do designers make mock-ups?
10. What are prototypes used for?

Correlations

SCIENCE

1. Suppose your friend sprained an arm and couldn't move it. Design a device to allow your friend to fasten buttons and open and close zippers without help.

MATH

1. Try some consumer research. Ask your friends and family to rate the following cereals on a 1 to 5 scale (5 points means "like it a lot"; 1 means "dislike it a lot"): corn flakes, oatmeal, raisin bran, shredded wheat. Draw a graph showing the survey results.

LANGUAGE ARTS

1. Select one of the following problems. Write a report about the basic problem-solving techniques you would use to find a solution. (A) Your community needs to reduce the amount of waste going into the landfill. (B) Many of the students in your school skip lunch. (C) Senior citizens need transportation to get to the grocery store, doctor's office, etc.

SOCIAL STUDIES

1. The date is 1875, and you and your family live in New York. As part of a settlement program, the government has offered your family 100 acres of free land in the Indiana Territory if you will move there. Use the problem-solving procedure to make the decision whether to accept the offer or remain in New York.

CHAPTER 

Designing Graphic Solutions

Introduction

How do you express an idea? Using words? Using gestures? People often draw their ideas on paper. Sometimes they even make a model. They do these things because it is often easier to see a concept than to describe it. Sketches and other types of drawings are primary means of communication in many professions and trades. Perhaps this is why it is often said, "A picture is worth a thousand words."

After reading this chapter, you should be able to

Identify three types of pictorial drawings.

Explain the usefulness of orthographic/multiview drawings.

Explain the value of using different line types.

Discuss the purpose of dimensioning.

Describe the use of scale in drawings.

Demonstrate good lettering techniques.

Give examples of the types of things that should be noted in a parts list.

Discuss the use of models in presenting graphic solutions.

Words you will need

pictorial
oblique
isometric
perspective
multiview
dimensioning
scale
bill of materials
computer modeling

Pictorial Sketching

A **pictorial** sketch gives us the most realistic view of what an object looks like. Some pictorial sketches are *rough sketches*—quick, freehand drawings. Rough sketches are usually the first type of pictorial we try. They are quick, and they allow us to get our ideas down on paper as fast as possible. When ideas or design solutions are being generated, it is best to be able to draw them rapidly and refine them later.

Refined, or finished, sketches are drawn more carefully. They are sometimes drawn with drafting instruments or a computer, perhaps even using color and shading.

There are three categories of pictorial drawings: oblique, isometric, and perspective. Each type has certain unique characteristics, but if you are new to sketching, just use the type you find easiest. Experience and practice will enable you to vary your choice as time goes on.

Oblique

Oblique drawings are probably the quickest to draw, but they appear the most distorted to the eye. The front view is always drawn without any changes. This is especially useful when the object you're drawing has a curved or round shape as seen from the front. Fig. 11–1.

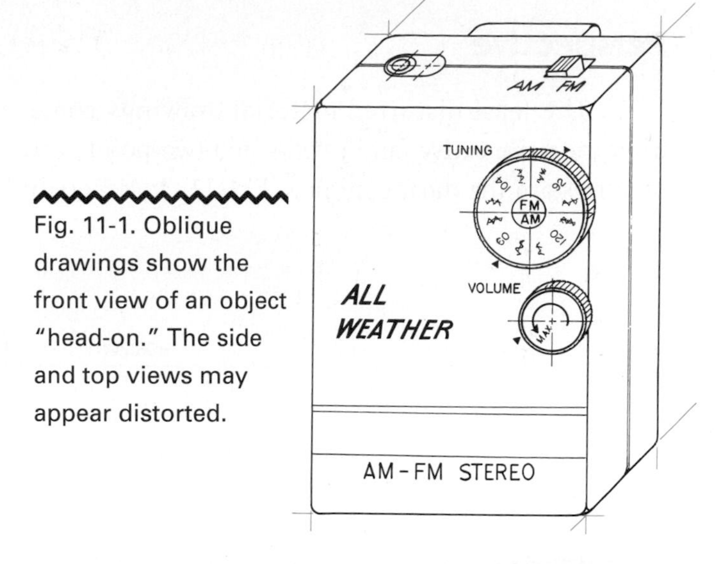

Fig. 11-1. Oblique drawings show the front view of an object "head-on." The side and top views may appear distorted.

Isometric

An **isometric** drawing can also be drawn reasonably fast, and you can manipulate it easily once you get used to it. Circles appear as ellipses in an isometric drawing. However, since this is often the way they appear in real life, they look fairly natural. Special isometric grid paper may be used to establish the depth angles on each side. The angles are usually 30 degrees from horizontal. Fig. 11–2.

Fig. 11-2. Isometric drawings allow people to make quick pictorial sketches with only moderate distortion.

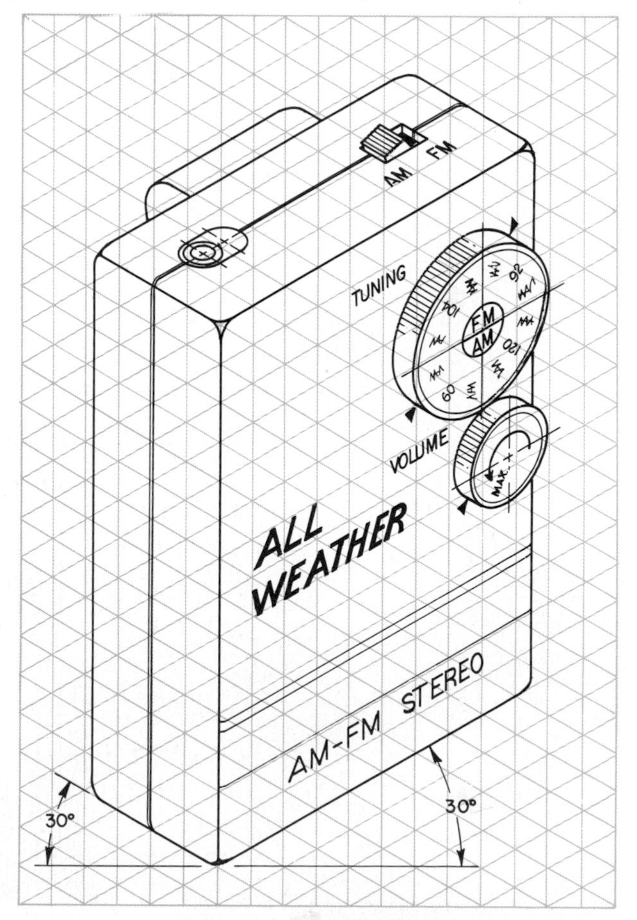

Perspective

The least distorted pictorial drawings come from the **perspective** family. One- and two-point perspectives are the most common. Fig. 11–3. Although an accurate perspective drawing done with instruments might take quite some time, it is not too difficult to make rapid perspective sketches.

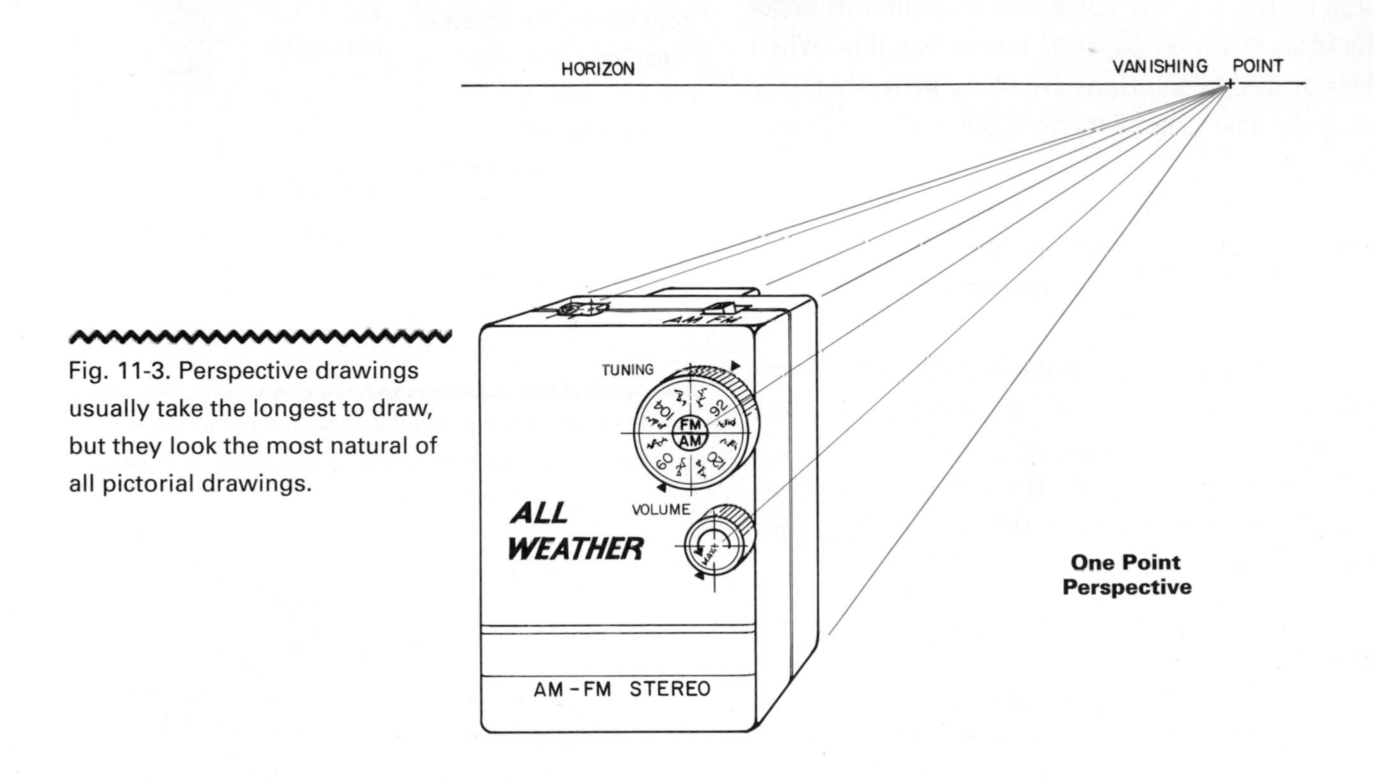

Fig. 11-3. Perspective drawings usually take the longest to draw, but they look the most natural of all pictorial drawings.

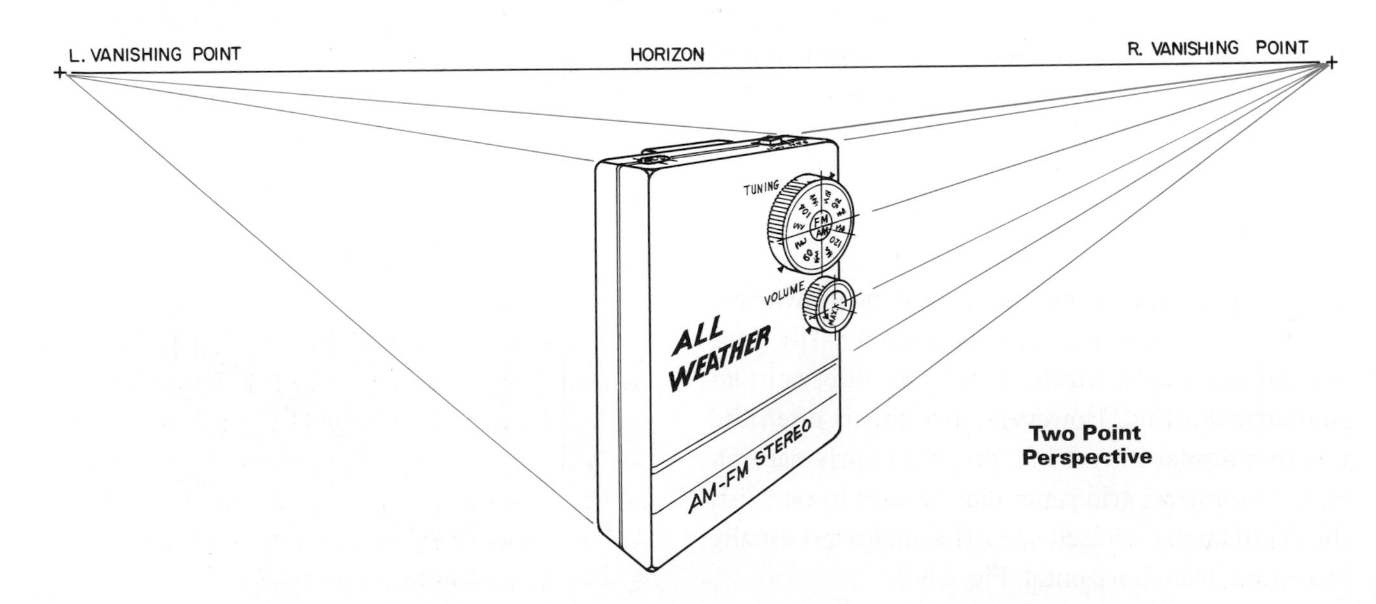

Renderings are typically perspective drawings that have color, shading, and perhaps even background material such as human figures or landscaping. Fig. 11–4. Renderings are often used by architects to show people what a building will look like when it is finished.

Fig. 11-4. Renderings are pictorial drawings that include details such as shading and backgrounds to make them more attractive. These drawings are often used to show potential customers what a building or product will look like.

►►► FOR DISCUSSION ◄◄◄

1. What is the difference between a rough sketch and a refined sketch?
2. Of the three types of pictorial drawings, why do you think architects usually use perspective drawings to show people how buildings will look?

■ Using Figs. 11–1, 2, and 3 as guides, make oblique, isometric, and perspective sketches of an object in the room. Which sketch was easiest for you to draw? Which looks the most realistic? Why?

Multiview Drawings

Pictorial drawings aren't the only kind of drawings drafters use. They also use **multiview drawings**, or *orthographic projections*, to relate ideas or solutions to problems. Multiview drawings allow you to show one or more sides of an item at a time. By aligning the selected views, the viewer can see exactly what the object looks like, or perhaps how it operates. Multiview drawings, like pictorial drawings, may be sketched roughly or drawn with great care and accuracy.

All objects can be viewed from six sides. Fig. 11–5. In the United States, the three most common views used in drafting are top, front, and right side. In most cases, all the necessary information about the object can be shown using these three views.

Since time and effort are always in demand, the rule is to draw as few views as necessary to describe the object adequately. Some objects, for example, require only one view. Fig. 11–6.

Line Types

Because technical drawing is a graphic language, drafters have agreed to use standard rules so that everyone can understand their drawings. Certain kinds of lines have special meanings. Although you can draw everything in a view using a single line type, a drawing that follows drafting standards will be easier to understand. Fig. 11–7.

Fig. 11-5. A multiview drawing, or orthographic projection, shows one or more of the six possible views of an object. The views are usually lined up as shown here.

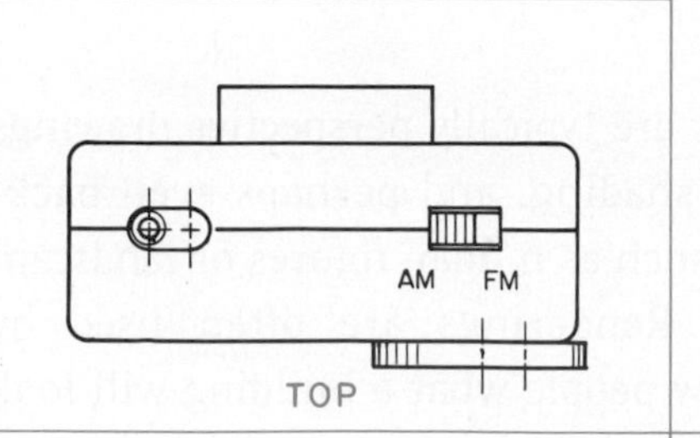

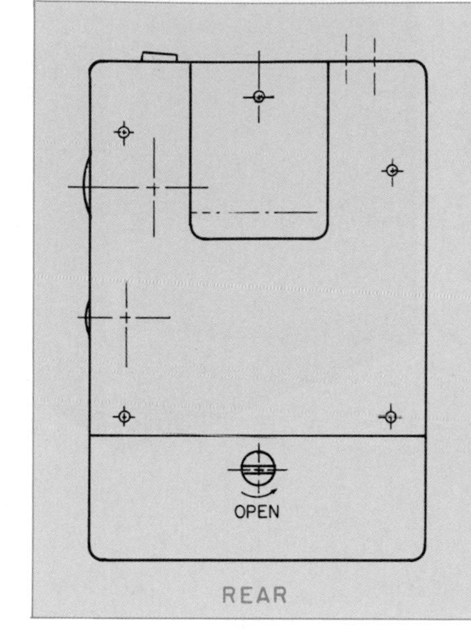

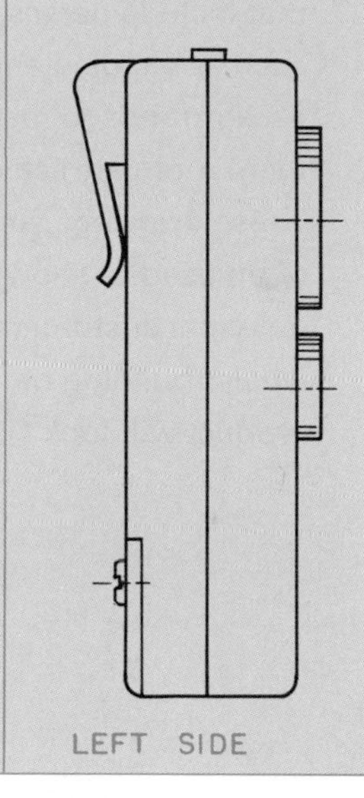

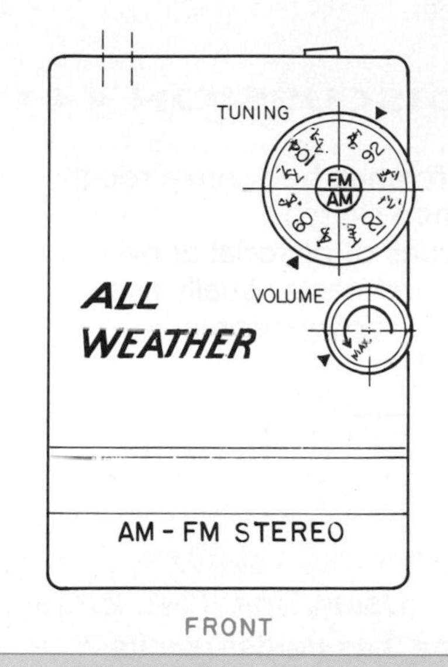

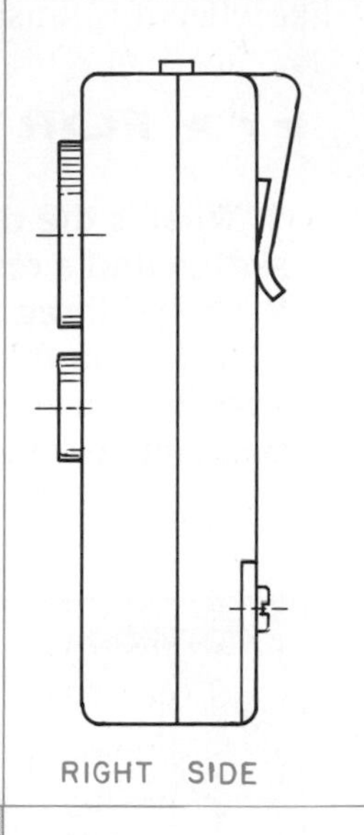

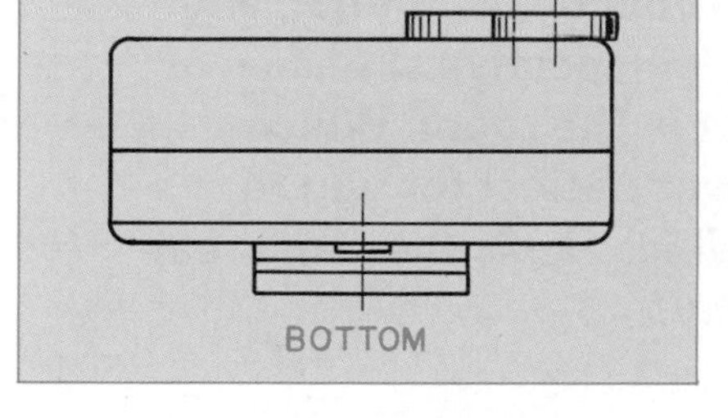

Fig. 11-6. This plastic mat, designed to allow chairs with casters to roll easily over a carpeted area, can be described using only one view. A side view could be drawn to show the mat's thickness, but it is simpler to write a note saying the stock is 1/8" thick plastic.

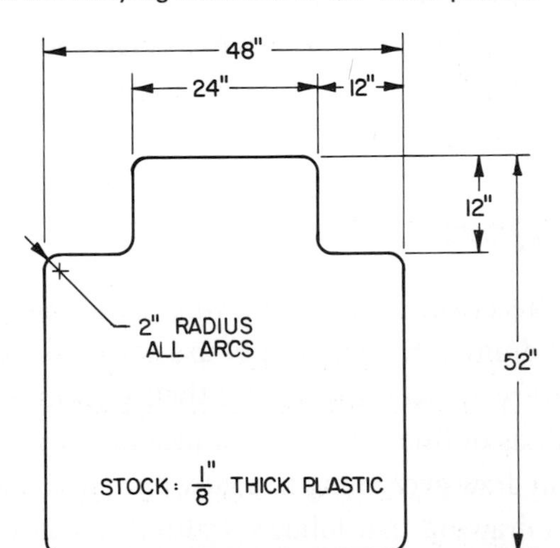

Fig. 11-7. Lines can have meaning, just like letters, words, and other symbols. The lines shown here have special meanings on technical drawings.

SOLID / OBJECT / VISIBLE LINE

HIDDEN / INVISIBLE LINE

CENTERLINE / LINE OF SYMMETRY

DIMENSION / EXTENSION LINE

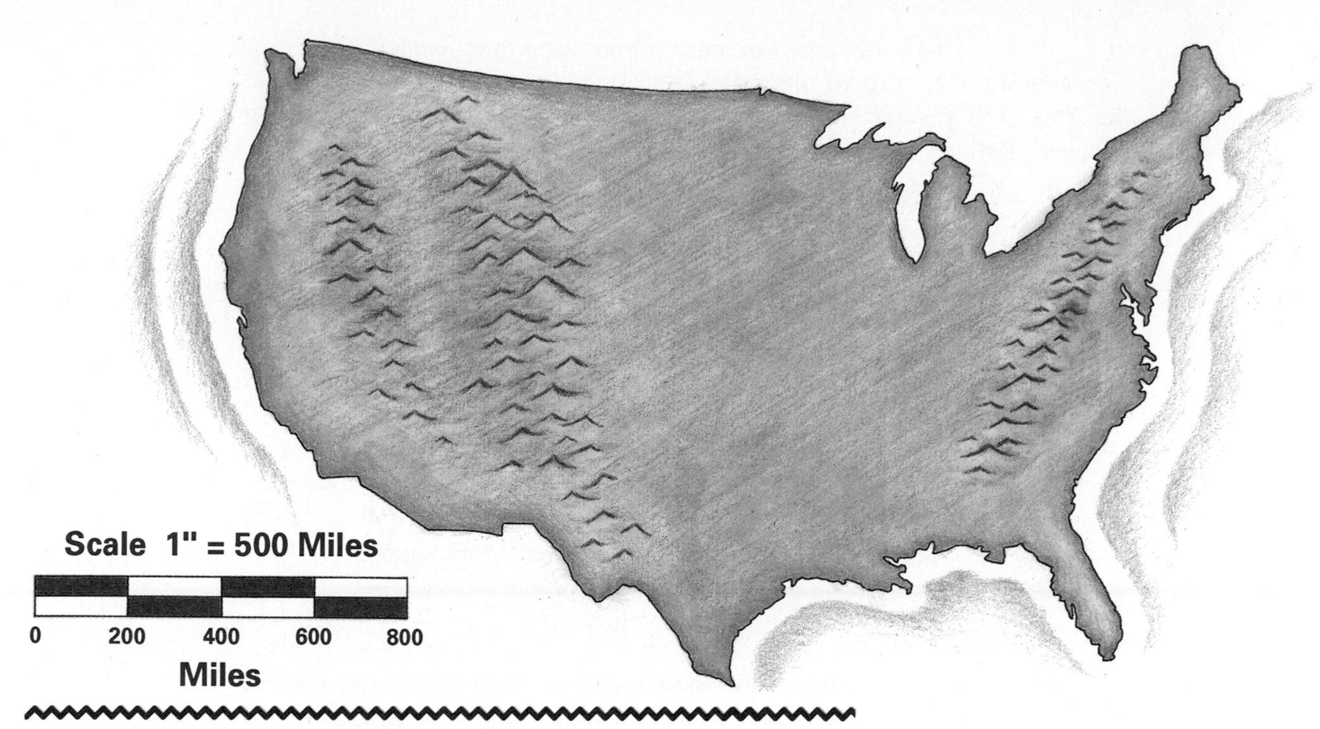

Fig. 11-8. A bar is a convenient way to show scale. You could also write a proportion, such as 1" = 500 miles. This means that one inch on the map as drawn equals 500 miles in actual distance.

Dimensioning and Scale

The purpose of all technical drawing is to communicate information. Since much of this information has to do with the measurements of objects, **dimensioning** is usually needed. Dimensions show things such as size, shape, and location. Figure 11–7 shows the type of line that is commonly used to indicate dimensions.

Some items, too large in real life to fit on a reasonably sized piece of paper, must be drawn to **scale**. This means everything is in the correct proportion, but not drawn full size. Fig. 11–8. Other drawings may show a small object or area larger than it really is for clarity. In fact, it is not unusual to see drawings that have views or parts of views drawn at different scales on the same piece of paper. The scale you select depends on the size of the actual object and the size of the paper you are using.

Lettering

To be readable, technical drawings must include acceptable lettering. Fig. 11–9. There are many rules concerning lettering. Here are a few basic ones.

- The lettering should not be too big or too small. For most drawings, letters and whole numbers should be 1/8" high. Fractions should be 1/4" high.
- The space *between* words should be equal to the width of the letter "O."
- Letters *within* a word should be spaced so that they seem to be equal. Some letters are narrower that others, so the actual space will vary.
- Use a sharp pencil. Keep your lettering neat and clean.

HOW TO FORM LETTERS IF YOU ARE RIGHT-HANDED

ARROWS ARE USED TO SHOW THE DIRECTION AND ORDER OF ALL STROKES.

A B C D E F G H I J K L M N O P

Q R S T U V W X Y Z &

HOW TO FORM LETTERS IF YOU ARE LEFT-HANDED

A B C D E F G H I J K L M N O P

Q R S T U V W X Y Z &

HOW TO FORM NUMBERS IF YOU ARE RIGHT-HANDED

1 2 3 4 5 6 7 8 9 0 $\frac{1}{2}$ $\frac{3}{4}$ $4\frac{11}{16}$

HOW TO FORM NUMBERS IF YOU ARE LEFT-HANDED

1 2 3 4 5 6 7 8 9 0 $\frac{1}{2}$ $\frac{3}{4}$ $4\frac{11}{16}$

Fig. 11-9. Most notes on drawings are written in capital letters. On a sheet of notebook paper, practice making letters and numbers. Make the strokes in the order shown here. The "boxes" around the letters and numbers are guidelines to help you see how they are proportioned.

Types of Drawings

Drafters make several different kinds of drawings to meet different needs. *Detail drawings* show individual parts or sections of objects in a larger scale to allow greater clarity. *Working drawings* are plans that show sufficient information for something to be built or manufactured. *Assembly drawings* show how certain devices fit together. Assembly drawings are often pictorial. They are used in repair manuals, do-it-yourself kits, and even advertisements. Fig. 11–10.

Bill of Materials

Some drawings, particularly working and assembly drawings, include a parts list, also known as a **bill of materials.** The bill of materials is a concise list of the sizes and characteristics of the actual parts that make up the device. Fig. 11–11.

▶▶▶ FOR DISCUSSION ◀◀◀

1. Name all the views possible in multiview drawing.

2. Why should you include a bill of materials as part of a set of drawings?

■ Graphically illustrate your solution to the community activity in Chapter 10, which involved crossing a river. You may use a CAD system or draw your solution by hand.

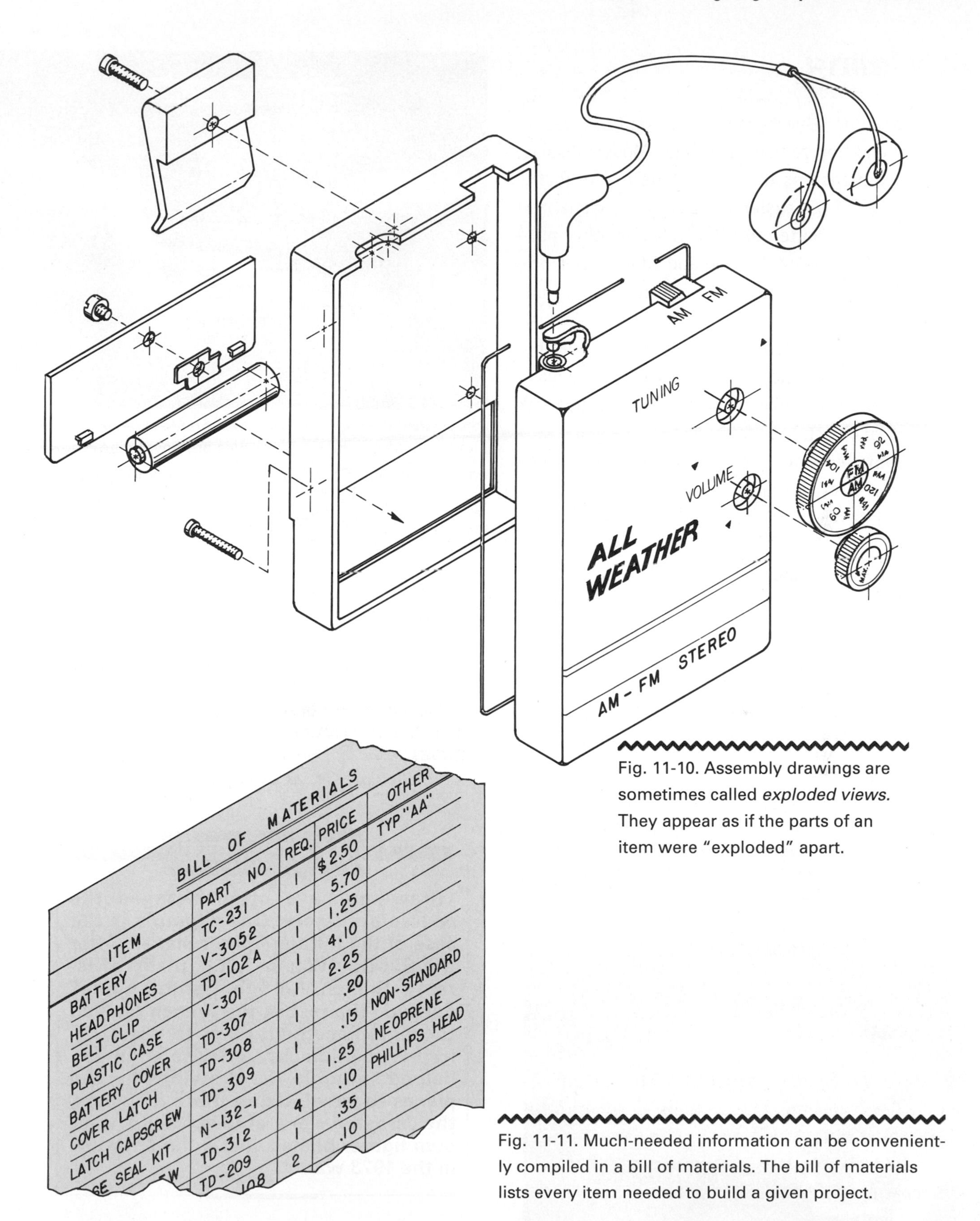

Fig. 11-10. Assembly drawings are sometimes called *exploded views.* They appear as if the parts of an item were "exploded" apart.

BILL OF MATERIALS

ITEM	PART NO.	REQ.	PRICE	OTHER
BATTERY	TC-231	1	$2.50	TYP "AA"
HEADPHONES	V-3052	1	5.70	
BELT CLIP	TD-102 A	1	1.25	
PLASTIC CASE	V-301	1	4.10	
BATTERY COVER	TD-307	1	2.25	
COVER LATCH	TD-308	1	.20	NON-STANDARD
LATCH CAPSCREW	TD-309	1	.15	NEOPRENE
SE SEAL KIT	N-132-1	1	1.25	PHILLIPS HEAD
EW	TD-312	4	.10	
	TD-209	1	.35	
	108	2	.10	

Fig. 11-11. Much-needed information can be conveniently compiled in a bill of materials. The bill of materials lists every item needed to build a given project.

Modeling

For some projects, you might want to create a model in addition to your drawings. Models display all three dimensions in a more true-to-life form, making the object easier for people to visualize. Thus, models are often used for sales or other presentations, advertising, testing, and education. Models may be the actual size of the object, or they may be scaled up or down. Fig. 11–12. Models built just for appearance are often made from inexpensive and easy-to-work items such as cardboard, clay, or plaster. Working models may use parts that will be used in the actual object.

Computer modeling creates a pictorial image of an object electronically. In addition to being able to manipulate the image displayed on the screen display, sophisticated computer programs also allow people to test and experiment with the object before it is built. Fig. 11–13. This means that devices can be designed, modeled, and tested on the computer before the first one is manufactured. Therefore, computer modeling can greatly speed up product development.

Fig. 11-13. More and more models are now being created electronically. Computer models save developers time and money they would otherwise have to spend testing and revising a real model.

▶▶▶ FOR DISCUSSION ◀◀◀

1. How might a product model be used in a sales presentation?
2. Explain why a teacher might wish to have a model of a device for demonstration purposes.

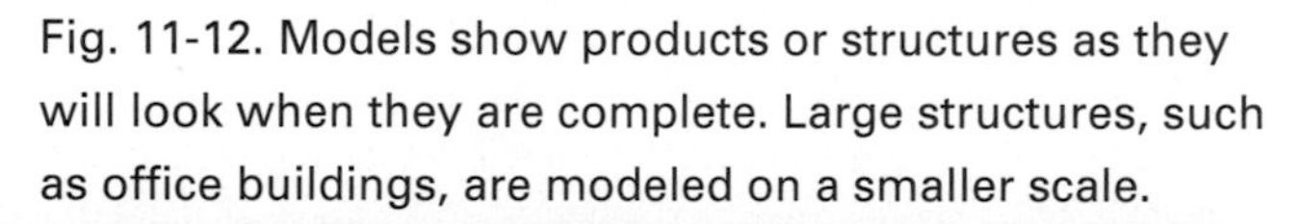

Fig. 11-12. Models show products or structures as they will look when they are complete. Large structures, such as office buildings, are modeled on a smaller scale.

IMPACT

Drawings have often changed the course of history. One example is the case of the Mirage fighter planes. After the Six-Day War in 1967, many of Israel's French Mirage fighter planes needed repairs, but the French refused to sell replacement parts to Israel. Alfred Frauenknecht, a Swiss engineer, helped smuggle the drawings for the planes to Israel. Using these drawings, the Israelis were able to build their own fighter planes, just in time for use in the 1973 war.

CHAPTER 11

Chapter Highlights

- Ideas are often expressed graphically or by making a model.
- Sketching is a convenient method for quick graphic expression.
- Graphic communication can be achieved using one of several types of pictorial or multiview drawings.
- To allow drawings to communicate clearly, numerous rules for line type, dimensioning, and lettering have been formed.
- Computers are now being used for both ordinary drawing and modeling purposes.

Test Your Knowledge

1. Why are rough sketches preferred as a method for communicating ideas?
2. Name three types of pictorial sketches.
3. How does a rendering vary from a simple perspective sketch?
4. What is a common reason for needing a multiview drawing?
5. Why is it usually necessary to draw only three sides of an object?
6. Identify three line types, and tell how each is used.
7. What is the purpose of dimensioning?
8. What scale might you use to draw an accurate sketch of your classroom?
9. Why do drawings often include notes?
10. What could a bill of materials be used for?

Correlations

SCIENCE

1. Have a friend place a carpenter's level on top of your head to mark your height on a wall. Measure your height in meters, centimeters, and millimeters.

MATH

1. Make a scale drawing of your classroom. Let 1 inch = 4 feet. What are the dimensions of your drawing?

LANGUAGE ARTS

1. Draw a floor plan of your home. How could you use written language (words and numbers) to add more information to the floor plan?

SOCIAL STUDIES

1. Research an invention—past or present. What was the step-by-step process the inventor used to develop his or her invention? Be specific. Were sketches and drawings used? Did the inventor rely on models to perfect the invention? Approximately how long did it take to complete the invention?

Make a Difference

Odyssey of the Mind

"Your mission. . . should you decide to accept it. . . will be to design, build, and drive a battery-powered, lightweight vehicle that performs various technical tasks and to create a theme for its presentation."

When the six-member team of sixth graders from Washington School in Pekin, Illinois, decided to take the Odyssey of the Mind challenge, they focused on OMer's Buggy Lite as their long-term problem. They did not even dream of reaching the world level of competition. Their goal was to cooperate to meet the long list of Odyssey of the Mind (OM) rules and regulations. Their hard work and enthusiasm paid off, and they finished first in the state of Illinois and fifteenth in the world.

The Odyssey of the Mind Program encourages students to think creatively to solve problems. OM competition consists of three areas: the long-term problem, style, and the spontaneous problem. High scores at the regional level allow students to compete at the state level. Top scores at the state level mean a chance to compete at the world competition.

Members of the award-winning team (Erin Elmore, Amy Kriegsman, Ross Schaefer, Molly Soldwedel, Andrew Thompson, and Matt Tibbs) were in the same sixth-grade class, but they did not really know each other very well. Each student wanted to work on the long-term problem of OMer's Buggy Lite, so they decided they would be a team!

Molly's father teaches cooperative learning in their school district. He helped the team members learn to brainstorm and support each other. Matt's father has a home workshop, which he opened to the group. He taught the entire team how to measure materials, operate hand tools and power tools, and follow safety rules. Amy's mother taught the members of the group how to use a sewing machine. Each student made a flag and costume to represent a country of the Middle East as part of the *style* of the long-term solution. Rosalie Wendelin, the coordinator of Pekin's OM program, drilled the team on possible spontaneous problems they might receive at the competition.

The OM team worked many hours to complete their solution to the long-term problem in time for the regional competition. The sponsors were not allowed to do any of the work or help with the brainstorming or decision-making processes. The team scored extremely well at the regional competition held in Bloomington, IL, and placed second. They lost thirty points because their vehicle weighed 75 pounds (30 pounds over the maximum weight to score any points).

Knowing that they could change their design before the state competition, the team observed the design of the other entries. They replaced the plywood sides of their vehicle with Styrofoam™. They also added cardboard scenery backdrops to their presentation. Again, their hard work paid off, and the group won first place at the state competition in April.

At the world competition, held at the University of Tennessee, the Pekin team competed against sixth, seventh, and eighth graders from all over the world. Their vehicle performed perfectly as their presentation—which showed peace between the nations involved in the Middle East conflict—brought forth tears and cheers from the audience.

When the lively sixth graders were asked which part of the OM competition they liked best, they each had a different answer. Erin enjoyed designing and painting the scenery. Amy liked working with the power tools. Ross found the spontaneous problems to be most challenging, and Molly liked brainstorming. Andrew discovered that he could run a sewing machine as well as a "nasty" table saw, and Matt was able to "drive" before turning sixteen.

All of the team members agreed immediately upon two things: (l) the BEST part of OM was making it to the world competition, and (2) they can't wait to start working together as a team on the next long-term problem!

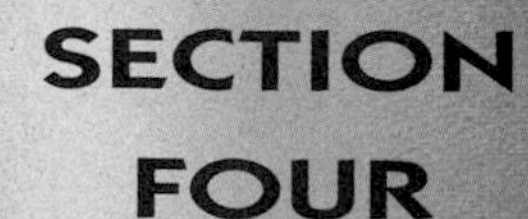

COMMUNICATION SYSTEMS

Safety!

Activity Brief

Communicating a Message

PART 1: Here's the Situation ....

A communication system is basically a way in which a message is sent from one point or person to another. Since the invention of movable metal type in the 1440s, we have depended a great deal on the written word, particularly in the form of books, magazines, and newspapers. As new technology has been developed, communication systems have become increasingly complex and have added a global perspective to our daily lives. Today, computers and video technologies have given us the means by which visual messages can be sent nearly anywhere in the world almost instantly.

As you do this activity and as you read Chapters 12 through 17, you'll learn more about the various ways in which messages are communicated. You'll also gain an understanding of the impacts communication has on our lives.

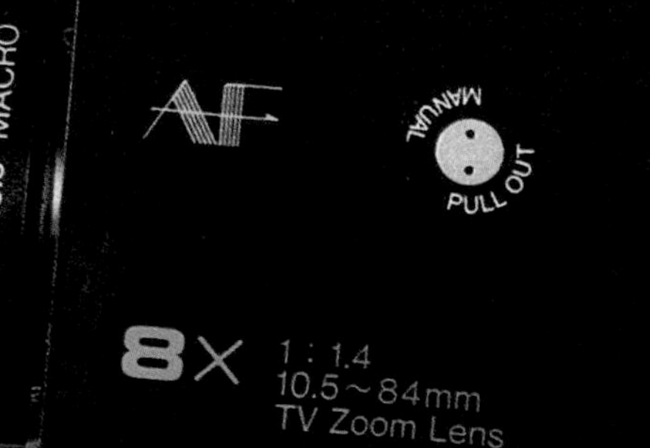

PART 2: Your Challenge...........

This activity will help you learn about video means of communication. Working with one or two classmates, you will make a videotape recording (a video) of a message that will either inform viewers or explain to them how to do something. Your presentation should not be one in which you simply stand in front of a camera and talk. Be creative and think of other ways in which you can effectively express your message. How can you capture and hold the attention of viewers? What are some of the various techniques you could use? Have fun. Try several ideas. In the end, however, make sure your intended message gets across to your viewers.

PART 3: Specifications and Limits...........

Your video and the way in which you produce it will need to meet certain standards. Read the following specifications and limits carefully before you begin.

1. Working in small groups of 2 or 3 students, you are to develop and produce a video that is 4–8 minutes in length.
2. As you work, the following must be approved by your teacher:
   - message
   - storyboards
   - script
   - completed video

PART 4: Materials...........

Equipment and supplies you might use include the following:

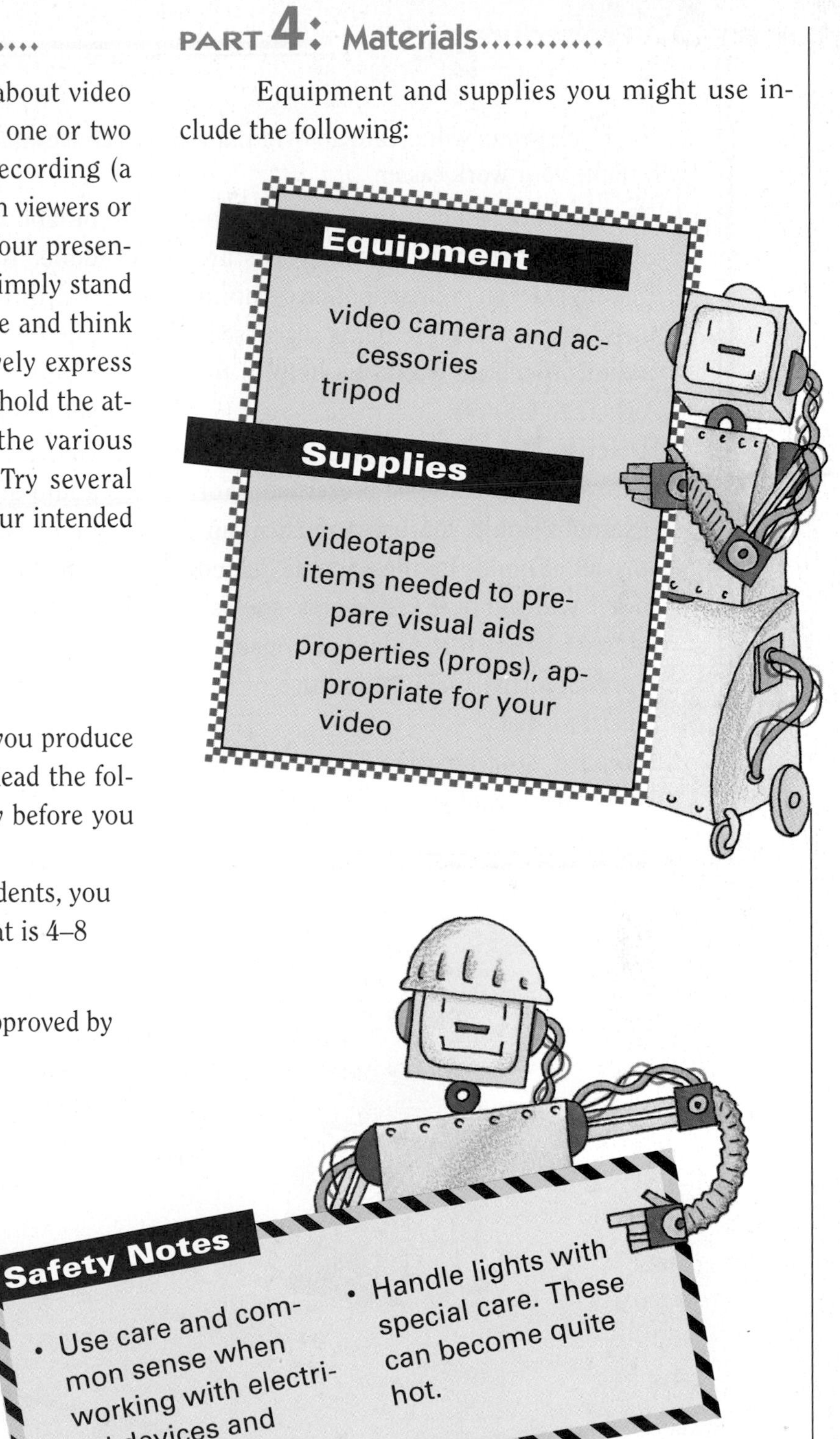

Equipment
video camera and accessories
tripod
Supplies
videotape
items needed to prepare visual aids
properties (props), appropriate for your video
Safety Notes
• Use care and common sense when working with electrical devices and equipment.
• Handle lights with special care. These can become quite hot.

PART 5: Procedure

Each group will create an original video to convey a message. Following these steps will make your work easier.

1. As a group, select a message to produce. (For example, what events are taking place in your school or community? What issues are being discussed? What procedure would be helpful for others to know?)
2. Discuss the various ways in which your message could be presented. (For example, could you use a musical approach? Could the message be "acted out"? Will you need visual aids, such as signs?) Brainstorm ideas. Choose an approach that makes good use of the visual format.
3. Prepare storyboards. These should summarize what your video is going to be about and how it is going to be done. You should prepare one storyboard for each shot, scene, or setup. Fig. IV-1.
4. Write a detailed script that follows your storyboard sequence. Fig. IV-2.
5. Gather the props and prepare the audio and/or visual aids you need to help convey the message. Mark instructions for their use on the script.
6. Your teacher will demonstrate how to set up and operate video equipment properly. Practice using the equipment before you tape your video.
7. Rehearse your production.
8. Record your message on videotape.

Storyboard using a 3" x 5" index card

STORYBOARD FOR: BASKETBALL TRYOUTS

Page

Scene: 1

Action/Treatment

BEGIN WITH A CLOSE UP OF OUR VARSITY SQUAD. PAN OUT TO A LONG SHOT OF THE PLAYERS SHOOTING THE BALL.

DESCRIBE THE FUN AND CHALLENGE OF PLAYING BASKETBALL.

Scene: 2

Action/Treatment

TIGHT SHOT OF A PLAYER SLAM DUNKING THE BALL.

DIALOGUE: "YOU TOO CAN JOIN THE TEAM!"

Scene: 3

Action/Treatment

MEDIUM SHOT OF THE COACH WITH THE TEAM BEHIND HER.

DESCRIBE WHEN AND WHERE TO TRY OUT.

Storyboard sheet

Fig. IV-1. Each storyboard should include a sequence number, a sketch of the scene, a description of the action to take place, and how the scene will be shot. Include other information, such as dialogue, if you consider it helpful to the production.

Research Tip:
Find out about various storyboarding styles and techniques.

Group: ___________ **Topic:** ___________ **Page:** ___________

Setting ____

____

Dialogue or Narrative	**Actions/Production Notes**
Speaker # _______ :	
Speaker # _______ :	
Speaker # _______ :	
Speaker # _______ :	

Research Tip:
Find out about various script formats.

Fig. IV-2. Possible script format.

PART 6: For Additional Help..........

For more help with this activity, look up the following terms. You'll find some of them in this book. (Check the index.) You'll find others in dictionaries, encyclopedias, and other resource materials.

brainstorm
dialogue
narrative
property (prop), as used in a play
public service message
scene
scenery
script
special effects
storyboard
video
video camera
videotape

PART 7: How Well Did You Meet the Challenge?..........

After you've finished taping your message, show it to the class. Discuss responses to the following questions:

1. Was the video enjoyable?
2. Could everyone in the video be seen and heard?
3. Were all props and audio and visual aids used effectively?
4. Was the intended message conveyed to the viewers?
5. In what ways could the video be improved?

PART 8: Extending Your Experience..........

This activity helps you learn the following:

- how to use visual techniques to effectively present a message
- how to operate video equipment
- the basic processes and "behind-the-scenes" steps that are involved with any type of video production

Think about the following questions and discuss them in class. You'll find more information about communication techniques and technologies in Chapters 12–17 in this section, "Communication Systems."

1. If you were to prepare another video, what other techniques would you like to try? Why?
2. How might the steps which you followed in your production help you as a worker in any occupation?
3. What are some positive and negative impacts that have resulted from the increased use of video technology?

CHAPTER 12

Types of Communication Systems

Introduction

Communication is the process by which people, animals, and machines send and receive information. Without communication, it would be almost impossible to carry on even the simplest routines.

After reading this chapter, you should be able to

Demonstrate oral and written communication.
Give an example of telecommunication.
Describe a basic model of communication.
Discuss the history of communication.
List three types of electronic communication.

Words you will need

signal
oral communication
graphic communication
transmit
telecommunication

What Is Communication?

One of the basic needs in life is to communicate. What do we mean by "communication"? The definitions may vary. Basically, though, communication is any means of sending a message, or **signal**, from one point or person to another.

Communication may take place between people, animals, machines, or some combination of these. For example, you are using person-to-person communication when you talk to a friend. When you play a computer game, you are communicating with a machine.

▶▶▶ FOR DISCUSSION ◀◀◀

1. Give at least one example of each of the following kinds of communication:
- **people-to-animal**
- **animal-to-animal**
- **machine-to-machine**

2. What do you think is meant by the term "foreign language"?

The Communication Process

How is communication accomplished? Let's find out. Suppose you wanted to know if the local public library rents or loans videotapes to its users or patrons. How could you communicate your question to someone who is likely to know the answer?

Basic Types of Communication

You could, of course, simply ask the first person you meet. This would be spoken, or **oral communication.**

Let's say you did just that, but the person answered in a strange language. You could not understand the answer. Would this be communication? Well, messages were certainly sent from one person to another. The messages, however, were not understood. You and the other person would need to restate your messages in different ways until you could understand each other. Fig. 12–1.

On the other hand, you could try some other way to get the answer to your question. How about sending a letter to the librarian asking about the videotapes? This would be written, or **graphic communication**.

Fig. 12-1. In the Biblical Tower of Babel, everyone spoke a different language. It became impossible to communicate. Today, language differences can still make communication difficult.

How would you send your letter? You would probably mail it at the nearest mailbox. The postal service would send, or **transmit**, your message by taking it to the library.

Are there any other ways to send your letter? If you and the library both had computers, you could send your message through a modem. A *modem* is a device that enables one computer to communicate with another computer through telephone lines. Fig. 12–2.

You could also communicate with the library in other ways. You could use the telephone, for example. All of these ways of communication have certain things in common. When we list these things, a pattern, or *model*, develops.

Communication Model

All communication involves some type of message. The message is transmitted from a sender to a receiver. Since communication is rarely perfect, there may be some need for adjustment, as shown in Fig. 12–3. The same idea in different words might be expressed as shown in Figure 12–4.

In all cases, what occurs is simply a message being sent and received, with occasional need for clarification. (This is the "adjustment," or feedback, part of the loop. Chapter 9 tells more about feedback.) As you read the next few chapters, keep this model in mind. It will help you understand how communication technology works.

▸▸▸ FOR DISCUSSION ◂◂◂

1. Suppose you were talking to a friend on the telephone. There was static on the line, and you could not hear what your friend was saying. In the communication model, what parts of your conversation would correspond to input, process, output, and feedback?

2. What advantages might someone gain by communicating using a modem?

Fig. 12-2. A modem allows people to communicate over long distances using computers and telephone lines. The modem may be a separate unit, or it may be built into the computer.

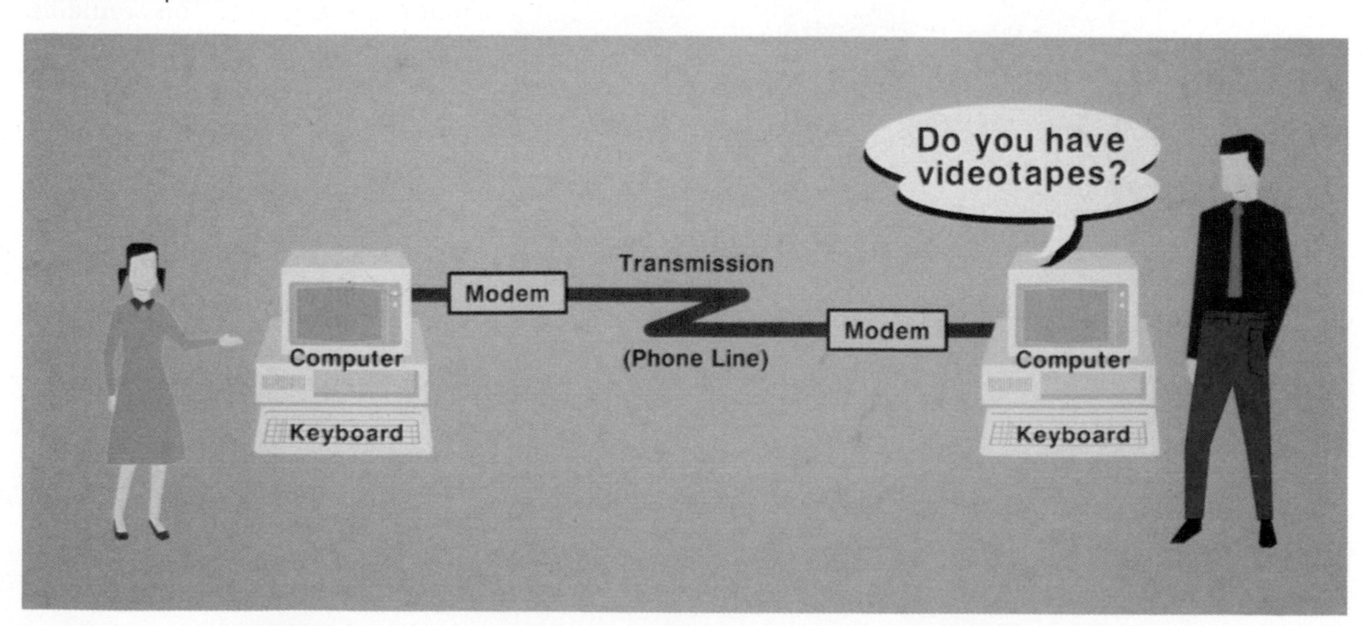

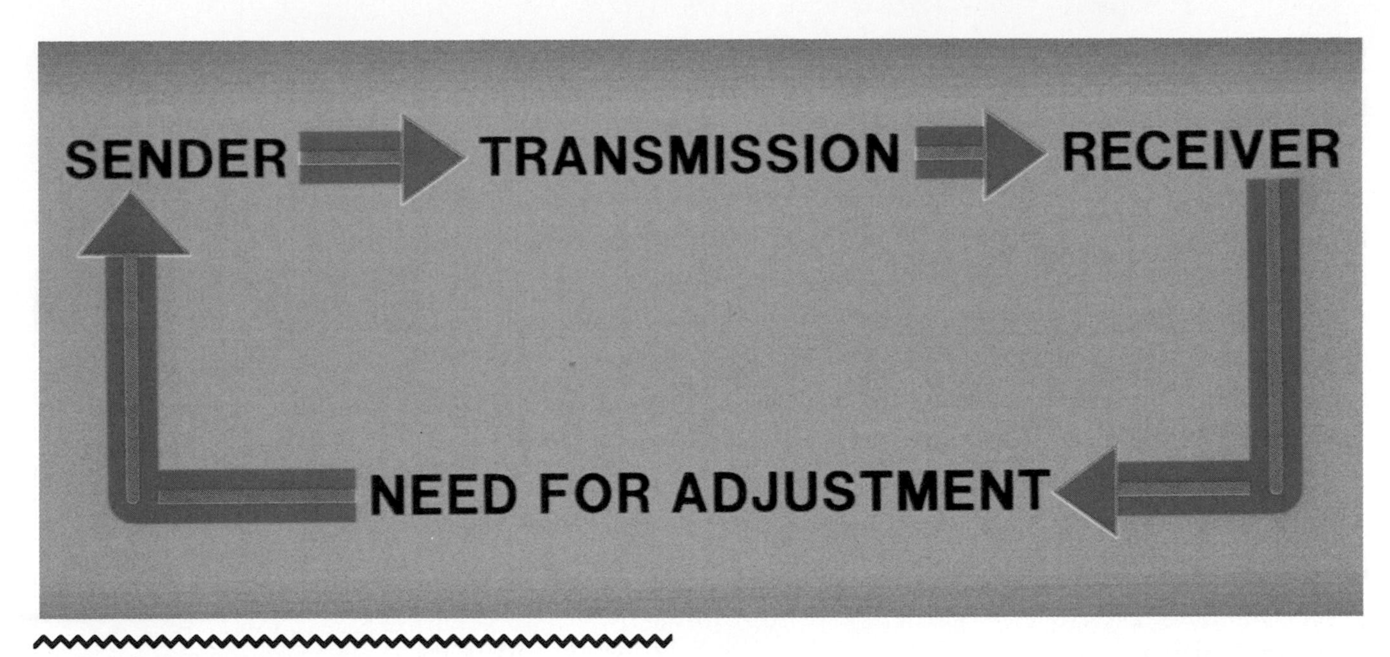

Fig. 12-3. One model for communication.

Fig. 12-4. The communication model can also be expressed this way.

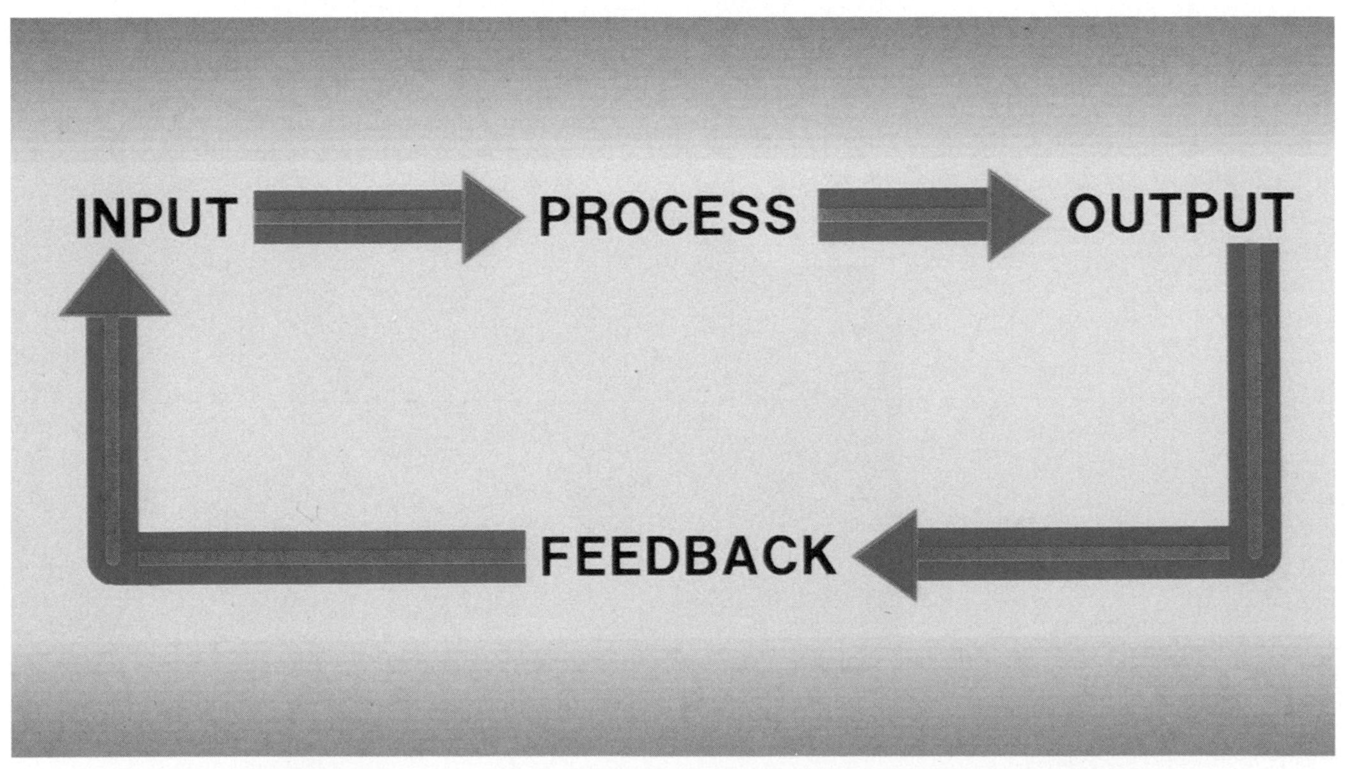

Early Forms of Communication

Before the development of spoken and written language, how do you think people conveyed their ideas? How could you communicate without words right now, if you needed to do so? Perhaps you could draw pictures, as we know the cave dwellers did. Perhaps you could use some sort of sign language, such as rubbing your stomach to indicate hunger. What about "body language" and facial expression? People still use these means of communication today to help express their meaning. Fig. 12-5.

Hieroglyphics

Early written communication began with Egyptian hieroglyphics around 3000 B.C. Hieroglyphics were a system of symbols that represented ideas and sounds. The hieroglyphics were typically written with ink on a paperlike substance made from papyrus reeds. Fig. 12-6.

Cuneiform Writing

Around the same time, the Mesopotamians developed a system of writing called *cuneiform*. Cuneiform writing consisted of wedge-shaped characters that represented things or ideas. The characters were scratched into soft clay surfaces, which later hardened to make permanent records. Fig. 12-7.

People have come a long way since these early forms of written language were used. Today people speak and write thousands of languages and dialects (similar versions of a single language).

TECHNOLOGY TRIVIA

Although stone, wood, or metal were sometimes used, cuneiform writers preferred soft clay. When the slanted edge of a stylus is pressed into soft clay, it leaves a wedge-shaped mark. Thus, the wedge-shaped appearance of cuneiform writing is the result of the means by which the writing was done—an early example of how the medium can influence the message.

Fig. 12-5. Communication is not limited to speech. How are these women communicating?

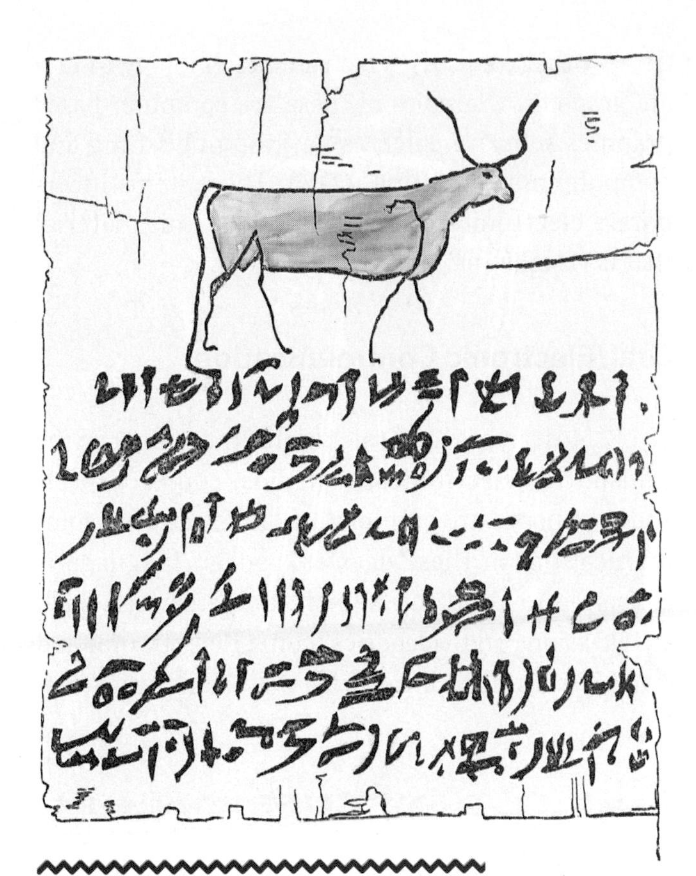

Fig. 12-6. Egyptian hieroglyphics were made up of pictures that symbolized ideas, actions, or sounds. Since the same symbol could have several different meanings, hieroglyphics are difficult to interpret.

Fig. 12-7. The characters in cuneiform writing began as picture symbols. As time passed, however, each symbol became associated with a single syllable. This syllable-writing was a vast improvement over the picture-symbol method because it was much more flexible.

■ Work with a classmate to develop a secret written code. Using the code, compose a message and have your teacher check it. Then write it on the chalkboard. See if other members of the class can decipher the message (make out its meaning).

▶▶▶ FOR DISCUSSION ◀◀◀

1. Why do you think some messages later require clarification?
2. What might be the result if you tried to start your own language?

Modern Communication

Using techniques both old and new, modern communication allows us great choice in selecting how to go about our daily routines. Some processes are purely visual, or graphic. Others are oral. However, much of modern communication is a combination of electronics with graphic communication, oral communication, or both. Although these techniques will be reviewed in detail in later chapters, a brief summary of each is given here.

Graphic Communication

We are surrounded by graphic media. Books, magazines, signs, and billboards communicate through printed words and pictures. Pictures appear in advertising and product labels. Graphic communication may involve photography, drafting, typesetting, printing, and even body or sign language. These topics will be reviewed in the next chapter.

Electronic Communication

Sometimes a signal or message is transmitted by an electronic device. The device changes the signal into a form that allows it to be sent through wires, fiber-optic cables, or even through the air. When the message arrives at the receiver's end, another device changes the signal back into a form the receiver can understand. Electronic devices allow **telecommunication**, or communication over long distances, to be an everyday event.

IMPACT

Electronic communication is helping to make the world more democratic. As more people are able to communicate in more ways, it becomes harder to hide the truth or promote lies. A dictator may control the newspapers or television stations, but computer networks, fax machines, and car phones are much harder to control.

Graphic/Electronic Communication

Several machines have been developed to transmit graphic messages electronically. Items such as computer modems, telegraphs, and facsimile (fax) machines fall into this category.

Other items use electronic technology to create graphics. Examples of these are computer-based graphics software, such as desktop publishing and computer-aided drafting (CAD). These items incorporate electronics to create and transmit material that is visual in nature.

Oral/Electronic Communication

Most types of oral communication can be transmitted electronically. Radios, compact discs, and telephones are examples of oral/electronic communication. In these devices, spoken language is converted to electronic signals. Some devices, such as televisions and videodiscs, combine both oral and graphic communication with electronic transmission.

▶▶▶ FOR DISCUSSION ◀◀◀

1. Describe two ways in which modern devices have made oral communication easier and faster.
2. Give an example of oral and graphic communication being used together without electronic transmission. Why is electronic transmission usually used?

■ **What types of communication systems will you have in your model community? Choose one and develop a plan for constructing it.**

CHAPTER 12 Review

Chapter Highlights

- Communication is an important process in our lives.
- Through various means, people, animals, and machines are able to send and receive information, both locally and over long distances.
- All communication consists of a sender transmitting a message to a receiver; sometimes an adjustment is needed to clarify the message.
- Communication has progressed a great deal in a technological sense, allowing much to be done electronically.

Test Your Knowledge

1. Define *communication.*
2. What is the difference between a spoken question and a written question?
3. What determines whether a language is considered foreign?
4. How are modems used in communication?
5. Identify two ways to find out whether your school library has a book you want to read.
6. Sketch your version of a communication model.
7. When would it be appropriate to utilize body language?
8. List five spoken languages.
9. Name five examples of graphic communication.
10. What are the differences between graphic and electronic communication?

Correlations

SCIENCE

1. A gorilla was taught sign language by her trainer. Find a magazine article about this controversial experiment and tell the class about it.

MATH

1. Suppose you are a Stone Age human scouting for mammoths. You find a herd of them, and you want to go back and tell the others in your tribe how many there are. Numbers haven't been invented yet. How will you count the mammoths?

LANGUAGE ARTS

1. Do you, or someone you know, speak a language other than English? If so, translate this sentence into the other language: "Let there be peace on Earth." Prepare a bulletin board display of all the different translations your class can provide.

SOCIAL STUDIES

1. Magazines are one form of graphic communication. What were the first magazines published in the United States and when were they published? What was their purpose? Write a report about your findings.
2. Are magazines still an important form of communication today? Tell why or why not.

CHAPTER 13

Graphic Communication Processes

Introduction

Communication through printing, photography, and technical drawing (drafting) is a major activity in the world today. Can you imagine what your daily life might be like if newspapers and magazines could not be mass-produced? Can you envision going to a library or bookstore that had just a handful of books because books could not be printed in sufficient quantities?

After reading this chapter, you should be able to

Describe the history of graphic communication.
List various printing techniques.
Describe the process of photography.
Explain what is meant by design, drawing, and computer-aided design and drawing.

Words you will need

movable metal type
printing press
offset printing
screen printing
photography
drafting
computer-aided drafting (CAD)
desktop publishing

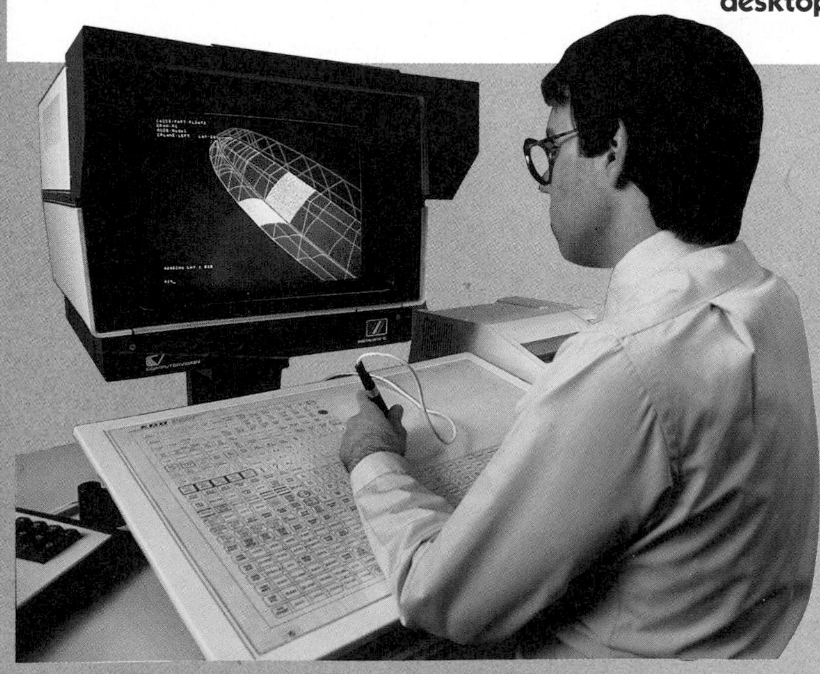

Printing

People use many different methods of communication. How do people decide which method to use? Let's suppose, for example, you wanted to start a newsletter for your classmates. How would you go about it?

Would you telephone all the students in your school to read them the stories in the newsletter? No, that wouldn't be very practical. Could you paint the news on a large bulletin board, and simply repaint it when more stories were written? Yes, perhaps, but that approach also seems quite awkward.

Most school newsletters are produced using printing processes that allow many copies to be made quickly. The reason for this will become clear when you examine printing technology more closely.

Early Printing Processes

Printing as a process began several hundred years ago when people wrote or copied whole documents by hand. In fact, hand copying has been a source of employment and income for large numbers of people through the ages.

Hand copying was an expensive and time-consuming process, as you might well imagine. Eventually, people developed the technique of carving pictures and individual letters out of wood. Then they coated the wood with ink and transferred the image onto paper. Fig. 13–1.

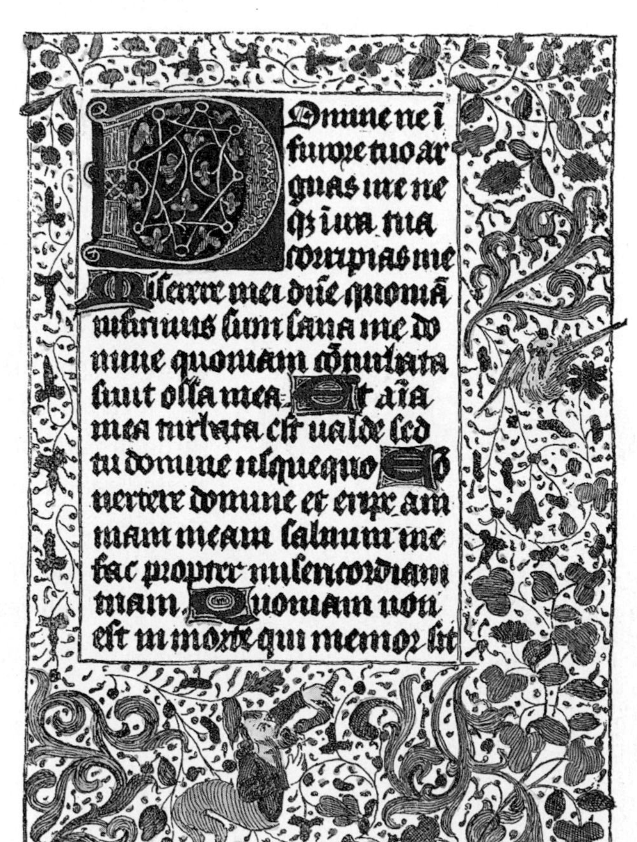

Fig. 13-1. Prior to the 1400s, documents were hand-copied or printed from individually carved letters. This page is from the Book of Hours, published in 1350. Books such as this one were rare and expensive because they required so much time and skill to create.

This remained quite a lengthy procedure, since each letter had to be carved individually. Still, it was an improvement. The same pieces could be used again and again.

In the mid-1400s, people began creating metal letters by pouring molten metal into molds. Since each mold could be used to create many copies of itself, the printing process became much easier and faster.

Until recently, this **movable metal type** was used by everyone who wanted to print many copies of a text. Fig. 13–2. The type was set up in a **printing press**, a machine that pressed the movable type against paper.

Early printing presses were fed one sheet at a time. They were powered entirely by humans or animals. When steam and electrical power became available, the presses became more automated.

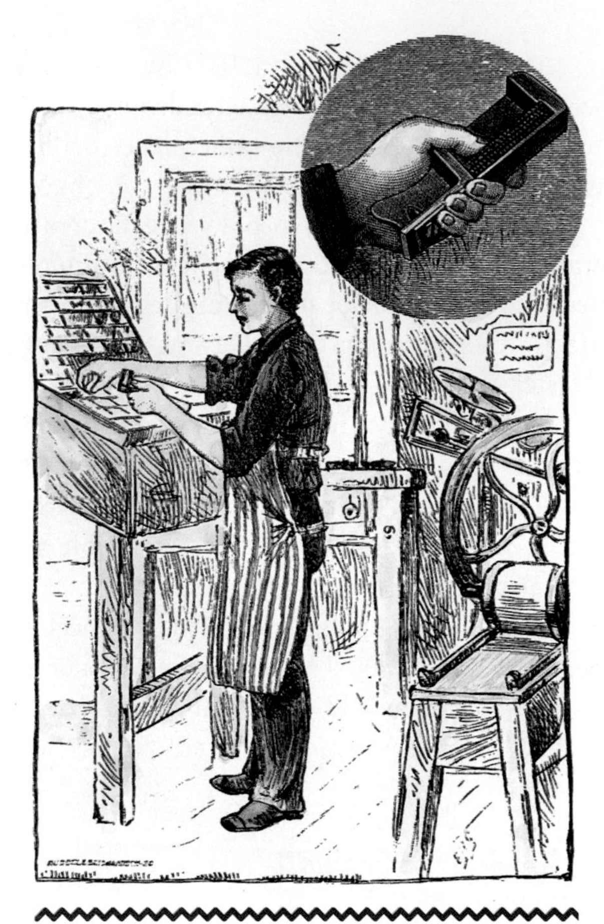

Fig. 13-2. The use of movable metal type greatly decreased the time and expense needed to typeset text.

Newer typesetting methods are electronic. They no longer require the physical placement of individual letters into a frame or machine. Modern high-speed presses use paper from huge rolls. The paper uncoils at a blurring rate as the paper is pulled through the press.

TECHNOLOGY TRIVIA

Johann Gutenberg, a German, invented the first successful printing press around the year 1440. The technology spread rapidly over Europe. By the end of the century, between 15 and 20 million books had been printed.

Offset Printing

The printing process itself has changed greatly. Today there are many ways to print an image (words and/or pictures). One of the most common is **offset printing**. In offset printing, the image to be printed is transferred from an inked form to a rubber blanket. From there it is transferred to the paper.

The most popular technique of offset printing is *offset lithography*. The image is first copied onto a printing plate. The plate is a thin, flat sheet, usually made of metal. The plate is chemically treated so that the image area will have a greasy surface. The plate is then curved around a cylinder on the press.

To print an image, the printer moistens the plate with water. When ink is applied, the water repels it. The ink sticks only to the greasy image area. The inked area then "offsets" its image to another cylinder. From this cylinder, the image is printed onto the paper. Fig. 13–3.

Screen Printing

There are several other ways of printing. Each method has its own best applications. To transfer a design onto a tee-shirt, for instance, the most common process is **screen printing**. This method was originally called *silk screening* because silk was used for the screen material.

In screen printing, the ink is squeezed through parts of the screen that are not blocked by a stencil material. Fig. 13–4. The characteristics of stencils vary greatly. In fact, some stencils can be "cut" by hand or made photographically using automated procedures.

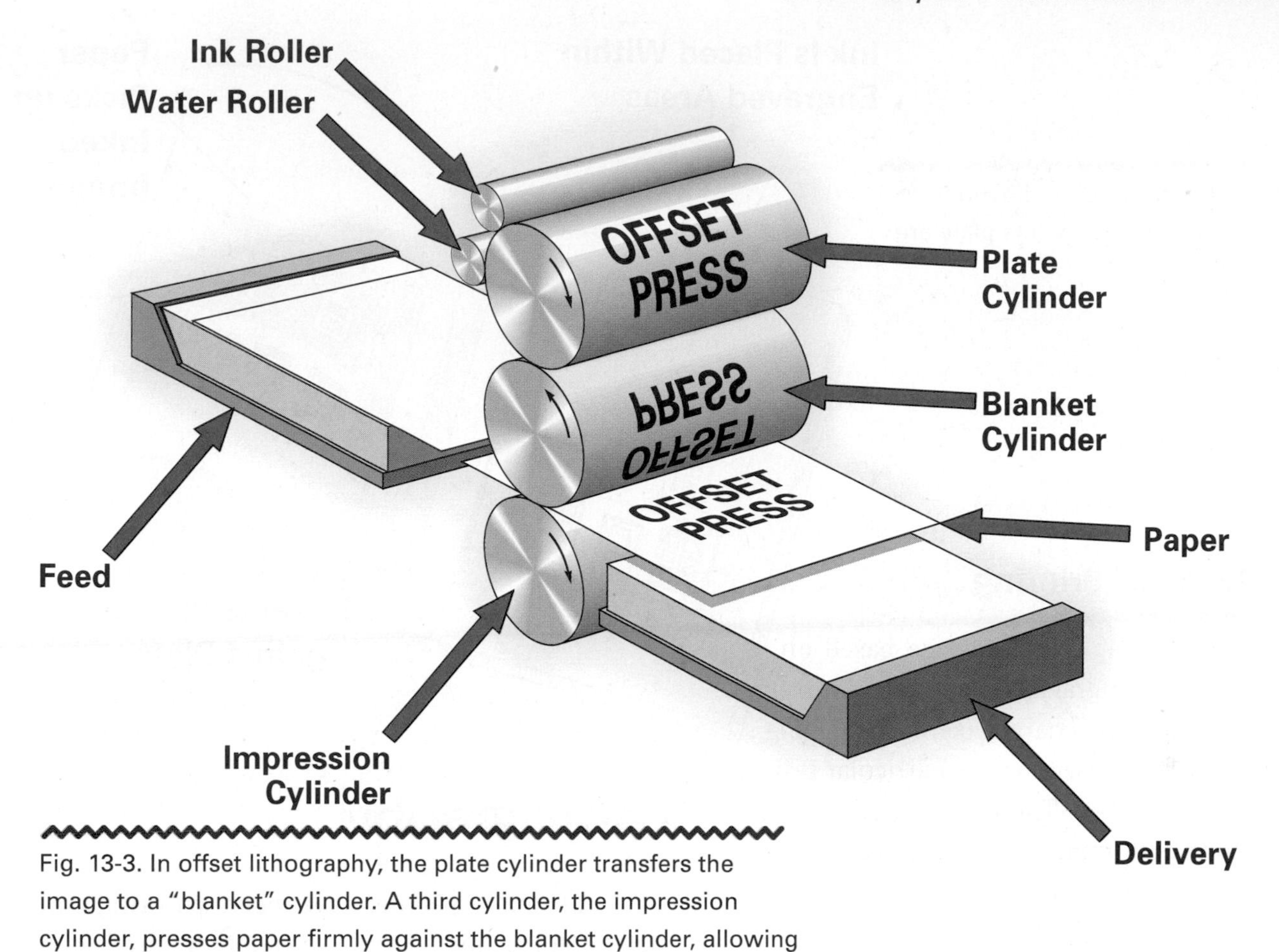

Fig. 13-3. In offset lithography, the plate cylinder transfers the image to a "blanket" cylinder. A third cylinder, the impression cylinder, presses paper firmly against the blanket cylinder, allowing the impression to be transferred to the paper.

Fig. 13-4. Many of the messages or pictures you see on tee-shirts were made by screen printing. The squeegee is used to force ink through the "open" parts of the screen.

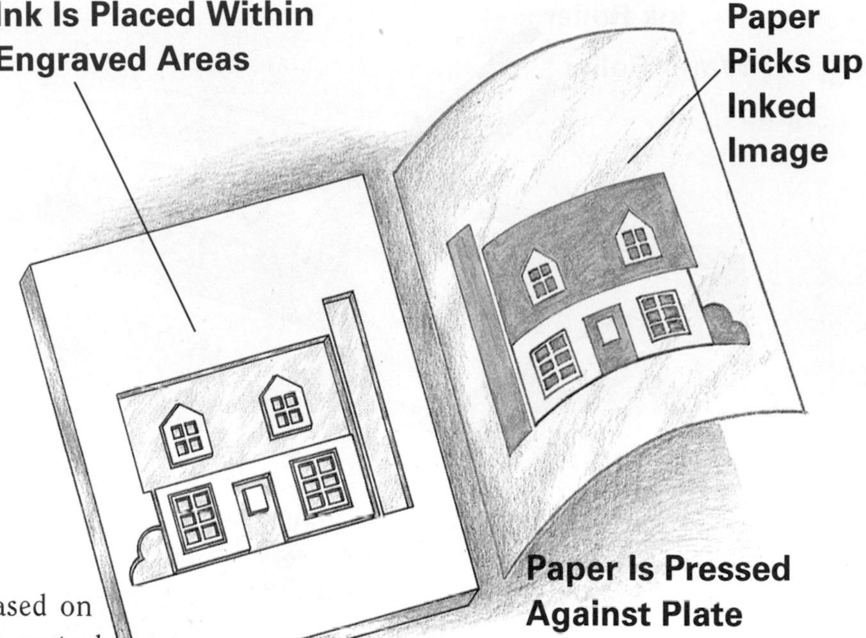

Fig. 13-5. In engraved, or gravure, printing, etched areas on the plate are loaded with ink.

Other Types of Printing

Other printing processes are based on principles such as engraving, electronic control (laser), or electrostatic charges (xerography). Fig. 13–5. The method chosen for any particular task depends on many factors. The printer must consider factors such as cost, number of copies, and the material on which the image will be printed.

▶▶▶ FOR DISCUSSION ◀◀◀

1. How did people spread or share news before the development of mass printing techniques?
2. What skills might a typesetter need today that would not have been needed before electronic typesetting techniques became available?

■ **Prepare a front page for the Glenville News announcing that MegaIndustries will soon be opening a new plant.**

Photography

Although people have been painting and drawing for many, many years, **photography** is only about 150 years old. People discovered that certain chemicals are sensitive to light. When these chemicals are manufactured into film and used with a camera, images can be captured.

A camera has only a few basic parts. In many ways, it resembles the human eye. Light enters the small opening at the lens and is focused on the film. The light-sensitive film captures the image. The image is developed into a *negative.* The negative can then be made into pictures of different sizes. Fig. 13–6.

Will you have pictures in your proposed newsletter? Would your classmates benefit by having some images other than straight text? There is a saying that a picture is worth a thousand words. That might explain why people consider pictures so important. Think what your local newspaper would look like if there were no pictures in it! Pictures, like text, may be printed using a variety of techniques.

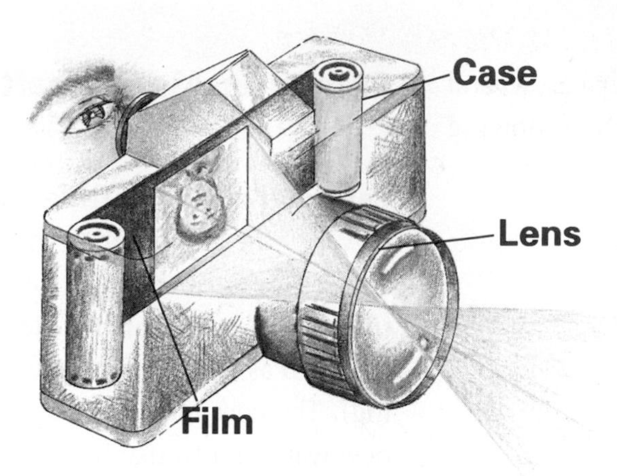

Fig. 13-6. When you take a picture, the image passes through the lens onto unexposed film.

▶▶▶ FOR DISCUSSION ◀◀◀

1. When photography first became popular, some people thought painting would decline as an art form. Why do you think they thought so?
2. Photography is now used for many important tasks, such as helping law enforcement officers identify criminals quickly. Name some other important uses of photography.

Drafting

Another method of communicating detailed information is called **drafting**, or technical drawing. Drafters draw different views of an object or device. Other people utilize the specific information in the drawings to build, use, or repair the object or device.

As an experiment, try to describe in words how a bicycle chain works. It's not very easy, is it? A drafter, however, could draw several views of such a chain. The views would give enough information to allow the viewer to see exactly how it worked.

Drafting and Design

As you read in Chapter 11, most technical drawings show the top, front, and right-side views of an object. Fig. 13–7. Other types of technical drawings show how the object might look in perspective, or "3-D."

Fig. 13-7. This drafter creates drawings at a drawing board, using special pens, inks, and templates to draw accurately.

With respect to design, the drafter has several choices. The designer must consider the following characteristics of the object:

- size
- shape
- proportion
- appearance
- purpose
- material
- weight
- texture
- color
- cost

Most designs are the result of a long list of features and priorities that change on a case-by-case basis. Can you think of reasons why every design makes different demands upon the designer?

Computer-Aided Drafting

Drafters can now use computers for drawing. **Computer-aided drafting (CAD)** has added the power of the computer to the skill of the drafter. Fig. 13–8. When the drafter uses the computer in both the design and drafting stages, the process is referred to as *CADD* (computer-aided design and drafting). The computer enables its operator to design in new ways and to use techniques never before available.

▶▶▶ FOR DISCUSSION ◀◀◀

1. Have computers changed the basic role of the drafter? If so, how?
2. Why do you think designers might choose to use CADD rather than traditional design techniques?

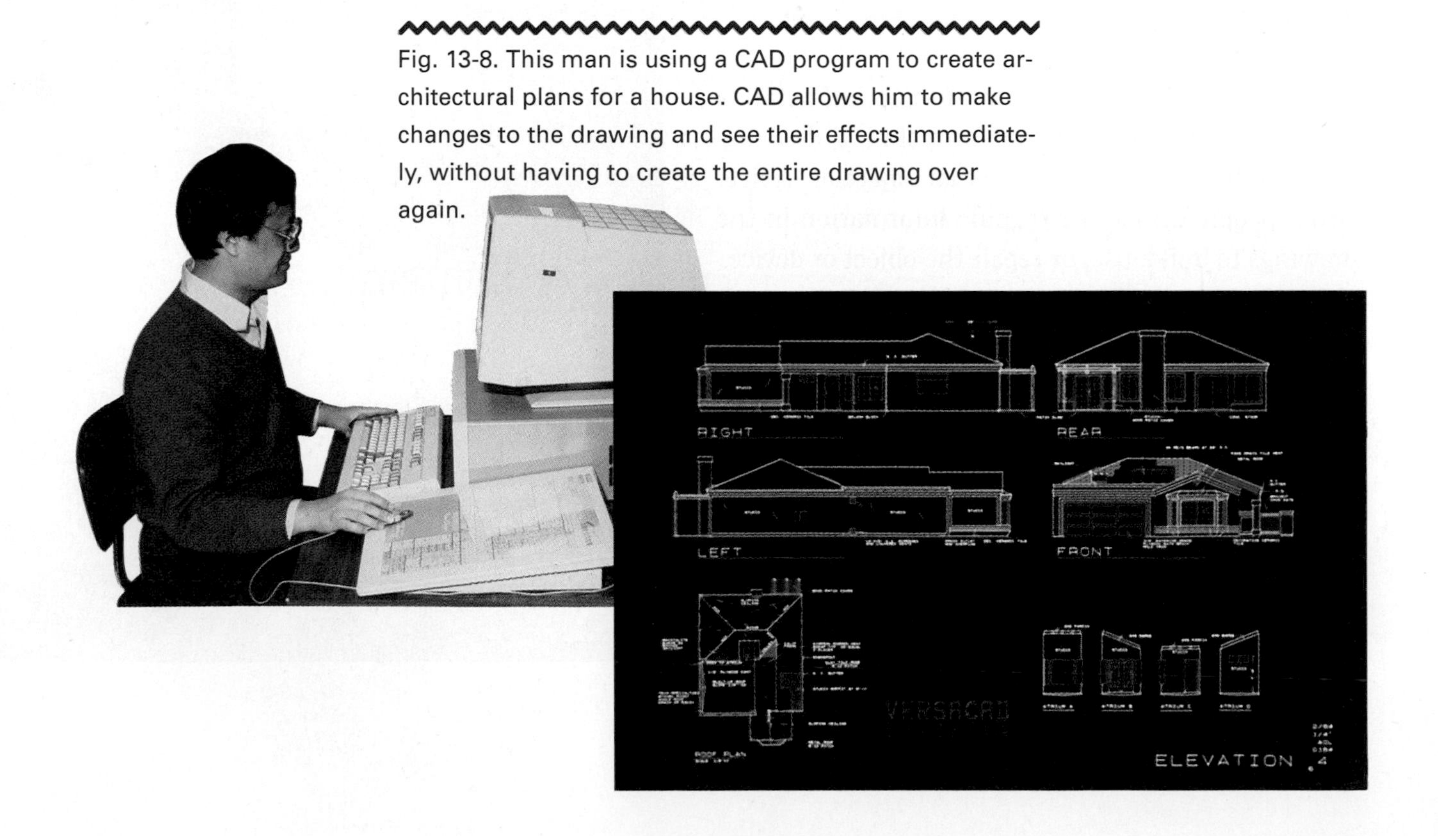

Fig. 13-8. This man is using a CAD program to create architectural plans for a house. CAD allows him to make changes to the drawing and see their effects immediately, without having to create the entire drawing over again.

IMPACT

Computer-aided drafting has changed industry, and it is affecting education, too. Ten years ago, few schools taught CAD. Today, most drafting classes include at least some training in CAD. Schools need to keep pace with technology in order to prepare students for the world of work.

Shaping the Message

Have you ever wondered how the printed message is "shaped"? How do people determine what advertisements, product packaging, or greeting cards will look like?

Often the answer lies in a "storyboard" layout approach. Fig. 13–9. In this process, several possible layouts are roughly sketched or boxed out. The different visual ideas are reviewed, and the person in charge decides which scheme is most effective or appealing. The designer then refines the preliminary work until it evolves into an acceptable final form.

Fig. 13-9. Several rough sketches may be done to give clients an idea of what can be done. After a client chooses the "rough" that he or she prefers, the drafter or artist creates increasingly detailed and refined versions of the chosen idea.

Rough Sketches

Refined Version

Final Version

One new technique that makes this refining process much more economical is **desktop publishing.** Recent advances in microcomputer technology now allow anyone who has a computer to do many of the tasks that used to be done by typesetters and printers. With software that is now available, people can create text and graphics and quickly lay them out in several ways to see which layout is best. When the layout has been finalized and approved, it can be sent electronically to a typesetter or printer. This reduces much of the time and expense involved in older forms of graphic communication design.

▶▶▶ FOR DISCUSSION ◀◀◀

1. How does the storyboard layout help designers?
2. By what criteria might a marketing manager make a final decision about the appearance of a book cover?

■ Collect at least three newspaper or magazine advertisements that include pictures. Cover the words and the product itself (if shown) and see if your classmates can identify the type of product being advertised just by looking at the pictures.

Fig. 13-10. A desktop publishing system includes hardware (the computer and related components) and software (usually several different programs). With this system, text and illustrations can be created and arranged into pages.

CHAPTER 13

Chapter Highlights

- Graphic communication has a significant impact on our everyday lives.
- Printing has evolved over the years into a variety of techniques to mass produce printed matter.
- Photography is a relatively new method of graphic communication. It was developed to capture images permanently.
- Drafters draw views of objects, sometimes using a computer, to enable us to build, use, or repair the objects accurately.

Test Your Knowledge

1. Name some examples of graphic communication.
2. Define the term *printing technology*.
3. Why do you think wood was the material used to carve letters for early printing endeavors?
4. Why was the invention of movable metal type significant?
5. Explain the technique of offset printing.
6. What is the function of the screen in screen printing?
7. What part of the body is most closely comparable to a camera? Explain.
8. What allows an image to be captured on a roll of film?
9. List several possible job descriptions for drafters.
10. Describe one approach to designing the graphics for the outside of a cereal box.

Correlations

SCIENCE

1. Build a pinhole camera. Make an exposure on a piece of sheet film.

MATH

1. Photography and drafting both are based on ratio and proportion. A tree in your photograph is 40 millimeters high. If the scale is 10 mm = 1 meter, how tall is the tree in the real world?

LANGUAGE ARTS

1. Bring a photograph from a magazine or newspaper to class to put on the bulletin board. Select a different photograph from the display and write a paragraph describing the situation that it might present.

SOCIAL STUDIES

1. Many newspapers are going out of business because people prefer to get their news from television or radio. In a report, explain why we need (or do not need) newspapers.

CHAPTER 14

Telecommunication

Introduction

Early telecommunication relied on the creativity of people attempting to be seen or heard from a distance using non-electronic methods. People developed various schemes over hundreds of years to achieve long-distance communication.

Modern telecommunication is most often electronic. This means that the message is somehow changed into a signal that can be relayed (passed along), or transmitted through cable or air. This electronic processing allows the signal to be sent near or far, even thousands of miles away.

After reading this chapter, you should be able to

Define telecommunication.

Describe the development of pre-electronic telecommunication.

Explain early telecommunication techniques, such as those that used drums, horns, smoke, fire, and flags.

Discuss the concepts of modern telecommunication.

Words you will need

megaphone
signal processing
optical telescopes
wigwag
semaphore signaling
electromagnetic pulses

Basic Telecommunication

How loud can you yell? Can you make yourself heard in the next classroom? Can you project your voice from one end of your school building to the other? At some distance, your voice eventually becomes too soft to be heard. What could you do to make your voice heard at a greater distance?

Perhaps you could use a megaphone of some type. A **megaphone** is a device that helps project sound in a chosen direction. The simplest megaphone is a cone-shaped device that focuses your voice in a single direction. Fig. 14–1. Another possibility is an electrical amplifying megaphone that makes your voice louder and focuses it in one direction. You might even use a microphone-amplifier-speaker system to make your voice louder and project it further in a more general direction. Fig. 14–2.

Even with the help of these devices, could you yell loud enough to be heard in the next city? Probably not. Your voice can only carry over a rather limited distance through the "open air."

Fig. 14-1. Megaphones are often used by cheerleaders and mascots at school sports events. (Credit: Manual High School Rams)

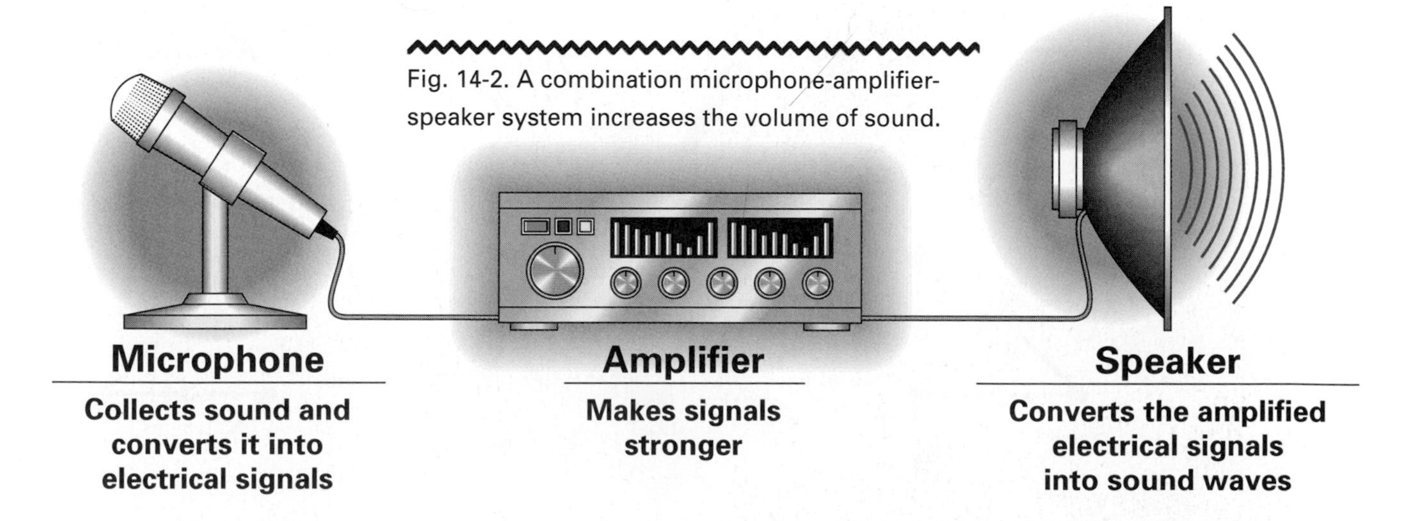
Microphone
Collects sound and converts it into electrical signals
Amplifier
Makes signals stronger
Speaker
Converts the amplified electrical signals into sound waves

Fig. 14-2. A combination microphone-amplifier-speaker system increases the volume of sound.

As you learned in Chapter 12, the ability to communicate over long distances is called telecommunication. The prefix *tele* means "distant." So how could you make your voice heard in a distant location? The answer lies in a technology called **signal processing**. The idea is to electronically convert or modify your voice (or any signal containing information) into a form that lends itself to long-distance transmission. There are many ways to do this, and we shall explore some of them in the next chapter.

▶▶▶ FOR DISCUSSION ◀◀◀

1. How does a megaphone help someone to be heard?
2. Define telecommunication.

■ **Design and construct a megaphone.**

Early Telecommunication

From the earliest times, people sought out ways to communicate over long distances. Among the first techniques were drum beating, horn blowing, and smoke signaling. Drums were made from hollow logs covered with tightly-stretched animal skins. Different sizes of drums produced different sounds, so messages could be sent by both rhythm and pitch.

Horns were typically made from materials at hand or from the horns of local animals. Rams' horns were used for communication in many parts of the world. Like the drums, the horn sounds varied by rhythm and pitch.

Fig. 14-3. Chappe's invention was a device mounted on a large vertical beam on top of a tower. The device could be moved into various shapes. Some of the shapes are shown below. Each shape had a different meaning. Since the device was large, it could be seen over long distances. The towers were placed from 3 to 6 miles apart.

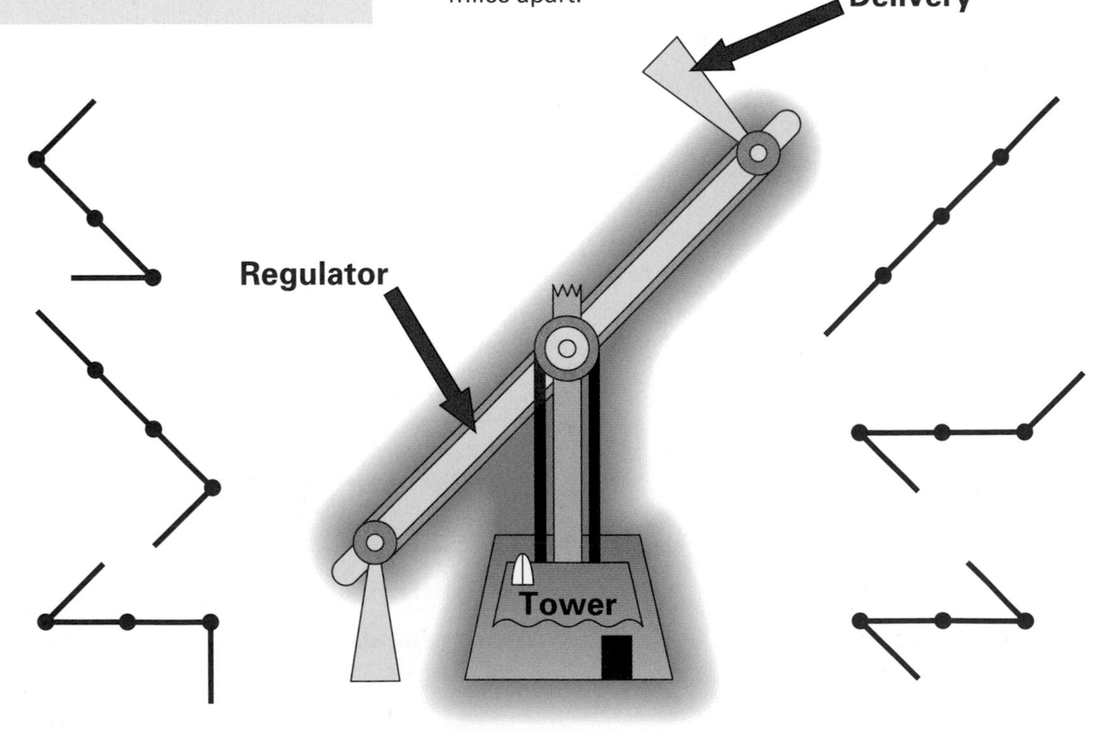

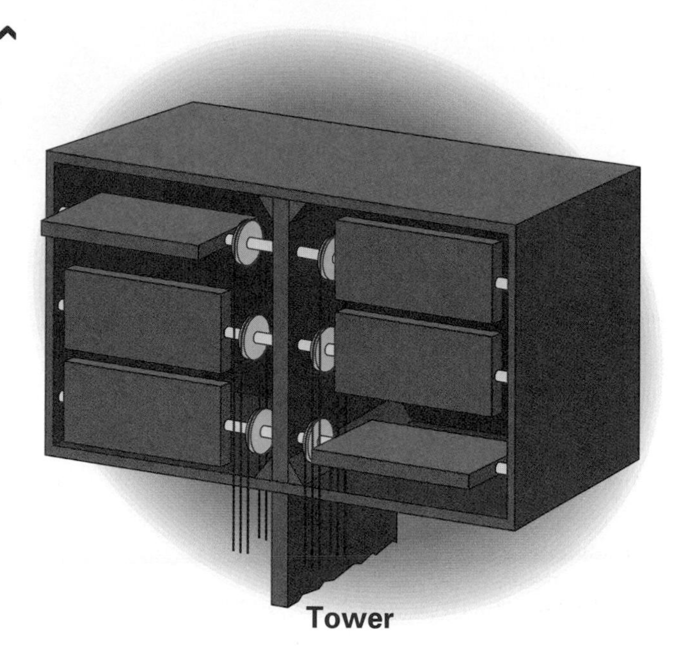

Fig. 14-4. The Murray shutter semaphore allowed signals to be sent by opening or closing one or more of six shutters. The semaphore was mounted on a tower, and people viewed the message through telescopes.

In some areas, people used green, or live, wood to create smoky fires. They used the smoke from the fires to send messages. The meaning of a message depended on the number and spacing of the puffs of smoke.

In the evening, the light from the fires was much more visible than smoke, so signals were sent by blocking the firelight in some meaningful pattern. People made these signal, or beacon, fires in places of high visibility so that they could be seen from as far away as possible.

Sophisticated visual systems, sometimes called **optical telegraphs**, were gradually developed over the years. From France came the Chappe semaphore in the late 1700s. Claude Chappe's idea was to have towers spaced every few miles, equipped with telescopes and his signaling invention. Fig. 14–3. The Murray shutter semaphore was a similar concept developed in England at about the same time. This device used light and shutters to convey information. Fig. 14–4.

TECHNOLOGY TRIVIA

When the Greeks defeated the Trojans around 1193 B.C., they used hilltop bonfires to signal news of their victory. With this system, the news reached the Greek city of Argos, 500 miles away, in just a few hours.

Relaying a signal from post to post was a vast improvement over sending a messenger, but people continued to seek out faster and more efficient means of telecommunication.

During the American Civil War (1861–1865), the newly established Signal Corps used an organized system of flags to transmit messages. The flags were large and square, with a red background surrounding a white square in the center (or the reverse, to maximize contrast with the background of the moment). Fig.14–5. In this system, known as **wigwag**, messages were sent by waving the flags in a series of predefined motions. For evening use, the same system of motions was repeated using torches.

Semaphore signaling is similar to wigwag in that two flags, or targets, are used. In this case, however, the flags are smaller. The message is interpreted by noting the position of the flags as they are brought to each of several possible arrangements. Fig. 14–6. The semaphore process has been used primarily on ships at sea and as warnings for oncoming trains at various points above the tracks.

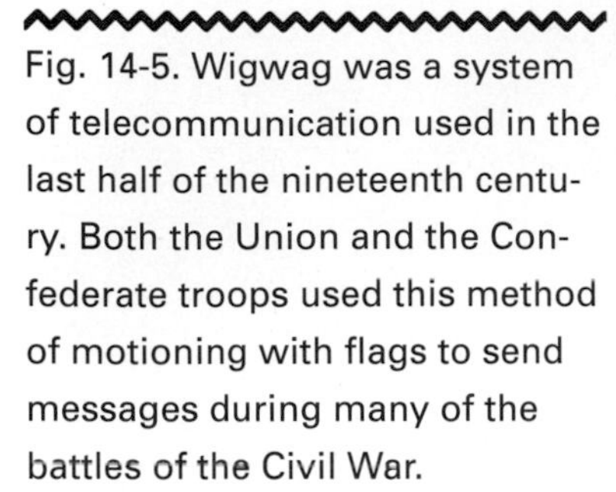

Fig. 14-5. Wigwag was a system of telecommunication used in the last half of the nineteenth century. Both the Union and the Confederate troops used this method of motioning with flags to send messages during many of the battles of the Civil War.

How well these methods worked depended on several factors, some of them beyond anyone's control. Precipitation, wind speed and direction, terrain, and available light affected their usefulness. In a high wind, for example, the sounds made by drums and horns were very difficult to direct. Smoke was difficult to use in high wind or rain, and it couldn't be used at all at night.

▶▶▶ FOR DISCUSSION ◀◀◀

1. Why do you think telescopes were used with the Chappe semaphore and the Murray shutter system?
2. Why would weather conditions be of concern for those attempting visual, line-of-sight telecommunication?

Fig. 14-6. Semaphore flags allow people to read messages by noting the positions of the two flags each time the sender pauses momentarily.

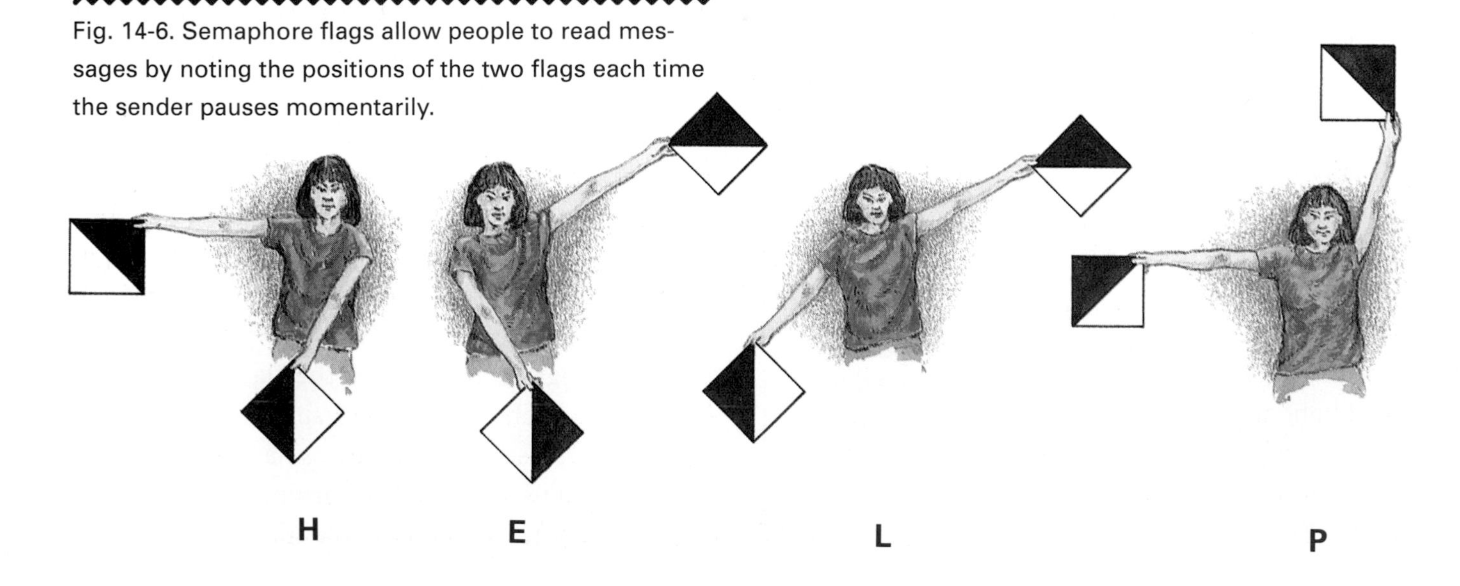

Modern Telecommunication

Telecommunication now occupies a very central role in our personal lives. How telecommunication devices work is therefore of importance to us.

Most modern telecommunication is electronic in nature. Electronic telecommunication signals are usually described as **electromagnetic pulses**, or waves. These signals can travel at the speed of light, thousands of miles per second. This speed makes them ideal for long-distance communication. Ordinary sound waves, by contrast, are much slower. In fact, they are rated in feet per second, rather than miles per second. Fig. 14–7.

Forms of Telecommunication

Almost everyone today is exposed to or uses telecommunication devices daily: radio, television, telephone, telegraph, facsimile (fax) machines—the list goes on and on. These devices, and others like them, allow us to send just about any type of signal over long distances.

For example, we can transmit written messages by fax or modem. Radios and telephones allow us to transmit oral messages. Satellite links enable us to send television broadcasts—audiovisual signals—around the world.

Fig. 14-7. Because light travels so much faster than sound, you can see lightning striking a mile away about 5 seconds before you hear the sound of thunder.

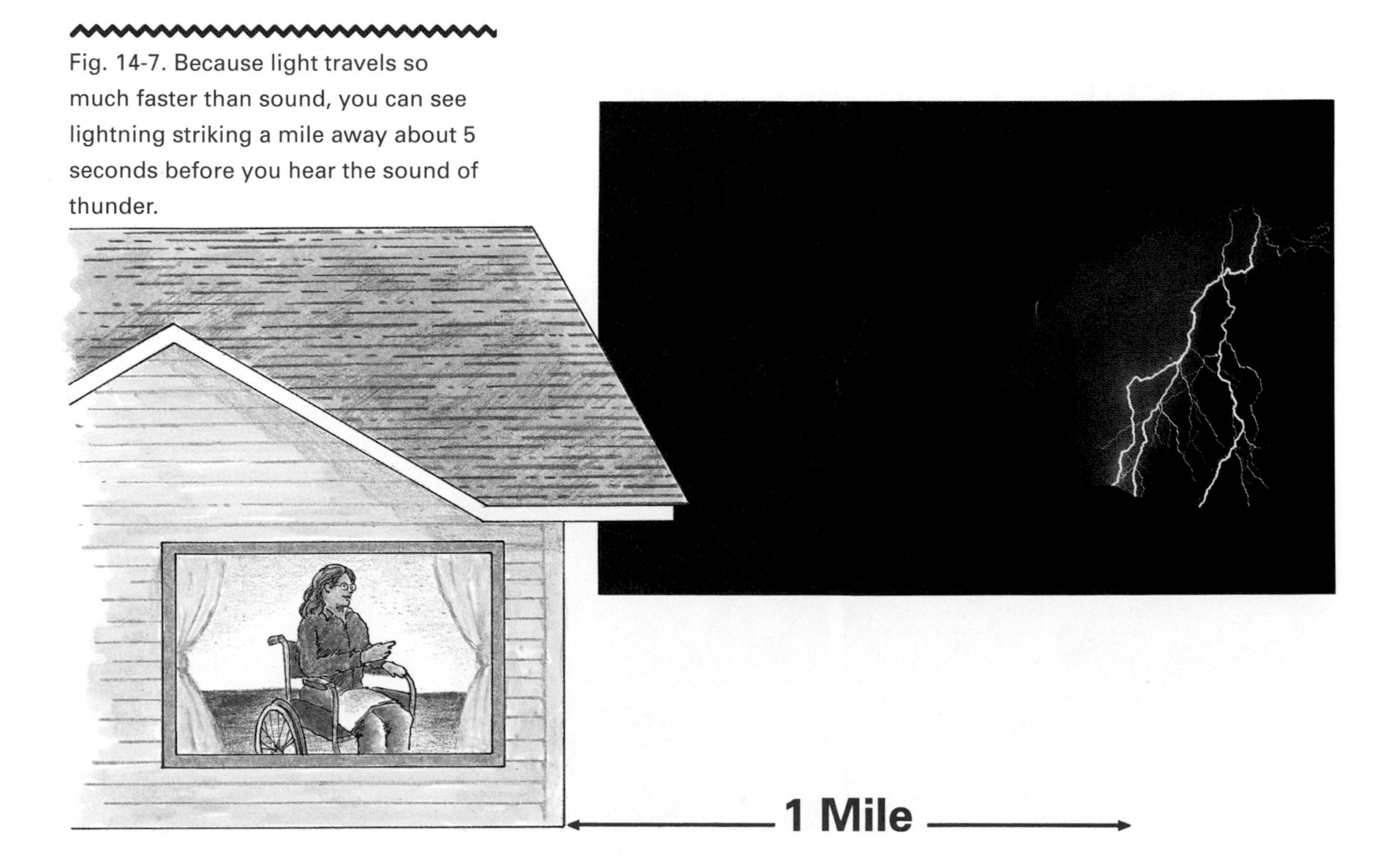

IMPACT

If something important happens halfway around the world this afternoon, we can learn all about it on the news this evening. When news can travel that fast, can any place on earth really seem far away? Can we still believe that what happens in other countries doesn't affect us?

Carriers of Telecommunication

Another aspect to consider in transmitting electronic signals is what carrier will be used. Many factors influence the choice of carrier, or medium, through which a signal is sent.

However, there are only two basic carriers to choose from: wire (or cable) and air. Thus some signals may be "hard-wired" through a cable, while others are radiated through the air or into space. Fig. 14–8. You might now sense what is meant by expressions such as "cable TV" or radio "airwaves." The cables may be above or below ground, or even under water.

▶▶▶ FOR DISCUSSION ◀◀◀

1. Name several modern telecommunication devices.
2. Why are electromagnetic pulses ideal for telecommunication purposes?

■ Construct the model communication system that you planned in the community activity for Chapter 12.

Fig. 14-8. Electronic signals may be sent to distant locations through the air from communication towers or through networks of cables.

CHAPTER 14 Review

Chapter Highlights

- Telecommunication, or communicating beyond the normal range of unaided sight and sound, has been going on for hundreds of years.
- Pre-electronic telecommunication involved systems of drums, horns, fire, smoke, and flags or targets.
- Modern telecommunication systems convert information into electronically processed signals which travel at the speed of light.

Test Your Knowledge

1. Why have people throughout history had a need or desire to telecommunicate?
2. Name three methods of telecommunication used before the advent of electricity.
3. Name two situations in which megaphones are used today.
4. Why do so many communication devices (telephone, telegraph, etc.) begin with the prefix *tele*?
5. What is meant by the term *signal processing*?
6. Why might sending a signal from station to station or from tower to tower by visual means be faster than using a human courier?
7. What is an electronic signal carrier?
8. Describe a pattern you could use to send messages by smoke signal.
9. Why do you think it is sometimes said that the earth is "getting smaller"?
10. What is meant by the term *cable television*?

Correlations

SCIENCE

1. In an encyclopedia, find a diagram of the electromagnetic spectrum. What wavelengths of the spectrum are used for communication?

MATH

1. In your block there are nine houses. Each is about 45 feet from the nearest utility pole. About how many feet of cable are needed to wire all nine houses for the newest cable television service?

LANGUAGE ARTS

1. Investigate the television cable system in your area. What advantages does the cable system offer? What disadvantages? Tell about your findings in an essay.

SOCIAL STUDIES

1. What inventor was responsible for the radio microphone? When did he invent it? What new forms of communication did that open up? Find out at least two other inventions created by this person.

CHAPTER 

Modern Communication Processes

Introduction

Although the purpose of communication never really changes, communication technology has evolved quite a bit. While "face-to-face" interaction might still be the most effective or desirable way to conduct some meetings, much of today's communication is electronic, at least in part.

After reading this chapter, you should be able to

Define electronic communication.

Give an example of communication that is entirely electronic.

Give an example of communication that is partially electronic.

Describe the development of modern methods of telecommunication.

Give examples of modern telecommunication devices.

Words you will need

synthesized speech
modem
modulation
demodulation
coaxial cable
fiber-optic cable
satellite communication
telegraph
Morse code

Electronic Communication

Some communication is partially electronic. The message can be understood directly by the sender, the receiver, or both. However, somewhere between sender and receiver, the actual message is changed into an electronic signal that allows it to reach its target. In some automobiles, for example, a sensor causes an internal computer to create speech artificially to send audible messages to the occupants. This **synthesized speech** is used to warn passengers of possible hazards by sending safety messages such as "your door is ajar."

Some communication is entirely electronic. When a computer "speaks" with another computer, it uses a special code which can be understood by both machines, but which is not readily understood by humans. This type of communication involves no sound and no pictures.

Instead, the computer utilizes electrical and/or light pulses. The pulses are sent rapidly through the air or through some sort of cable, or sometimes both. When one computer is wired directly to another, the requirements for successful communication are minimal—the computers must simply share a compatible format and protocol (called "handshaking") for the data being transferred. In other words, the receiving computer must know how to interpret the signals it receives.

When the second computer is being accessed through a phone line, these pulses are sent using a modem. A **modem** is a device that allows electronic messages to be sent by the processes known as ***mod*ulation** and ***dem*odulation**. The original signal is modulated, or incorporated into another frequency signal called a *carrier signal*. The modulated signal passes through the phone line. When it reaches the receiver, the original signal is demodulated, or separated from the carrier signal so that it can be read by the receiving computer. Fig. 15–1.

▶▶▶ FOR DISCUSSION ◀◀◀

1. What is synthesized speech?
2. Why would you hook up a computer to a phone line?

Fig. 15-1. A modem modifies a computer's signals for transmission over phone lines. Another modem is needed at the receiving computer to change the signal back to its original form.

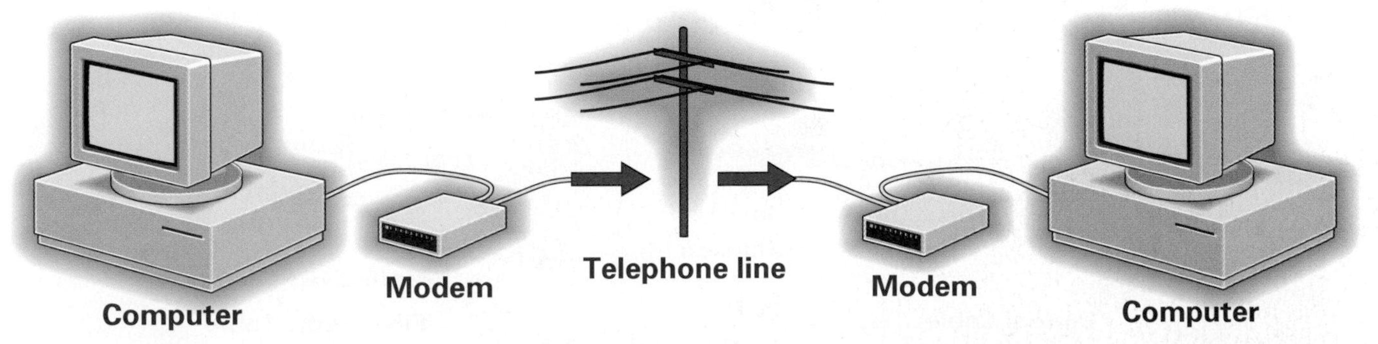

Electronic Telecommunication

Modern telecommunication consists primarily of electronic signals that can be sent over almost unlimited distances. Transmission is accomplished through the air or through electrically conductive wire or fiber optic cable.

Coaxial cable is a type of cable in which an electrically conductive wire, such as copper, is surrounded by another conductor and wrapped in an insulating material. This type of cable has been used for many years to transmit long-distance messages, but it is no longer the most efficient means available.

Many companies now use **fiber-optic cable** to transmit long-distance signals. This cable is made of thin glass or plastic fibers. Messages in the form of light signals can be transmitted through the transparent fibers. Fiber-optic cable weighs less than coaxial cable and is less likely to be affected by electrical interference. It also needs less power to carry a signal and has a much greater information-carrying capacity than standard copper wire. Fig. 15–2.

Transmission through the air, of course, needs no wire at all. Since these signals tend to travel in a straight line, however, they are sometimes beamed to satellites in the sky and bounced back to a different location on earth. This technique, called **satellite communication**, allows people to send and receive messages very quickly around the world. Fig. 15–3.

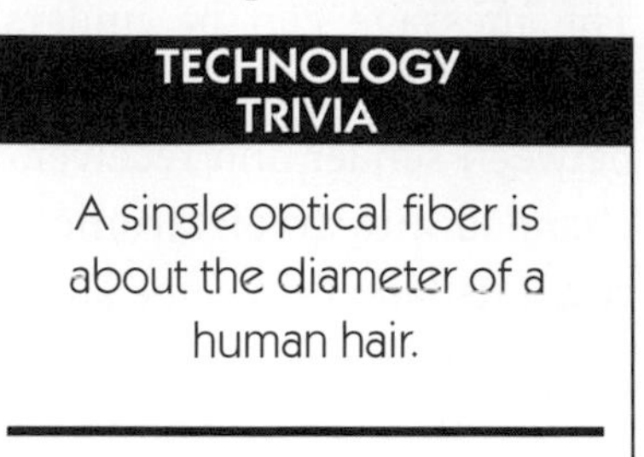

TECHNOLOGY TRIVIA

A single optical fiber is about the diameter of a human hair.

Telegraph

The first electrical device to be used for telecommunication was the **telegraph**, which was developed in the mid-1800s. Using a switch called a *telegraph key*, the operator "makes and breaks" (opens and closes) a circuit. This creates pulses of electricity that can be sent through wires.

To send messages by telegraph, the operator uses the telegraph key to create combinations of long and short pulses that stand for the letters of the alphabet and the numerals 0–9. Such a procedure, as devised by Samuel Morse, is ideal for transmitting messages in what is still known as **Morse code**. Fig. 15–4.

Fig. 15-2. Coaxial cables carry electrical signals. Fiber-optic cables, a more recent development, are designed to carry pulses of light.

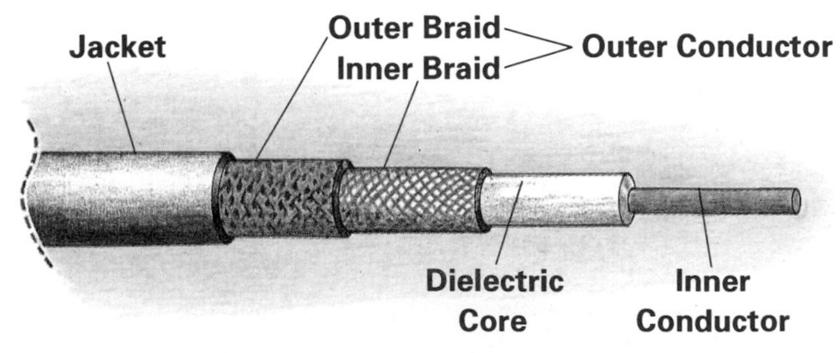

Coaxial Cable

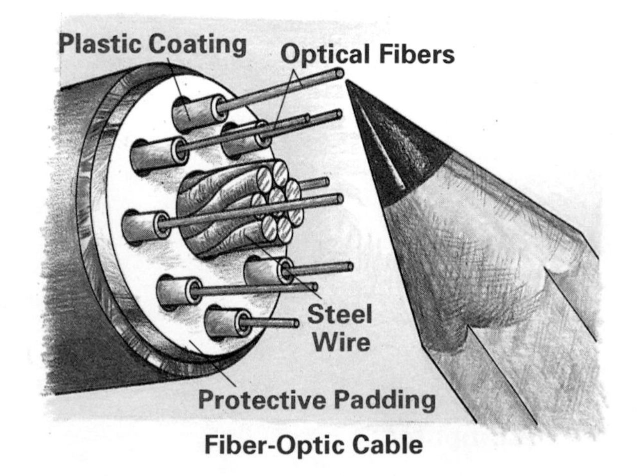

Fiber-Optic Cable

■ **Work with a small group of your classmates. Using a flashlight, practice sending and receiving messages in Morse code.**

Telephone

Soon after the telegraph came the invention of the telephone. The telephone became very popular and useful to the general public because no skill or code was necessary to operate it. The telephone converts speech into electrical pulses that can pass through wires. When the signal reaches the phone at the other end, the pulses are converted back into sounds again.

Much of the credit for the invention of the telephone is given to Alexander Graham Bell, but as with most complex devices, many other people made contributions to Bell's efforts. In fact, the development of the telephone continues into the present, with pocket-sized portable units now being marketed. Fig. 15–5.

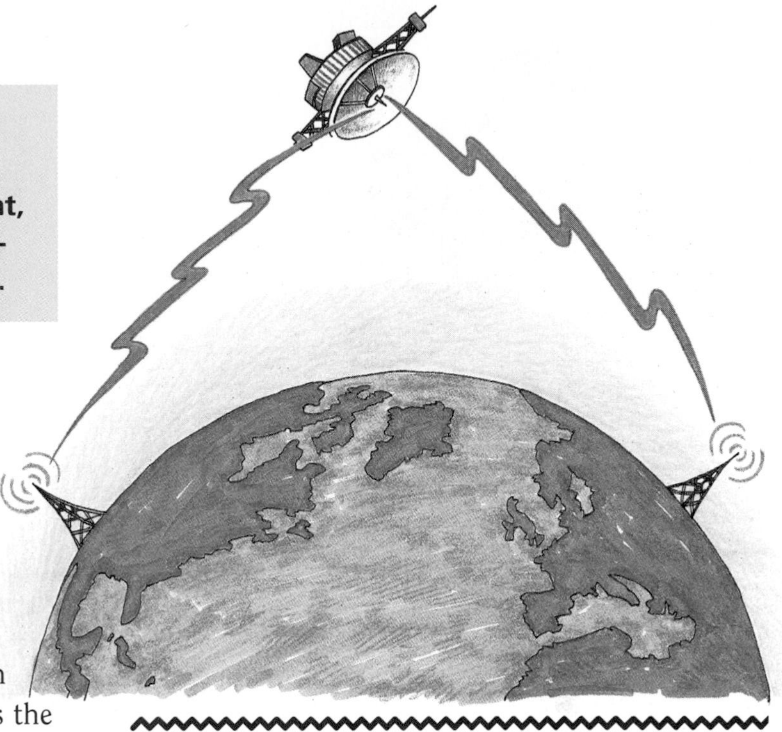

Fig. 15-3. Satellite transmission allows messages to be sent beyond the horizon. The satellites are designed to travel at the same speed as the earth rotates, so they seem to stay in one place over the earth.

Fig. 15-4. Using Morse code and a telegraph key, an operator can send messages through wires to distant locations. The telegraph was the first successful electrical telecommunication device.

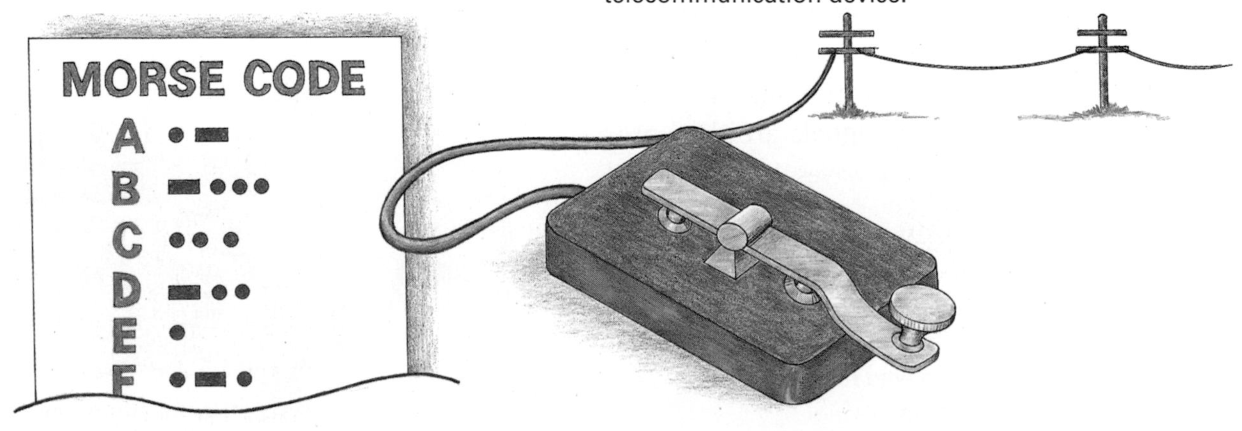

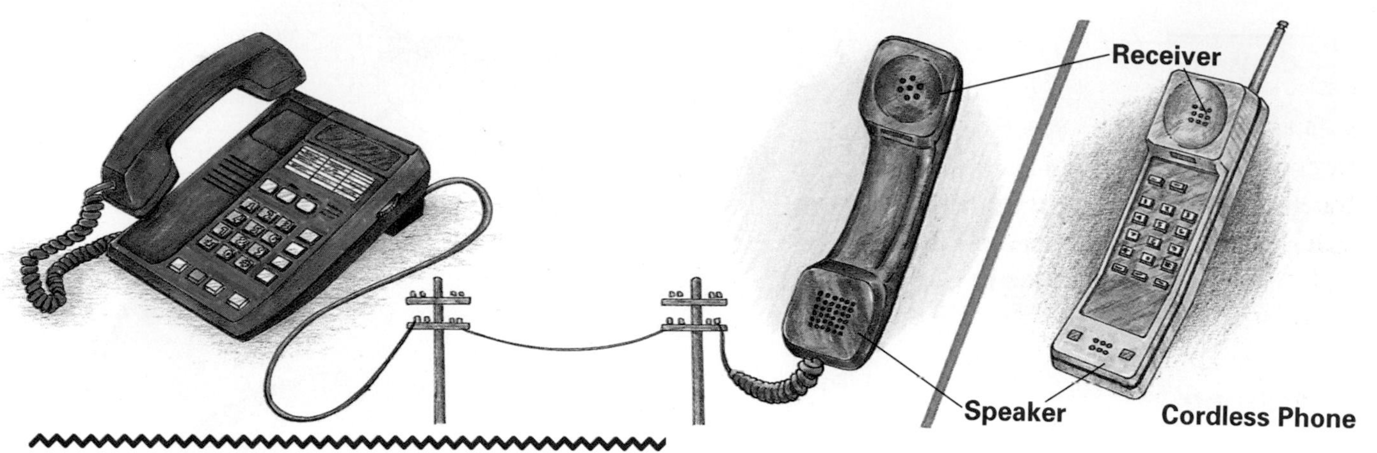

Fig. 15-5. Telephones convert sound into electrical signals that can be transmitted to other telephones, either by wire or through the air.

Radio

The next step in the development of communication technology was the creation of a telecommunication system that was not restricted to wires—something that could pass through the air. By 1900, Guglielmo Marconi and others had invented the "wireless telegraph"—the radio. Limited at first to Morse code, radio rapidly developed to the point of transmitting speech and even music through the air.

Basically, the sound being broadcast is modulated onto an electromagnetic carrier signal (radio wave). The radio waves spread out through the air in all directions from the antenna being used as the source of transmission. Other antennas act as receivers for the signals. The signals are then processed back into audible sound waves. Fig. 15–6.

Television

While radio transmission won raves from an appreciative public, it lacked one major asset—pictures. Working industriously toward that end, scientists developed the black-and-white television by the mid-1920s. Within a few years, the field of television grew so much that by 1950, the television had become a commonplace appliance in the home. Color models appeared around that time and made television programs even more popular.

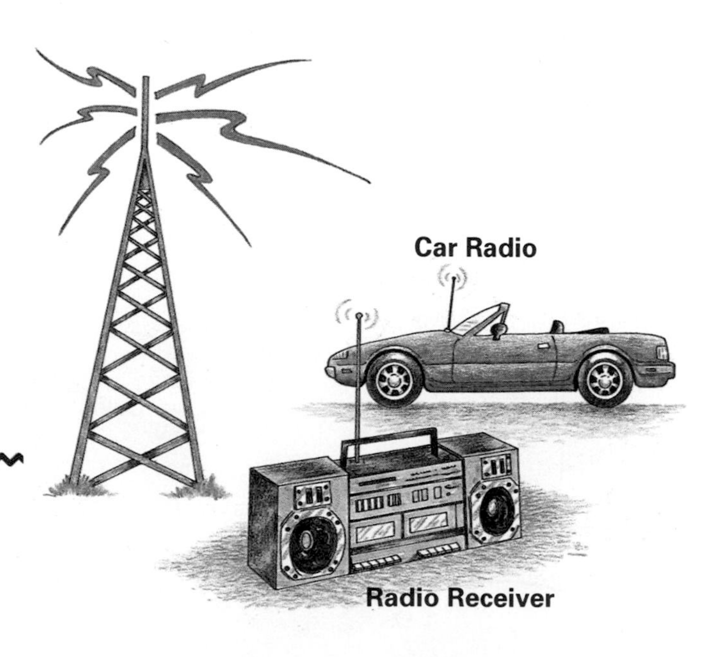

Fig. 15-6. Radio transmission allows sound to be broadcast in all directions. The radio signals can be heard by anyone who has a receiver.

The transmission of television signals involves both audio (sound) and video (picture) input. Microphones and television cameras process the information, combine it into a single complex signal, and transmit it from an antenna much as is done with radio. Within the receiving television, the images and sound are decoded so that the viewer can hear the sounds and see the images. Fig. 15–7.

▶▶▶ FOR DISCUSSION ◀◀◀

1. How does electronic telecommunication differ from non-electronic telecommunication?
2. Why do you think the radio was originally known as "wireless telegraphy"?

Fig. 15-7. Television pictures are "painted" on the screen one line at a time. The entire screen is refreshed ("repainted") 30 times every second. High-definition television (HDTV) provides a sharper picture by using more lines.

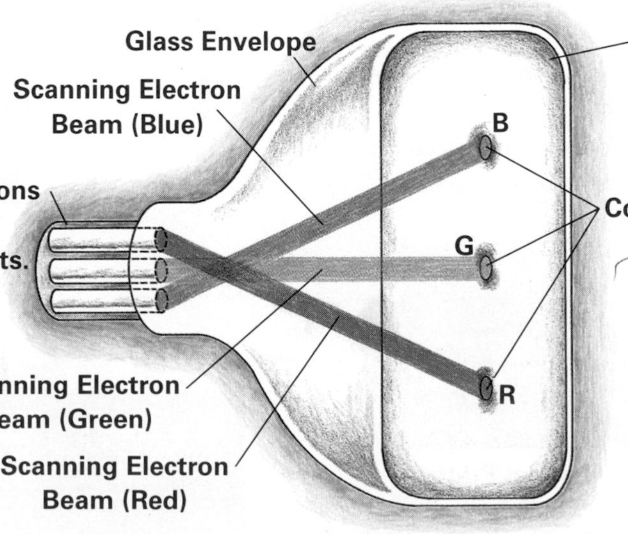

After separation and decoding, the video signal is sent to the picture tube.
Glass Envelope
Scanning Electron Beam (Blue)
Electron guns shoot electrons at the screen, which is covered with phosphor salts. There is one gun for each primary color: red, green, and blue.
Scanning Electron Beam (Green)
Scanning Electron Beam (Red)
B
G
R
Phosphorescent Screen
When electrons hit the phosphor salts, they glow.
Color Phosphors on Screen (Red, Green, Blue)

IMPACT

Television is part of our personal and national history. Ask your parents what they remember about John Kennedy's assassination, the first landing on the moon, or the Watergate scandal. What do you recall about the Gulf War of 1991? Chances are, the memories of these events will be the images that were seen on television.

Other Electronic Communication Devices

All communication devices convey information, or intelligence. That information may contain sound, graphic data, or a combination of both.

Graphic Electronic Communication

Some electronic communication is primarily graphic. Fax (facsimile) machines and electronic scanners, for example, can view an image of a picture or text, convert that image into electronic signals, and transmit the signals to another machine, which reproduces a copy of the original. Computer printers convert electronic signals into images on paper ("hard copy"). Fig. 15–8.

Audio-Electronic Communication

Other electronic devices are primarily audio. Audio-cassette players and CD-ROM players convert pre-recorded magnetic or digital information into sound. Fig.15–9. Walkie-talkies, CB (civilian band) and shortwave radios all send and receive airborne signals and convert them into sound.

Combined Audio/Graphic Electronic Communication

Several other devices have been invented which communicate both audio and graphic information electronically. Videodiscs (sometimes called *laser discs*), video tape recorders (VCRs) and picture phones, relying on various technologies, all transmit and/or receive motion pictures accompanied by a sound track.

▶▶▶ FOR DISCUSSION ◀◀◀

1. What is meant by the term "hard copy"?
2. Give an example of pre-recorded magnetic information.

■ The Glenville city council has decided to update the communication system used in the various city services offices. Do research to find out about the most recent communication systems and products. Organize the information you gather, and place it in a notebook for council members to review.

Fig. 15-8. Several different types of computer printers are now available. Non-impact printers do not actually strike the paper. Impact printers do strike the paper, much like the traditional typewriter.

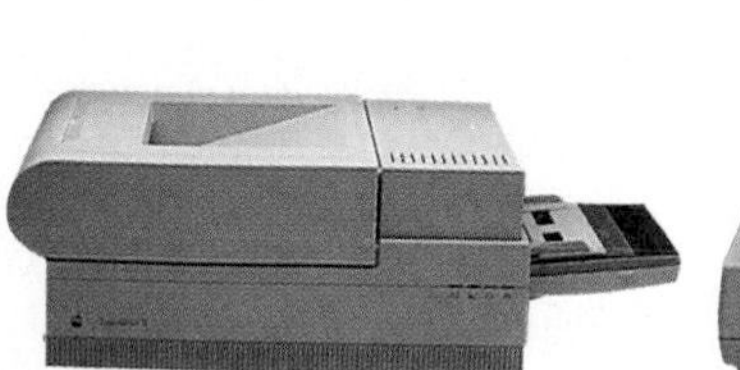

Non-Impact Printer

Impact Printer

Fig. 15-9. Small pits are burned into the surface of CD-ROMs. The pits are "read" by a fine beam of laser light.

CHAPTER 15 Review

Chapter Highlights

- Much of today's telecommunication is accomplished electronically.
- Electronic devices have changed the ways in which we communicate.
- Some electronic communication takes place through cables, and some is transmitted through the air.
- Portable electronic communication devices are now very popular and in widespread use.

Test Your Knowledge

1. What is the purpose of communication?
2. What types of technology are involved with the production of synthesized speech for an automobile?
3. What are the requirements for successful communication by modem?
4. Why might you want one computer to send messages to another?
5. Why is fiber-optic cable being used more and more instead of coaxial cable for many applications?
6. What is the function of a telegraph key?
7. What big advance in communication technology occurred with the invention of the radio?
8. About when were the first televisions introduced?
9. How does an electronic scanner work?
10. What is a videodisc?

Correlations

SCIENCE

1. Fiber optic transmissions use a property called internal reflection to make a beam of light travel along the fiber. Simulate this property using a water-filled aquarium and a bright flashlight. Tape the lens until only a tiny beam of light shines through. Shine the beam through the side of the aquarium up towards the water's surface. Move the flashlight up until the beam reflects back down into the water.

MATH

1. Suppose you have a VCR tape that can record for 2, 4, or 6 hours, depending on the speed at which you record. What is the greatest number of half-hour programs you could record on the tape?

LANGUAGE ARTS

1. Look through a television directory and find programming that will fit in these five categories: education, news, comedy, drama, and music. List the time and length of each program. In a brief speech, relate your findings to the class.

SOCIAL STUDIES

1. Find out more about Samuel Morse and the telegraph. Why did he invent it? What effect did it have on communication in the 1800s?

CHAPTER 16

Impacts of Communication Systems

Introduction

Because electronic communication and telecommunication are a basic part of modern life, people often take them for granted. You may not realize all the things you can do using something as common as the telephone. Fig. 16–1.

Consider the fact that it took about a month for news to travel from Europe to North America when ocean-going ships were the fastest means of communication. You might say that what took a month in the 1800s takes just a moment in the late 1900s.

After reading this chapter, you should be able to

Describe how modern communication affects people, society, and the environment.

Explain how electronic communication affects people's privacy.

Describe changes electronic communication has caused in job descriptions.

Explain how electronic communication may cause stress for some people.

Words you will need

credit history
teleconferencing
videoconferencing
telecommuting
techostress

Fig. 16-1. Early telephones had no dialing mechanism. The caller simply told the operator the name of the person he or she wanted to call.

Impacts on People

Was there ever a time when people did not use communication? It's unlikely, since people can communicate even by the looks on their faces. However, when we think of communication systems, we usually think of more technical and organized arrangements. Such systems allow information to be transferred from place to place efficiently. These systems have come a long way since people relied entirely on facial expressions or body language.

Because our communication systems are more complicated, people are influenced by them more than ever. However, the degree to which people are affected by communication varies.

In general, for example, younger people watch more television than adults. Does this have an impact? The answer is "yes!" Many youngsters prefer to watch television, for instance, instead of playing sports or doing other creative things. What is the result as they grow into adults? Does violence viewed on television shape the viewer? This is still being debated.

Before television was invented, were young people more likely to play creative and physical games? The answer to that depends on who you ask. Many people talk about how the whole family used to sit and listen to certain radio programs.

TECHNOLOGY TRIVIA

By the time you are 18, you will have watched more than 17,000 hours of television, and you will have seen about 360,000 commercials.

Well, OK, you say, how about before radio—what did kids do then? Once again, the answer is not the same from all sources. It was common, for example, to invite company over for sing-alongs or to play musical instruments together. Today's communication systems have a major influence on people's lives, but it is not always clear exactly what effect that influence has.

One way to find out how communication systems affect your life is to see what would happen without them. Let's say you are interested in sports—perhaps professional baseball. How do you learn of the outcome of the most recent game your favorite team played? Newspaper? Television? Radio?

These are all communication systems. If all of them were cut off one day, how could you find out the score of the game in question? You can see that we depend on these systems a great deal—perhaps more than you thought. Fig. 16–2.

■ **For two days, keep a log of the communication devices you use and how often you use them. How much do you rely on these devices? What would you do without them?**

Fig. 16-2. How do you communicate? Information and entertainment are available to us in many forms.

Privacy

What about privacy? You may have physical privacy in your own room from snooping eyes, but there are other types of privacy. For example, did you know that everyone who applies for a credit card may have his or her financial history checked? Many people have had such a file made about them, perhaps without knowing it. There are businesses who put together computerized credit files. They provide this information to banks and credit card companies. These credit files have information about loans, mortgages, or legal judgments a person may have had in the past.

Credit files can lead to all sorts of problems if an error occurs. Someone you never met, for example, might use your name and social security number (or some other identification document) to get a loan or credit while pretending to be you. If that person then fails to make the payments on time, your credit is affected, but you probably won't know it! Later, you may be rejected for a credit card with the explanation that your **credit history** is unsatisfactory.

This type of problem occurred often enough that lawmakers passed the Fair Credit Reporting Act of 1970. Among other things, this federal law allows people the right to review their credit files.

New Employment

Communication systems have also created new jobs. Many people are needed to install, maintain, and operate something like a cable television network. Computer systems require people to design methods of hooking up two or more computers, write software programs, and develop improved parts. Telephone companies have had to create and update electronic switching systems to meet increasing demand. The job descriptions for this new telephone technology did not exist just a few years ago.

Many communication companies have begun to use fiber-optic technology. Fig. 16–3. This is a whole new approach, which uses pulses of light instead of electronic pulses to transmit information. Fiber-optic technology requires companies to retrain their present staff or hire new employees who have the needed skills.

Fig. 16-3. Fiber-optic cables can carry thousands of times more information than standard telephone cables.

■ Council members need to know how city employees will be affected by a new communication system. Interview a number of employed adults to find out how their jobs have been affected by modern communication systems.

Changing Job Descriptions

Communication systems have also changed the way people do some jobs. Most businesses require communication between employer and employee. But what if the employer has several places of business, such as factories in several states? Modern communication allows people to communicate daily by phone and fax machine.

Some businesses take communication one step further with teleconferencing or videoconferencing. **Teleconferencing** makes use of the "conference call" feature offered by telephone companies. It allows several people, all in different locations, to speak together. **Videoconferencing** is similar to teleconferencing, but participants are also able to see as well as hear each other. Fig. 16–4.

Telecommuting

Some people "go to work" without even leaving their homes. These people do "office work" on their computers at home. **Telecommuting**, as it is sometimes called, allows workers to "commute to work" using the computer, a modem, and a phone line. Fig. 16–5.

Telecommuting is not for everyone. Some people have stopped working from their homes because they missed the friendships and the unofficial communication at their traditional offices. Others, however, express great pleasure in being released from the daily rush-hour traffic. They also enjoy the flexibility of working hours at home.

Techostress

How do people adapt to the ever-changing methods of modern communication? Not easily, it seems. The pressure that people face from technology has been described as **techostress** in some publications. People describe a feeling of uneasiness, of not knowing what is going on, of being uncomfortable with new technologies.

Fig. 16-4. This woman is participating in a videoconference. From her chair, she can talk to and see people in many different parts of the world.

Fig. 16-5. Telecommuters like this woman do all or some of their work at home. Many telecommuters send their work to a central office by modem or computer disk.

The fact that present communication systems allow people to communicate almost instantly 24 hours a day adds stress to people's lives. People are also stressed by the amount of information available on any particular topic. There is often not enough time to sift through it all. How do they decide when their research is complete?

▶▶▶ FOR DISCUSSION ◀◀◀

1. **How would you attempt to find out the score of any particular sports event?**
2. **Why do companies want to find out the credit history of people with whom they will do business?**

Fig. 16-6. People and businesses in almost every part of the world are now very dependent on electronic communication systems. What might be the effects in the office shown above if a power outage occurred?

Impacts of Communication Systems on Society

Would life in the U.S. be the same if our communication systems were rolled back overnight to what they were 100 years ago? It's very doubtful. Years ago, communication was just a part of doing business, working a farm, or governing a nation. Today, in many respects, electronic communication *is* the business. All kinds of organized groups, from bowling leagues to the federal government, depend on modern communication. If our communication networks failed, much of our society would be thrown into near chaos. Everyday activities would practically stop if our electronic communication systems were suspended or cut off. Fig. 16–6.

Think of all the electronic transactions that take place, and you will see how much we depend on our current communication systems. New York Telephone, for example, handles more than 100 million telephone calls a day in its regional calling area. Every day, people buy, sell, or trade millions of shares of stock electronically on the various exchanges. Every day, people pay for millions of items and services by credit card. Almost every transaction is somehow checked or recorded using computers and telephone lines. Fig. 16–7.

Fig. 16-7. Most restaurants and stores now use electronic devices similar to this one to check credit card transactions.

▶▶▶ FOR DISCUSSION ◀◀◀

1. Why would today's banking industry be in serious trouble without electronic communication?
2. Name at least five kinds of businesses that depend on telecommunication to transact much of their business.

Impacts of Communication Systems on the Environment

Today's communication systems also have an impact on our environment. Many of the things that people do affect the environment, and modern communication definitely affects what people do. Satellite communication, for example, provides important information that helps people treat the environment in a safe manner. For example, mineral exploration, forest fire fighting, crop selection and rotation, canal dredging, and many other activities rely on satellite communication. Fig. 16–8.

▶▶▶ FOR DISCUSSION ◀◀◀

1. How are satellites used to predict weather and assist in agricultural planning?
2. How might today's communication systems help people recover from an international disaster such as a large oil spill?

Fig. 16-8. Satellite communication photos can reveal much about the earth. A picture such as this one, for example, provides information about weather patterns.

CHAPTER 16 Review

Chapter Highlights

- Modern communication systems allow instant communication throughout the world.
- There are some drawbacks to all progress, even electronic communication.
- As electronic communication systems have advanced, privacy has become a real concern.
- Improvements in communication have caused some jobs to change and have created new jobs.
- Electronic communication has created a new type of job situation called *telecommuting*.

Test Your Knowledge

1. Name one way in which you are influenced by communication systems.
2. What did young people do for amusement before television was invented?
3. Why do people feel that their privacy is threatened by modern communication systems?
4. Discuss how an employer might manage several remote business locations.
5. What is teleconferencing?
6. What is the Fair Credit Reporting Act of 1970 and why was it passed?
7. Describe some jobs in the communication field that might not have existed 25 years ago.
8. What are some of the pros and cons of telecommuting?
9. Could telephone companies manually process or connect all the calls people make on a typical day? Explain your answer.
10. Discuss two ways in which communication technology can help people do things to help the environment.

Correlations

SCIENCE

1. Use fiber optic strands to move light around corners and through loose knots in the strand.
2. What practical uses might there be for moving light around corners?

MATH

1. If you watch TV from 10:05 a.m. until 2:35 p.m., how long have you been watching TV?
2. At that rate, how many hours would you watch per week?

LANGUAGE ARTS

1. Interview a cellular phone user. In a brief essay describe the user's satisfaction or dissatisfaction with the cellular phone.

SOCIAL STUDIES

1. Your book tells about some of television's negative aspects. What are some of the positive aspects of TV as a communication system? For example, how might news broadcasts, documentaries, and after-school specials help to inform TV viewers?

CHAPTER 17

Communication in the Future

Introduction

As electronic technology continues to improve, it influences much of today's society. Electronic power is increasing and, at the same time, the size and expense of circuitry are decreasing. These facts ensure that the influence of electronic technology can only expand.

How will improvements in electronics affect communication in the future? No one can say for sure, but this chapter may open your eyes to possibilities you may not have considered.

After reading this chapter, you should be able to

Discuss trends in modern communication.

Describe recent developments in telephone technology.

Explain the characteristics of the picturephone.

Discuss E-mail and voice mail.

Describe high-definition television (HDTV).

List several uses of a bulletin board system (BBS).

Discuss possible uses of two-way cable TV.

Words you will need

futurist
trend
cellular phone network
picturephone
pager
electronic mail (E-mail)
voice mail
high-definition television (HDTV)
bulletin board system (BBS)
electronic banking

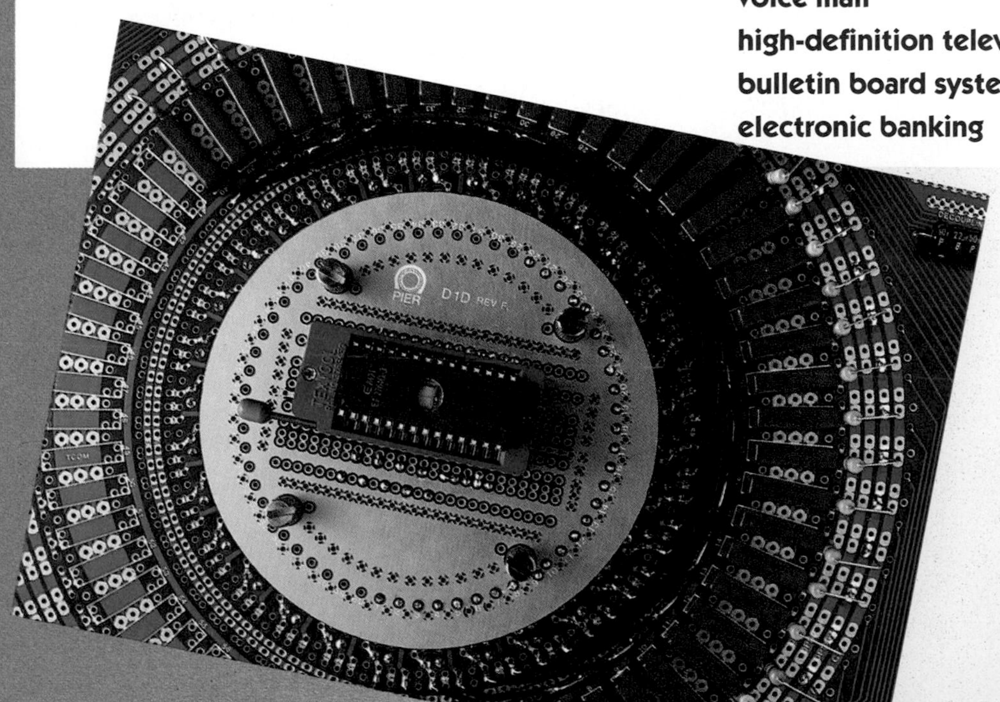

Predicting the Future

Have you ever dreamed about how life might be in the future? Thinking about the future is serious business for many people. In fact, some people make a living by predicting what might happen tomorrow.

Many people are concerned about the future for a wide range of reasons. Among these are economic interests, social planning, construction projects, and educational and legal issues.

People who try to predict the future are sometimes called **futurists**. Futurists appear to have a difficult task. How do you predict the future? Many futurists believe that the past is the key to the future. Studying history gives them a good starting point toward speculating about what might follow.

Another technique is to watch current **trends**, or the choices that most people seem to be making. Some futurists rely on trends, assuming that people will continue to make the same kinds of choices. For example, records show a trend toward the use of economical cars. The economy of cars has improved significantly during the last 20 years.

When you plot the average economy of cars on a graph, you can see an obvious trend. Fig. 17–1. Can you predict anything about the economy of cars of the future? It appears that, *if the trend continues*, gas mileage in cars will continue to improve.

▶▶▶ FOR DISCUSSION ◀◀◀

1. Why is it a good idea to make a graph when you are looking for a trend?
2. Why do people need to be able to predict the future?

■ **Design a bulletin board with the theme "communication in the future." This chapter will give you some ideas about what to include.**

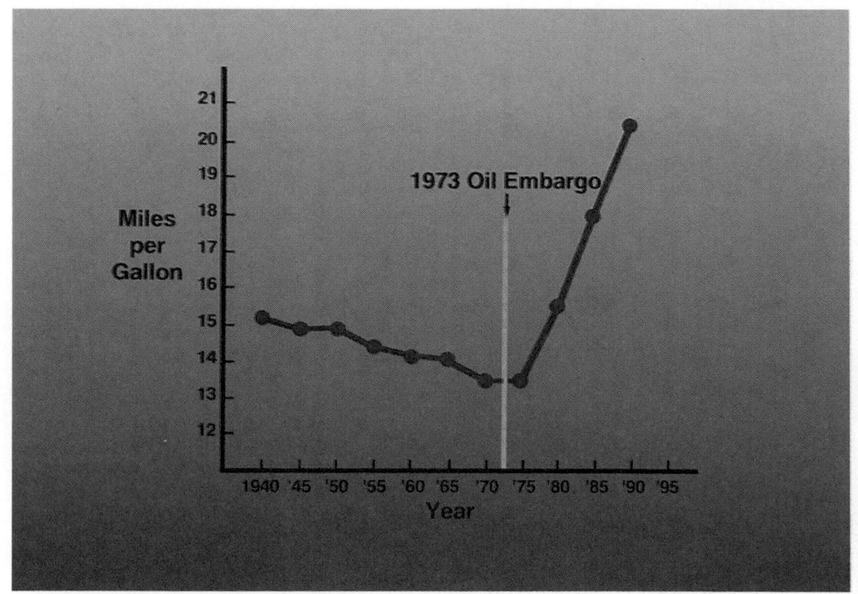

Fig. 17-1. This graph shows the average fuel economy of cars from 1940 to 1989. What trend does the graph indicate?

Current Trends in Communication

Let's look at some trends that are now developing in communication technology. Some of these trends have been developing since the early '60s. Others are much newer. They all give us important information about what might happen in the future.

Telephone

The portable phone is an example of a current trend in communication technology to make devices more portable—not tied down by wires. A portable phone allows people to make or receive calls within a certain range (usually up to 1,000 feet) of the "base station." (The base station is connected to a telephone jack.)

Fig. 17-2. Cellular phone networks divide a city or area into "cells." As a phone moves, the signal is passed from cell to cell to reach the phone.

The new **cellular phone network** is another type of portable phone. It allows people to make calls almost anywhere. Most major cities now support cellular phones. Fig. 17–2. Similar technology allows people to make telephone calls from airplanes in flight.

What will follow? The next generation of telephones will probably be small enough to fit inside an average shirt pocket. The unit might unfold, like a wallet, so it can reach the user's ear and mouth. People may be able to carry the pocket-phone around, placing and receiving calls from wherever they happen to be.

With the ability to call people regardless of their location, a new concept has arisen: the personal calling number (PCN), or personal communication network (PCN). Currently, telephone numbers are assigned to locations. In the future, people may be able to request a PCN, a number for their personal phone—whether they are at home or at work or even traveling. Furthermore—for Dick Tracy fans—yes, it is true, the wrist-phone is in the offing. Fig. 17–3.

Fig. 17-3. Telephone technology is improving so rapidly that wrist-phones will be practical in the near future.

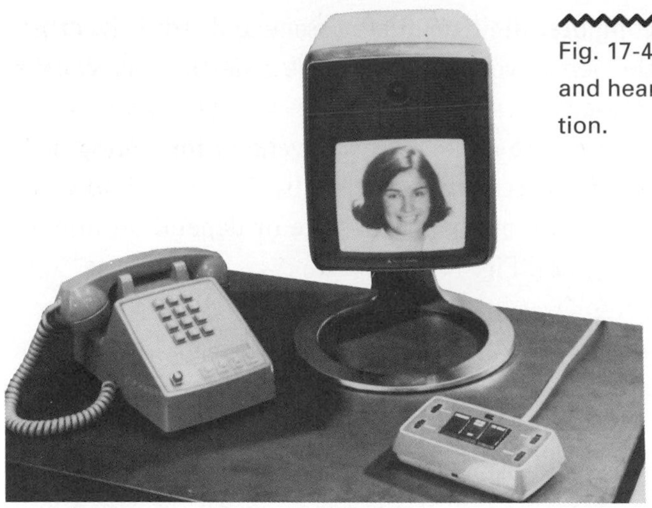

Fig. 17-4. Picturephones allow people both to see and hear each other during a telephone conversation.

Picturephone

At the 1964–65 World's Fair held in New York, AT&T introduced the **picturephone**. Heralded as the phone of the future, this invention enabled people to both see and hear each other. Fig. 17–4. For some reason, it was never a commercial success, but people are still predicting that its time has come.

Perhaps the best sign that these people are right is the increasing popularity of videoconferencing. The use of videoconferencing and picturephones will probably increase as the process improves and the cost decreases.

IMPACT

If most people had picturephones, what effects do you think that might have on our society? Would we feel more secure, being able to see the person who is calling us? Would we feel our privacy was invaded because the caller could see *us*? Do you think the advantages of a picturephone would outweigh the disadvantages?

Pagers

Did you ever see an electronic **pager** or "beeper"? It is a device that "beeps" when its number is dialed, getting the attention of the person carrying it. There are many types of pagers, ranging from simple sounding devices to more elaborate models that show messages on a miniature computer-like display. Fig. 17–5.

In the future, pagers will probably be more powerful. They will be more reliable and will have larger ranges. Perhaps they will even be able to transmit messages back to the sender. These functions might be included in the portable/laptop computers that are now becoming so popular.

TECHNOLOGY TRIVIA

With the help of communication satellites, it is already possible to page someone in another country.

Fig. 17-5. This physician is calling someone in response to a paged message. The message or telephone number appears in a liquid crystal display (LCD) on the pager.

E-Mail

Some people are now sending and receiving **electronic mail (E-mail)**. Electronic mail is sent by one computer to another. The messages are typed using a keyboard and sent by modem to a "mailbox" in a computer host that serves as an electronic bulletin board. This **bulletin board system (BBS)** stores the message files. The person to whom the E mail is addressed dials up the BBS to collect his or her mail. Some messages may even be "general," or public, so that anyone may read them. E–mail has been increasing in popularity, and the future promises more to come.

Voice Mail

Voice mail is also becoming increasingly popular. **Voice mail** is a combination of E-mail and the telephone. You dial the number of the computerized mail service using a touchtone phone. A message explains the options available, such as the voice mail codes for various people. You key in the desired code and if no one answers, you can leave a message. The computer digitizes the message and stores it. Later, the person you called dials the code and receives the message.

Voice mail is very convenient for calling people in different time zones. Also, voice mail does not require an answering service or depend on human assistance. Fig. 17–6.

Television

American television transmission presently uses a picture made up of 525 horizontal lines that are refreshed 30 times each second. In Japan and parts of Europe, however, a new system called **high-definition television (HDTV)**, has been introduced. HDTV uses different techniques to form pictures made up of 1000 or more lines per screen. The final product is a much cleaner, sharper image. Fig. 17–7. Since the HDTV picture is so much clearer, most people predict that it will be introduced in the United States in the near future.

Other changes in television will include the new flat-screen display. Present televisions use a *cathode ray tube* (CRT), which makes televisions

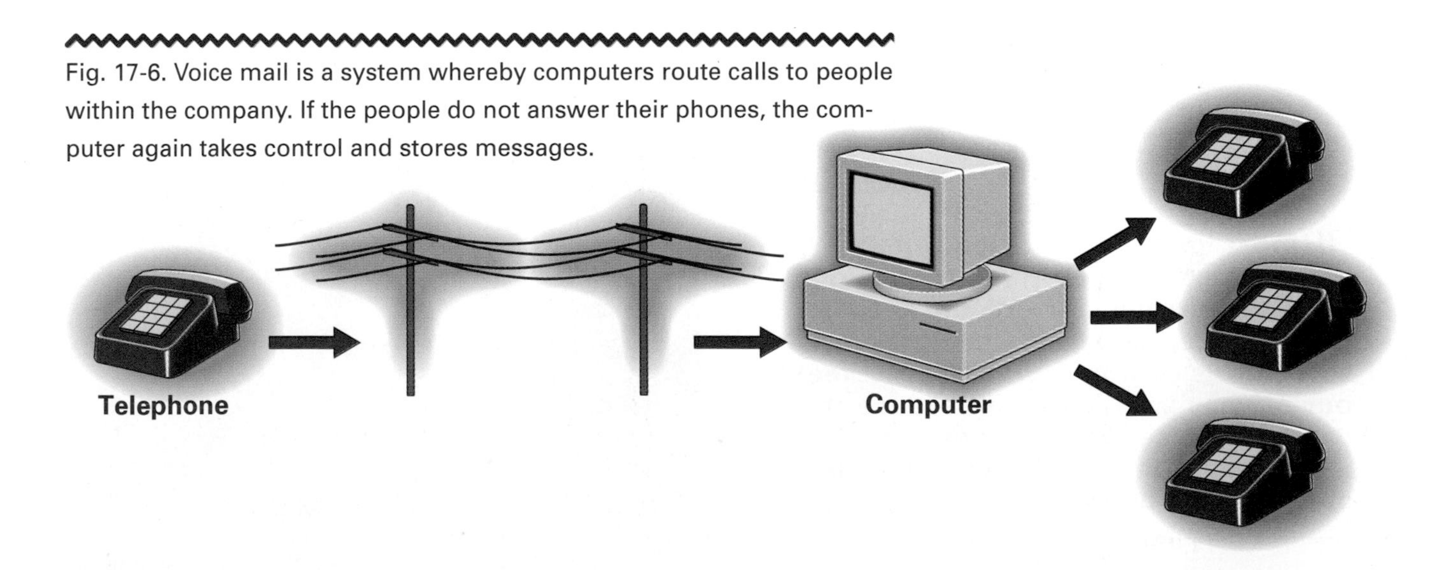

Fig. 17-6. Voice mail is a system whereby computers route calls to people within the company. If the people do not answer their phones, the computer again takes control and stores messages.

Fig. 17-7. By using many more horizontal lines per screen, HDTV provides much clearer images than standard televisions in use today.

rather bulky. However, televisions that have liquid crystal displays (LCDs) are now being developed. These televisions do not need a cathode ray tube. They may therefore be relatively flat, perhaps hanging on the wall like a poster or oil painting.

Electronic Communities

How would you describe your community? Most people think of a community as a group that lives within a certain area. But there are other types of communities, too. For example, "electronic communities" are made up of people who use computers to communicate on a regular basis.

Going "on-line" through a modem, people contact others who have similar interests. They meet through a bulletin board system. Members of this "community" are grouped by interest, not location. As more people begin using computers, more of these electronic communities are likely to sprout up, making this a movement of the future.

Electronic Banking

Just as automatic transaction machines (ATMs) have enabled people to conduct some banking activities when the bank is closed, **electronic banking** will free people from other banking limitations. Electronic banking, using touchtone phones and computers, allows people the luxury of banking even during the evenings or on holidays. Paying bills, transferring funds from one account to another, checking an account balance—all of these functions and more can now be done from your home or office. Although this trend has started slowly, the consumer's ability to bank by computer will expand in the '90s. Fig. 17–8.

Fig. 17-8. This woman is making a bank transaction by using her computer. The transaction information travels back and forth from the bank via modem.

Cable Television

Cable TV was originally designed to bring television signals to people who couldn't receive them using ordinary antennas. The system has grown since then and now includes special programming not available on the standard broadcast channels.

Some universities are now experimenting with using cable TV as a two-way medium. Students at distant locations use an interactive electronic device to respond to instructors at the universities. When it has been perfected, two-way cable may also be used for surveys, to request that certain programs be shown, or to streamline shop-at-home programs. Fig 17–9.

▶▶▶ FOR DISCUSSION ◀◀◀

1. Why is it that we can make electronic devices smaller today than we could in the past?
2. Electronic communities focus on similar interests, but they can be useful in other ways, too. Name some possible uses for an electronic community.

■ MegaIndustries owns a company that produces communication devices. This company is looking for ideas for new products. Using your imagination, design a new communication device.

Fig. 17-9. In the future, cable television may become a two-way interactive communication device. Even now cable TV is being used to allow personalized shopping from people's homes in some areas.

CHAPTER 17

Chapter Highlights

- Trends help us predict the future.
- Many electronic communication devices are becoming more portable.
- People may be members of electronic communities as well as geographical communities.
- Two-way communication, based on cable television, is now being developed.

Test Your Knowledge

1. What is the role of a futurist?
2. What is a trend? Give an example.
3. What is the difference between a standard telephone and a portable one?
4. Name at least one advantage of using a cellular phone.
5. How is a personal calling number (PCN) different from today's telephone numbers? Why might this be an advantage?
6. Think about the time and expense of attending a distant meeting. Why might a business manager suggest a teleconference instead?
7. Of what value is an electronic pager?
8. Compare E-mail and traditional mail.
9. Why might voice mail be valuable when you are trying to contact someone in a different time zone?
10. Why does HDTV yield a sharper picture than the televisions currently used in the United States?

Correlations

SCIENCE

1. Cordless phones transmit signals between the handset and the base unit. If you have a cordless phone, find out how far away you can take the handset from the base unit and still hear clearly. Do walls, trees, or metal objects interfere with the signal?

MATH

1. If a television picture is refreshed (repeated) 30 times per second, how many times is it refreshed per minute?

LANGUAGE ARTS

1. Imagine yourself as a futurist. In a creative essay, describe the type of telephone you would design for teenagers in the 21st century.

SOCIAL STUDIES

1. This chapter discusses various changes in telephone communication in the last twenty years. Many of these changes were made by the AT&T corporation. What invention preceded the formation of AT&T? When was the company formed and who established it?

You Can Make a Difference

When Are We Going to Use This in the Real World?

In 1966, ***foxfire*** was the name of a mountain plant that glows in the dark. Today, ***Foxfire*** is also the name of a magazine and a series of books that are written and published by students. In addition, it refers to a distinct style of teaching.

Eliot Wigginton began teaching school at Rabun Gap-Nacoochee School in the Appalachian Mountains of Georgia in 1966. His students were a mix of troubled students from Atlanta who lived in a dormitory and "townies," kids who lived too far from the crowded public school. His students had two things in common: they did not like school, and they had no desire to continue their education by attending college.

Wig, as the students called Wigginton, was a first-year teacher who knew he had to get the students interested if they were going to learn. His students were interested in putting together a literary magazine. Thus, ***Foxfire*** was born.

The first edition contained haiku (a Japanese form of verse) because most literary magazines at the time contained haiku. However, it also contained interviews with people who lived in the mountains and local citizens who knew Appalachian folklore and crafts. The students soon learned that their readers wanted to know more about the social customs of the area. They were not interested in haiku.

Putting together ***Foxfire*** magazine over the years has been a study in advances in publishing technology. In the 1960s, everything had to be typed letter-perfect because the text was photographed and printed photo-offset. If the typist made a mistake, the section had to be retyped. Now, students use desktop publishing computer technology to typeset, design, and lay out pages for the magazine themselves. They also do all the magazine photography themselves, printing pictures in a photo lab in the back of their classroom.

The first magazines were printed by the man who printed the weekly newspaper. The students ordered 600 copies of the first issue. Later, when ***Foxfire*** ran 5,000 copies per issue, the students had to find a new printer. ***Foxfire*** did not receive any money from the school for printing expenses. Students had to raise funds from other sources.

The students began to travel with Wig to explain to people all over the country what was different about the Foxfire program. In a Foxfire classroom, the teacher is a leader and guide rather than "the boss." While they are encouraged to make choices in *how* they want to learn material, students are responsible for learning the material and must demonstrate this. Students are encouraged to work and learn together, rather than compete against each other.

Just how successful has the Foxfire program been? Here are a few examples:

- The term ***Foxfire*** now is used to describe a democratic, student-centered, cooperative approach to education.
- The magazine is still in publication and has subscribers across the entire United States and abroad.
- Foxfire has also published a series of books that contain collections of articles from the magazines. The first one, ***The Foxfire Book***, has sold 4 million copies. The series has sold more than 7 million copies.
- Book royalties have made Foxfire a separate program, now located at the public Rabun County High School.
- Information about a way of life has been preserved.
- Foxfire has trained more than a thousand teachers around the United States in the Foxfire style of teaching.

Proceeds from book sales have been put to use in several ways—all of which benefit the Foxfire students at Rabun County High School, where the program moved in the 1980s. Proceeds have also paid for college scholarships for Foxfire students who never dreamed of attending college. Finally, proceeds have established Foxfire Center—log cabin offices on 110 acres in Mountain City, Georgia, where staff members organize Foxfire teacher training courses at colleges and universities around the United States.

A recent book the teenagers wrote, *Foxfire: 25 Years*, is about the experiences of Foxfire students themselves over the history of the program. It shows that many graduates of the Foxfire program have learned to make the connection between school and the real world. Most of them have found success in both worlds.

SECTION FIVE

PRODUCTION SYSTEMS: MANUFACTURING

Activity Brief

Technology in Manufacturing

PART 1: Here's the Situation ....

The development of new technology has greatly affected the ways in which products are manufactured. Robots, for example, are playing an increasingly important role in modern manufacturing. A robot is a machine that performs tasks usually done by people.

Today's typical industrial robot is a computer-controlled machine with a mechanical arm. In manufacturing, robots are used to pick up and move objects. Some weld, tighten bolts, and spray paint. Others sort items according to color and size. Objects handled by robots may be as large and heavy as certain aircraft parts or as small and fragile as the filaments used in light bulbs.

In manufacturing, technology is used both in the development and in the production of products. As you do this activity and as you read Chapters 18 through 23 in this section, you'll learn for yourself how this production system works.

PART 2: Your Challenge..........

Because the main robot used in manufacturing today performs like a person's arm, it is commonly called a *robotic arm*. Another name for it is **manipulator**.

Sometimes a robot arm is referred to as an *articulated arm*. This is because it has a series of pivot points, or joints, around which it moves. The pivot points are called **axes**. See Fig. V-1. The number of axes a robot arm has depends on the type of work it is designed to do. In manufacturing, robots commonly have six axes. However, the three basic axes can be identified as the joints in your arm are identified—shoulder, elbow, and wrist.

The "hand" of the robot is called a **gripper** or *end effector*. There is a wide variety of these devices because of the many different types of work that robots now do. A common type consists of two parts that open and close like pincers. These actions enable the robot to pick up, hold, and release parts or products being processed. See Fig. V-2.

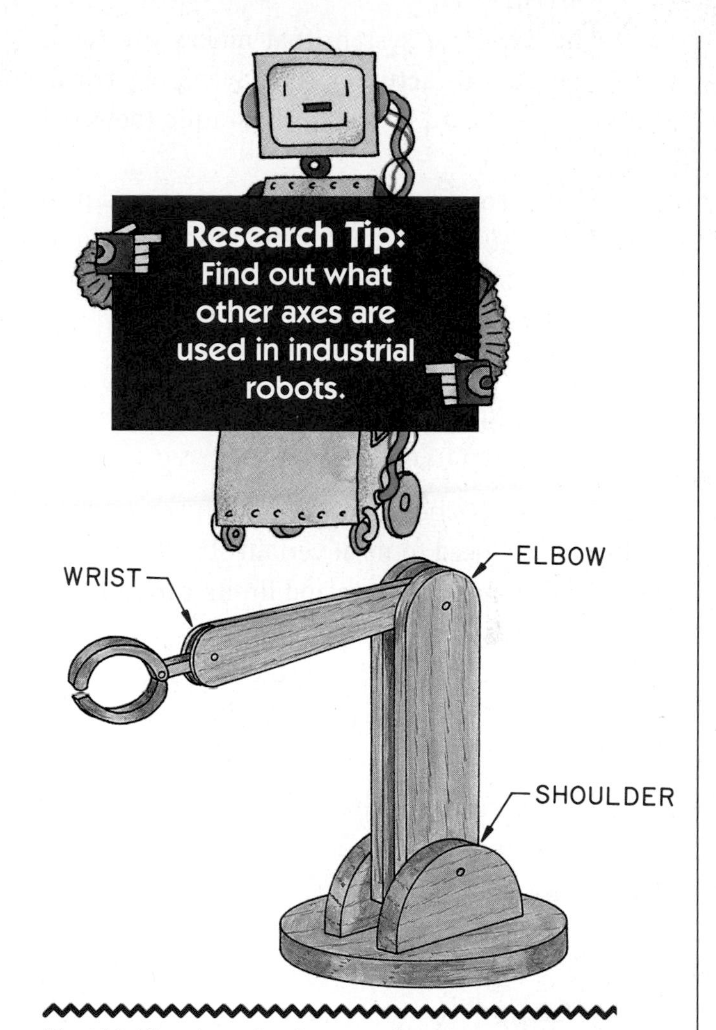

Research Tip:
Find out what other axes are used in industrial robots.
WRIST
ELBOW
SHOULDER

Fig. V-2. The three basic axes of a robot arm.

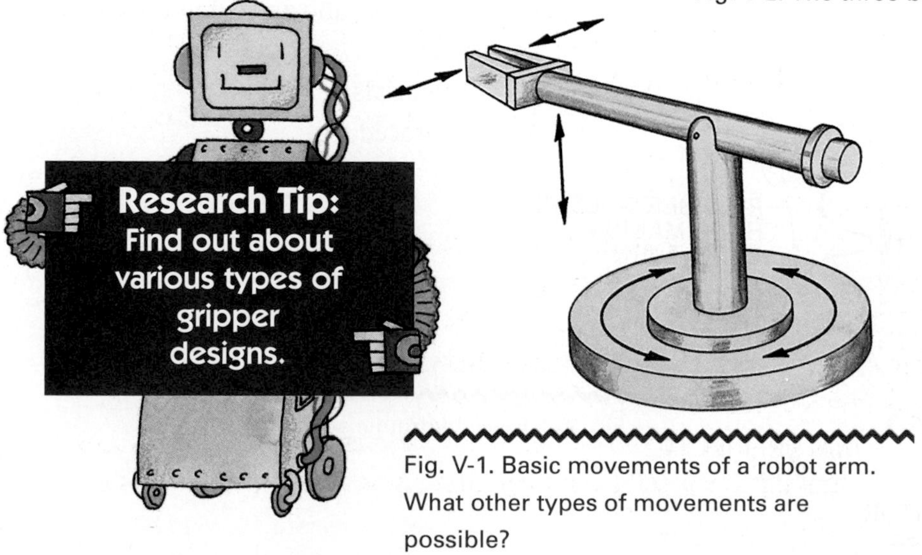

Research Tip:
Find out about various types of gripper designs.

Fig. V-1. Basic movements of a robot arm. What other types of movements are possible?

The device or system that makes the robot move is called an **actuator**. It may be electrical, pneumatic (powered by air), or hydraulic (powered by fluid).

In this activity, you will help design and produce a hydraulic robotic manipulator—a fluid-powered robotic arm—using common, everyday materials and creative thinking.

PART 3: Specifications and Limits

Your robot and the way in which you work to produce it will need to meet certain standards. Read the following specifications and limits carefully before you begin.

1. Working in groups of 2 or 3 students, you are to research and develop a prototype industrial robot.
2. The robot must be hydraulically controlled and must have two or more axes.
3. The robot must be able to pick up, move, and put down an object. This means that your design must allow for both horizontal (sideways) and vertical (up-and-down) movement.
4. A key element in the design of the robot is the power/control system. In this model, you will use syringes and ⅛″ plastic tubing to create a hydraulic system. Movement will require a water-filled piece of plastic tubing with a syringe on each end. Basically, this system works as follows:

   One syringe will be part of the actuator. The other syringe will be mounted appropriately, usually on the part of the arm you wish to move. Pushing in the plunger on the actuator syringe causes the other syringe to extend or "go out," moving the part on which it is mounted. Pulling out the plunger of the actuator syringe will cause the mounted plunger to retract or "go in," creating movement in the opposite direction. See Fig. V-3.

5. As you work, your teacher must approve the following:
   - rough sketch
   - working drawings, including front, top, and right side views and any detail drawings that are needed
   - bill of materials
   - plans of procedure

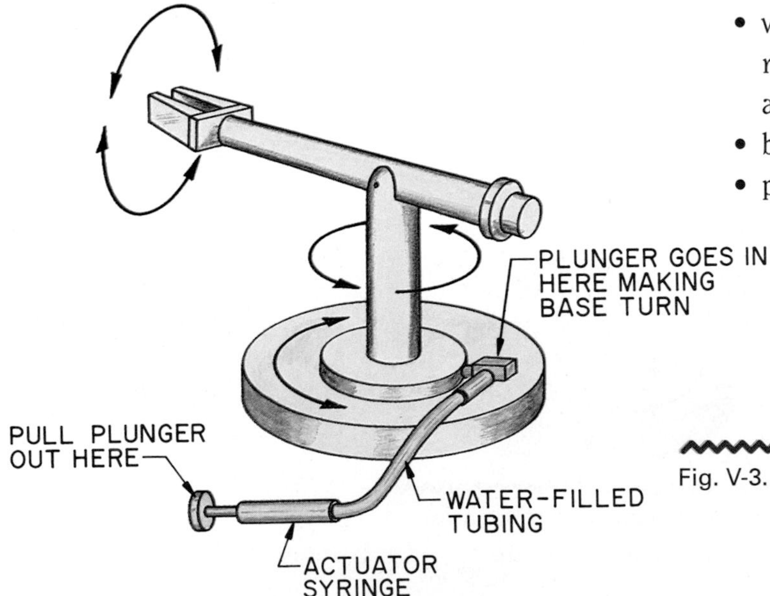

Fig. V-3. Basic hydraulic action.

Materials

plywood, assorted sizes
scrap lumber
dowel rods, various sizes
PVC plastic pipe
sheet metal
metal cans

Supplies

clear plastic tubing, ⅛ diameter
syringes, 6–20 cc
ball bearings
springs
adhesives
fasteners
abrasive paper

Equipment

drafting equipment or a CAD system
hand tools
power tools, hand-held
band saw
belt sander
drill press
safety glasses and other safety devices as needed

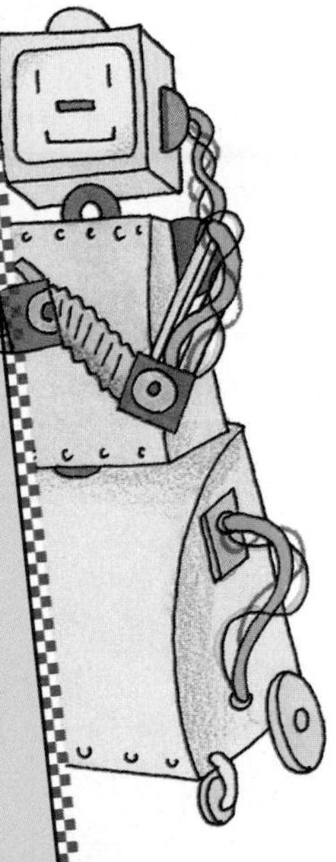

PART 4: Materials, Supplies, Equipment...........

Materials, supplies, and equipment you might use include the list at the left.

Safety Notes

- Observe all laboratory safety rules.
- Wear safety glasses and use other safety items as needed.
- Before using any power tools, be sure you understand how to operate them and always get your teacher's permission.

PART 5: Procedure...........

Each group will design a different robot and build it in different ways. Following these steps will make your work easier.

1. Examples of robots are shown in Fig. V-4. These are presented just to start you thinking. Remember your robot is to be a *prototype*. As a group, discuss various possibilities.
2. Make several sketches of possible robot designs and then combine your group's best ideas into a rough sketch. Modify your design as needed.
3. Prepare working drawings.
4. Create a bill of materials for making your robot.
5. As in the manufacturing industry, your production must be well-planned and organized. Prepare a plan of procedure for making your robot. Following a plan will help you work more efficiently.
6. Gather the materials and supplies that you will need.
7. Following your teacher's instructions for working in the lab, build your robot. Test and modify it as needed.

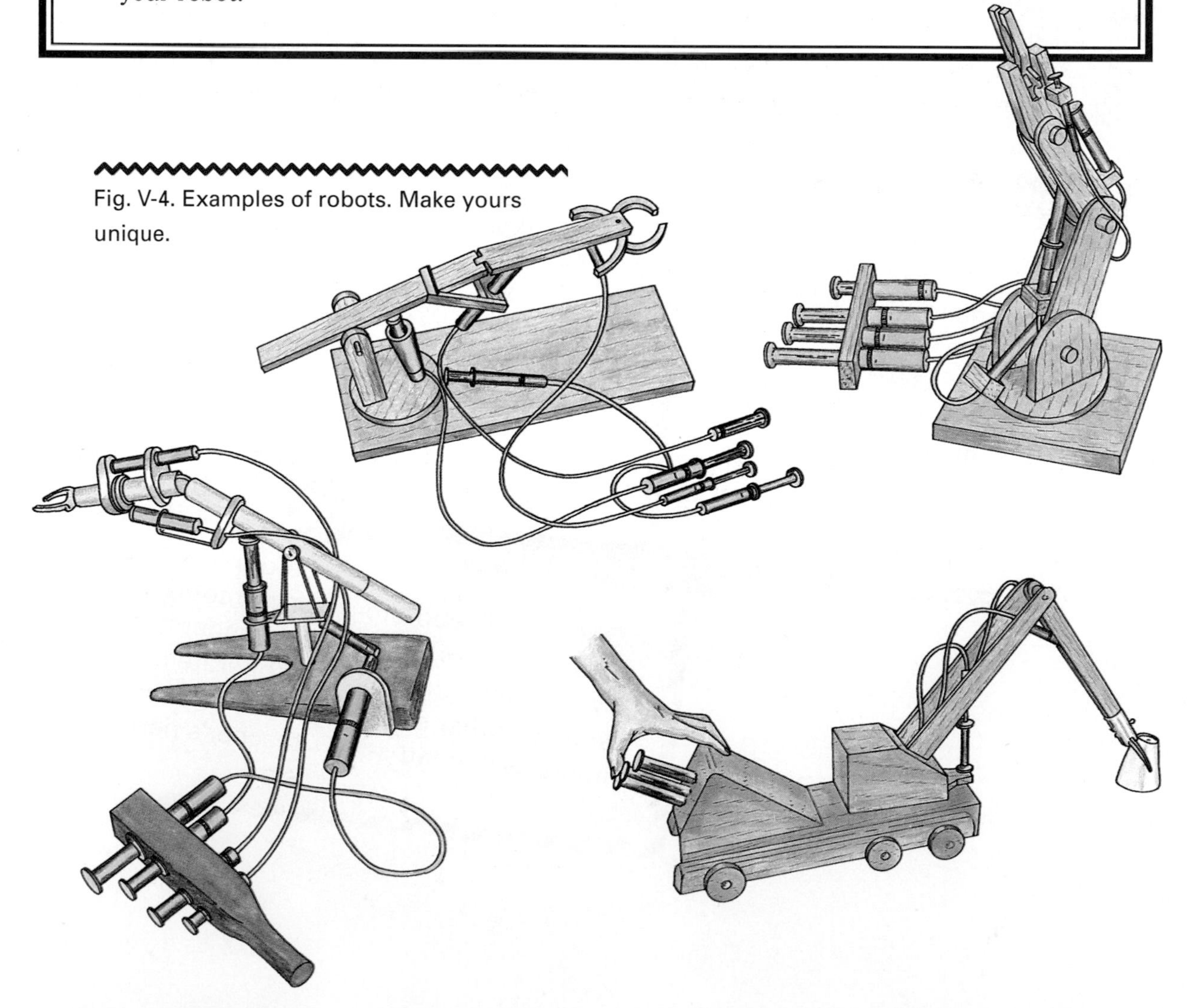

Fig. V-4. Examples of robots. Make yours unique.

PART 6: For Additional Help..........

For more help with this activity, look up the following terms. You'll find some of them in this book. (Check the index.) You'll find others in dictionaries, encyclopedias, and other resource materials.

drafting
hydraulics
leverage
plan of procedure
production line
prototype
research and development
robotics
thumbnail sketches
working drawings

PART 7: How Well Did You Meet the Challenge?..........

When your robot is completed, evaluate what you have done.

1. Prepare and give a presentation for the class. Explain your design. Relate problems you encountered and describe how you solved them. Demonstrate what your robot can do.
2. On a separate sheet of paper, write a brief evaluation of your robot and the work you did on it. If you had it all to do over again, what would you do differently?

PART 8: Extending Your Experience..........

This activity helps you learn about robots, the use of hydraulics, and procedures used in manufacturing. Think about the following questions and discuss them in class. You'll find more about manufacturing in Chapters 18–23 in this section, "Production Systems: Manufacturing."

1. If you were to make another robot, what new or additional features would you try to include?
2. How might your robot be used on a production line in your lab?
3. How could a motor be used to power the robot?
4. How could a computer be used to control robot operations?
5. What are some positive and negative impacts that result from using robots?
6. What uses do you foresee in the future for robots in routine daily life?

CHAPTER 18

What Is Manufacturing?

Introduction

How big would a system have to be to supply all the products we might ever need? Would the word *huge* or *gigantic* express the size of such a system? The system that performs this remarkable task is called a *production system*. Production systems change materials into more useful forms called products. Production systems are responsible for producing all the products we use each day.

The products of production supply communication, transportation, and biotechnology systems with all the things they need to produce products and services. Telephones, automobiles, medicines, and even buildings make up a tiny example of the work accomplished by production systems.

Production systems create these products through two activities: manufacturing and construction. This chapter and the chapters that follow will help you learn more about these systems.

After reading this chapter, you should be able to

Define production.
Define manufacturing.
Discuss the interdependence of systems.
Discuss the development of manufacturing.
Describe primary manufacturing.
Define secondary manufacturing.
Describe primary manufacturing processes.

Words you will need

production
manufacturing
construction
mass production
primary manufacturing
secondary manufacturing

What Is Production?

Look around your room. Each and every thing you see was produced by a production system. **Production** is the manufacture or construction of products. Books, furniture, clothing, windows, even the room itself is a product of this complex system. Fig. 18–1.

The goal of any production system is to create products. When products are produced in a factory, the process is called **manufacturing**. Books, furniture, windows and clothing are manufactured items.

Construction is the process of building something on a site. The room you are sitting in is a product of construction; so are houses, bridges, and tunnels. Fig. 18–2.

PRODUCTION SYSTEM

Input: Raw Materials → Process: Manufacturing, Construction → Output: Products

Fig. 18-1. Production systems supply us with every product that makes up our people-made world.

Fig. 18-2. Bridges are a type of construction that make it possible for people to cross rivers, gorges, and other natural barriers easily.

IMPACT

The manufacturing and construction industries have a huge impact on the economy. Together, they have an income of more than $900 billion per year.

Production Systems Are Interdependent

Manufacturing and construction systems depend on each other to produce products. That is, they are *interdependent*. Each system is vital to the success of the other. For example, a factory building is the output of a construction system. The people who assembled the building relied on manufactured materials such as bricks, pipes, and wire for its construction. Fig. 18–3. In fact, you will find that all technologies depend on each other. Let's see how this dependency works.

A factory that manufactures cosmetics relies heavily on transportation systems to deliver products. Most factories own or rent fleets of trucks from manufacturers to accomplish this goal. The trucks travel over roads created by construction systems. Communication systems within the factory store and transfer the information necessary for the production of millions of products each year. Fig. 18–4.

▶▶▶ FOR DISCUSSION ◀◀◀

1. Describe how a hospital is dependent on transportation, communication and production systems.
2. Make a list of ten manufactured materials that are used in house construction.

Fig. 18-3. Manufacturing and construction systems depend on each other. Factories are needed to produce building materials such as brick, concrete, steel, wire, and pipe. These same materials can then be used to build more factories.

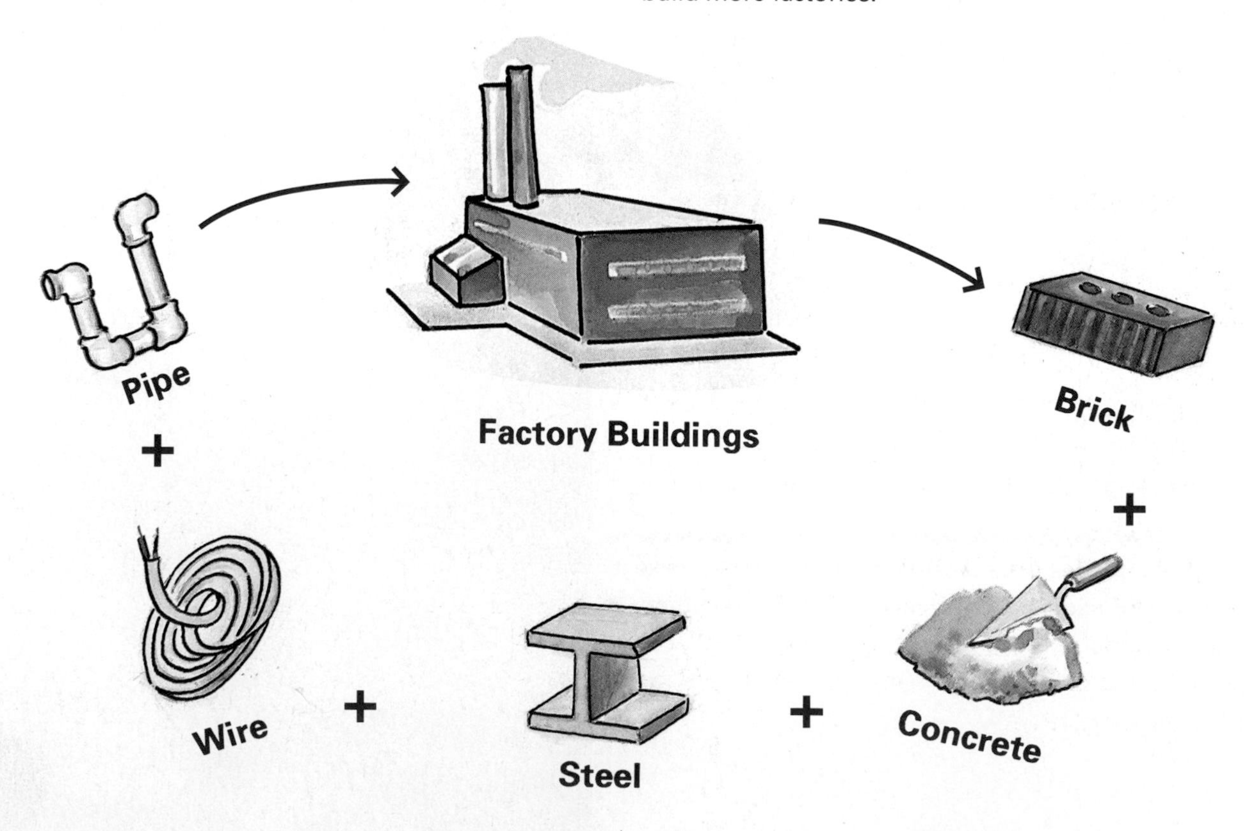

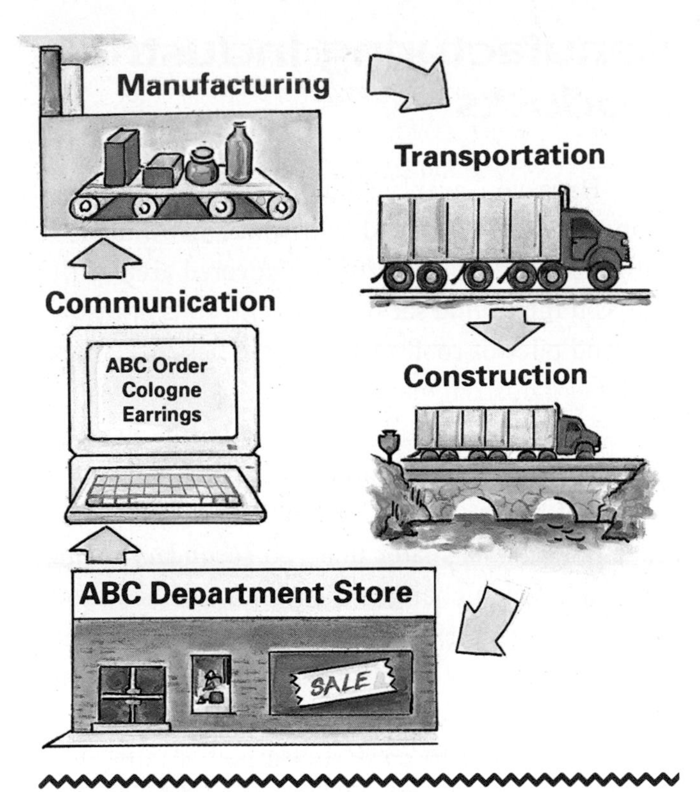

Fig. 18-4. All technological systems are dependent on each other.

Development of Manufacturing

Modern manufacturing is the result of many changes in how products are made. We can look back at some of these changes by studying the people who created them.

Early craftspeople were responsible for providing communities with everyday products. Fig. 18–5. If you lived during the 1800s, you might have been one of these craftspeople. You might have been a tailor, for example.

Tailors provided people with custom-made clothing. Working in a small shop, the tailor fitted, cut, and sewed by hand garments such as shirts and dresses. The tailor worked on one garment at a time until it was completed.

The Industrial Revolution brought new machines and techniques to production. This changed forever the methods used to produce products. Factories were soon equipped with sewing machines and hundreds of people to operate them. These factories could mass-produce hundreds of garments at a time. The impact of the Industrial Revolution on the tailor's role as a producer of garments was disastrous.

Mass production is the process of producing large quantities of products in factories. In a clothing factory, fabric travels from one machine to another on an assembly line. One machine operator performs one process, such as sewing the front and back together. The operator then

TECHNOLOGY TRIVIA

The first practical sewing machine was constructed by Elias Howe, an American inventor. He patented it in 1846. After Howe's invention, Isaac Singer and others also manufactured sewing machines. In 1854, Howe established his right to collect royalties on ALL sewing machines manufactured after his invention.

Fig. 18-5. Before the Industrial Revolution, families made much of what they needed. They depended on skilled craftspeople for things they couldn't make for themselves.

passes the garment on to the next operator, who sews on a sleeve. Each operator does exactly the same part again and again. Each garment is exactly the same. However, each garment is produced by many people instead of just one. Fig. 18–6.

Tailors could not compete with the speed of mass production and the low prices of mass-produced products. They, like many other craftspeople, were replaced by the factory system of manufacture.

Manufacturing today still reflects the factory system of production. Later, we will look at modern manufacturing techniques and how they differ from earlier production techniques.

►►► FOR DISCUSSION ◄◄◄

1. Before craftspeople, how did families obtain the everyday products they needed?
2. Describe how the factory system affected the work of the blacksmith.

Fig. 18-6. Each factory worker performs only one small part in the mass production of products, whereas craftspeople are responsible for entire products. Mass production made it possible for people with fewer skills to find work.

Manufacturing Industrial Products

Have you ever visited a lumberyard? Most lumberyards are stacked with piles of lumber in many sizes and shapes. Under a covered area in the yard, you might find sacks of cement, rolls of insulation, and piles of roofing shingles. These are all examples of industrial materials.

Industrial materials are created by primary manufacturing. **Primary manufacturing** is the conversion of raw materials into industrial materials. Industrial materials later undergo **secondary manufacturing**, which changes them into finished products. Most factories produce either industrial materials through primary manufacturing or finished products through secondary manufacturing.

If you were to construct a picnic table, you might make it with redwood lumber. The redwood you purchased from the lumberyard would be an example of an industrial material that has undergone primary manufacturing. The raw material (trees) has been manufactured into board lumber. You then begin secondary manufacturing to change the lumber into the table as you cut, assemble, sand, and apply a finish. Fig. 18–7.

Primary Manufacturing Processes

All raw materials must be processed or changed into more useful forms before they can be made into finished products. This is the goal of primary manufacturing. Primary manufacturing involves three processes: obtaining the material, refining the material, and creating industrial materials.

TECHNOLOGY TRIVIA

The world's tallest tree, a redwood, is in Redwood Creek Grove, California. Its tip, which is slowly dying back, reaches 362 feet.

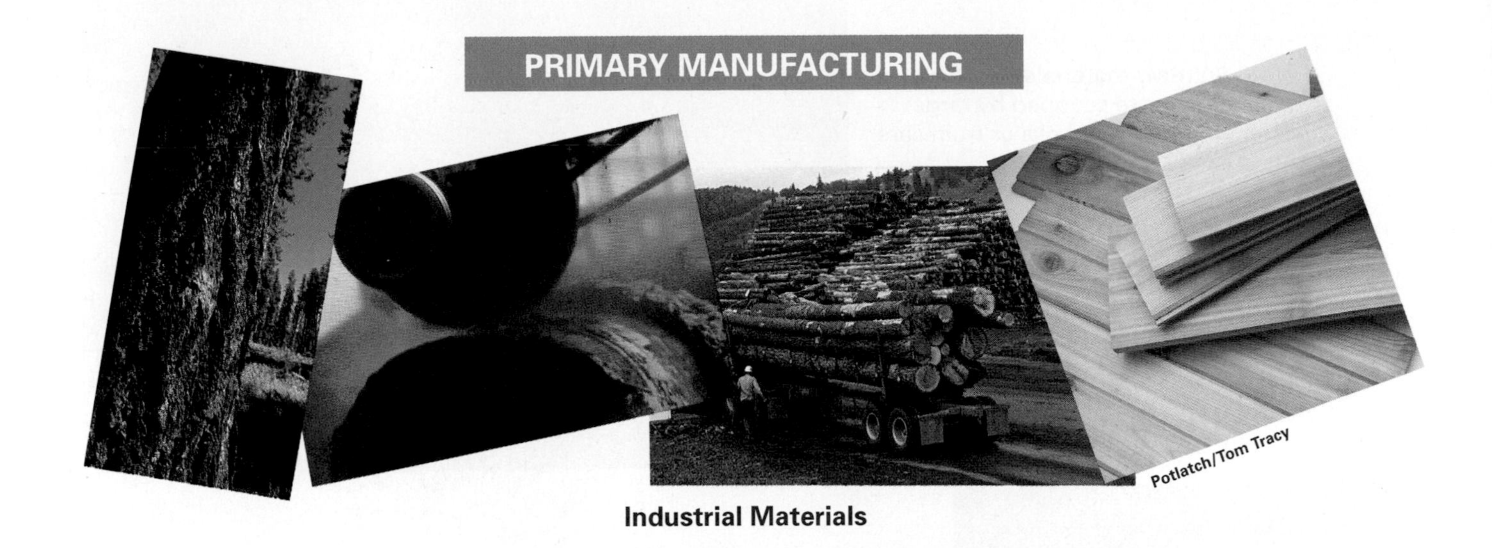

Fig. 18-7. Wood is a raw material that must be processed into industrial materials before it can be used for products such as picnic tables.

Obtaining Raw Materials

Materials are gathered by mining, drilling, or harvesting. Fig. 18–8. Materials that are mined are dug from the earth. Coal, iron ore, and copper are mined materials.

Liquids and gases can be collected by drilling. Large drilling rigs cut holes deep into the ground and pump out materials such as oil and natural gas.

Harvesting is the method used to gather renewable resources. For example, trees are harvested by cutting forested areas.

Fig. 18-8A. Raw materials found close to the surface are scooped by large power shovels into trucks or train cars. This is called *strip mining.*

Fig. 18-8B. *Drilling* for raw materials can be done on land or under water. Drilling stops when an underground pool of the material is reached.

Fig. 18-8C. Gathering renewable resources, such as trees, is called *harvesting.* Selective cutting harvests only the mature trees.

Fig. 18-8. Ways to obtain raw materials.

Fig. 18-9. Giant machines are used to refine raw materials and get them ready for manufacturing into industrial products. Here, bark is being removed from a log.

Refining Materials

Raw materials are very rarely used in their raw form. Primary manufacturing processes are used to refine them (clean them up).

Logs harvested from trees must have their branches and bark removed. Minerals that have been mined must have soil, rocks, and other impurities removed. Refining gets the materials ready for manufacturing. Fig. 18–9.

Manufacturing Industrial Materials

Many different processes are used to manufacture refined materials into metal, wood, ceramic, plastic, and composite industrial materials.

Steel, for example, is manufactured into sheets, plates, and bars. Logs are cut into boards and veneers of various—but standard—shapes and sizes. Silicates are combined into cement. Fig. 18–10. Industrial materials are manufactured into standard sizes and shapes to insure that consumers will be able to purchase identical materials each time a job requires them.

■ **Collect samples of various types of materials and make a display.**

▶▶▶ FOR DISCUSSION ◀◀◀

1. Make a list of the many industrial materials used in the manufacture of automobiles.
2. Describe how the materials used to make applesauce are harvested, refined, and produced.

■ **Investigate the processing of metal, wood, ceramic, plastic, and composite materials. Choose one or two material processing plants that possibly could be established in the community.**

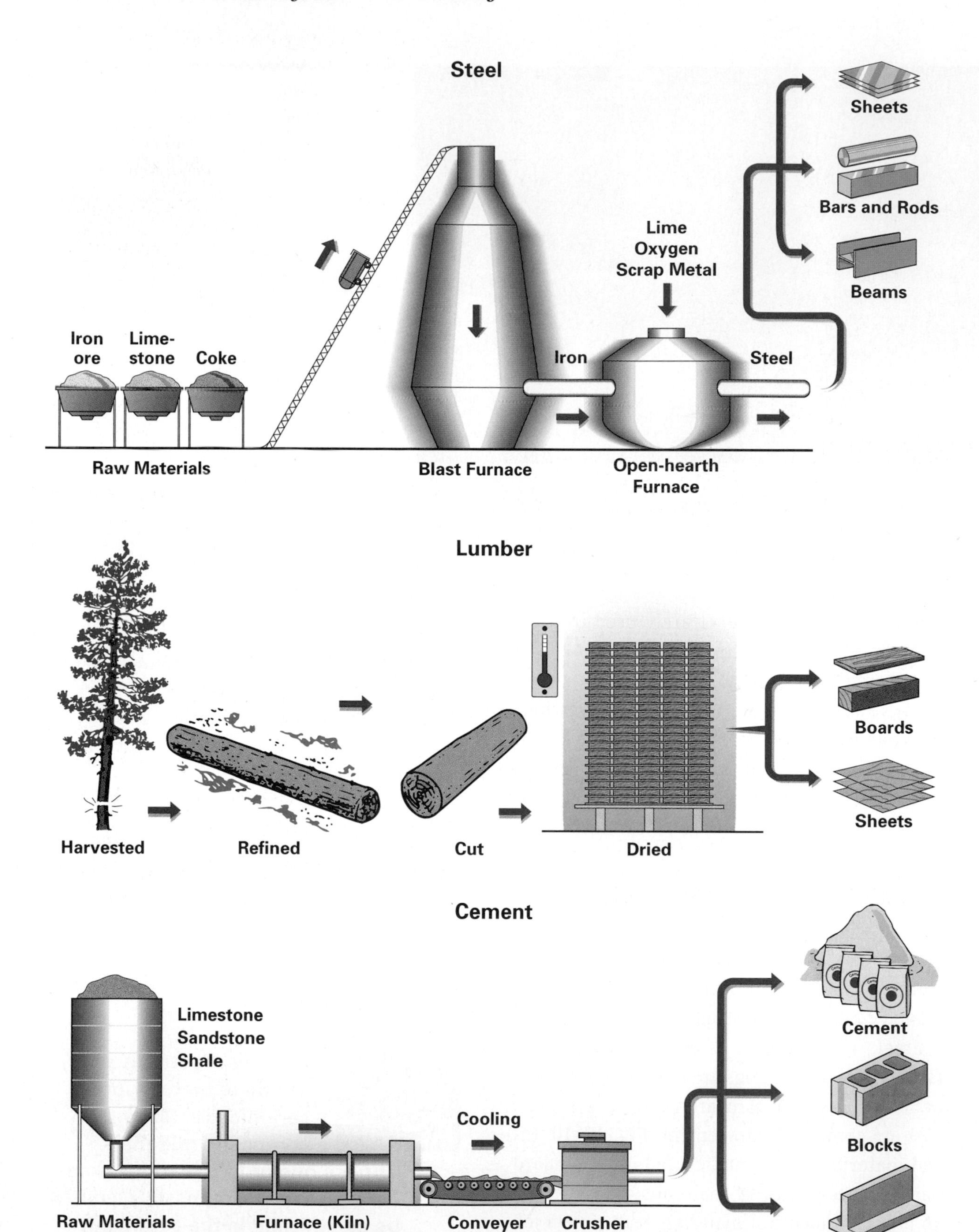

Fig. 18-10. Many different processes are used to manufacture industrial materials.

CHAPTER 18 Review

Chapter Highlights

- Production system activities include manufacturing and construction.
- All technological systems are dependent on each other.
- Today's manufacturing is a result of changes in how products are produced.
- Primary manufacturing produces industrial materials.
- Secondary manufacturing produces finished products.
- Primary manufacturing processes include gathering materials, refining materials, and manufacturing industrial products.

Test Your Knowledge

1. What are the two activities performed by production systems?
2. How are manufacturing and construction different from each other?
3. Describe how manufacturing and construction systems are interdependent.
4. Describe the factory system of production.
5. What impact did the factory system have on craftspeople?
6. List four primary manufactured products and describe how they are used in secondary manufacturing.
7. What are three methods used to gather raw materials?
8. What is the goal of refining materials?
9. Wood is processed into many different shapes and sizes. How do people know what to buy?
10. List three shapes in which steel products can be purchased.

Correlations

SCIENCE

1. Cement is a manufactured product used for construction. Have your teacher help you mix a small batch of cement. Pour the cement around a test tube placed in the center of a container. Put a thermometer in the test tube. Record the temperature at regular intervals until the cement has hardened.

MATH

1. You need 84 board feet of redwood lumber to construct a picnic table. (One board foot is the amount of lumber in a board $12 \times 12 \times 1$ inches.) If the wood costs $2.00 per board foot, how much will it cost you to buy enough lumber for the table?

LANGUAGE ARTS

1. Handcrafted clothing and furniture are becoming more popular. Write a paragraph comparing and contrasting a manufactured item with a similar item made by hand.

SOCIAL STUDIES

1. The chapter describes the effect of the Industrial Revolution on the tailor's trade. Name five other occupations that were common before industrialization. How did machines change those occupations?

CHAPTER 19

Manufacturing Processes

Introduction

How is a beautiful piece of furniture made? How was the engine in your family car produced? Both of these items are the result of material processing activities. **Material processing** is a series of steps or operations used to change materials into finished products. Fig. 19–1.

People process all kinds of things. Each morning, you probably process your hair—you use a series of steps to produce the hair style you like. You may first wash your hair with shampoo. Then you may rinse it, use a conditioner, and blow it dry. You might even add other products, such as gel or hair spray.

Most materials also undergo a combination of processes on their way to becoming products. In this chapter, we will study many of these manufacturing processes, including forming, separating, combining, and conditioning.

After reading this chapter, you should be able to

Describe forming, separating, combining, and conditioning processes.

Give examples of products created using the above processes.

Words you will need

material processing
forming processes
casting
pressing
forging
molding
bending
extrusion
separating processes
shearing
combining processes
adhesives
cohesion
conditioning processes

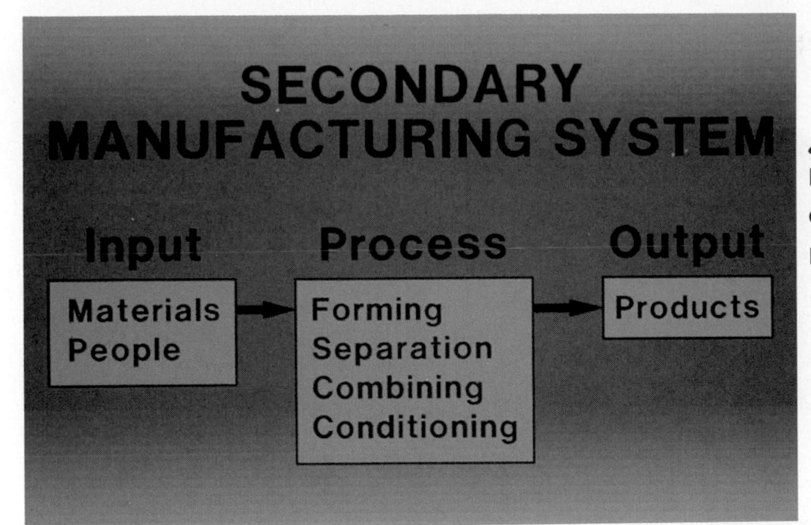

Fig. 19-1. People use tools and machines to transform materials into products.

Forming Processes

In Chapter 18, you reviewed some of the processes used in primary manufacturing. The results of these processes are industrial materials that are ready to be changed into finished products. Material processing activities also take place in secondary manufacturing.

Forming processes shape liquids and soft materials into finished products. When materials are formed, they are shaped without removing or adding material. If you have ever molded clay in your hands, you have performed a forming process. Common forming techniques are casting, pressing, forging, molding, bending, and extruding.

Casting

Casting is a technique in which a material is poured into a mold. The casting material takes the shape of the mold, much like water takes the shape of an ice cube tray. Fig. 19–2.

Pressing

Pressing is similar to casting, except that in pressing, a plunger forces the material into the mold. Powdered materials are often formed by pressing. Fig. 19–3.

Fig. 19-2. Molten (liquid) metal, liquid clay (slip), and some plastics can be formed by casting. First, a pattern is made in the shape of the object. Then a mold is created using the pattern. Liquid poured into the mold takes the shape of the cavity.

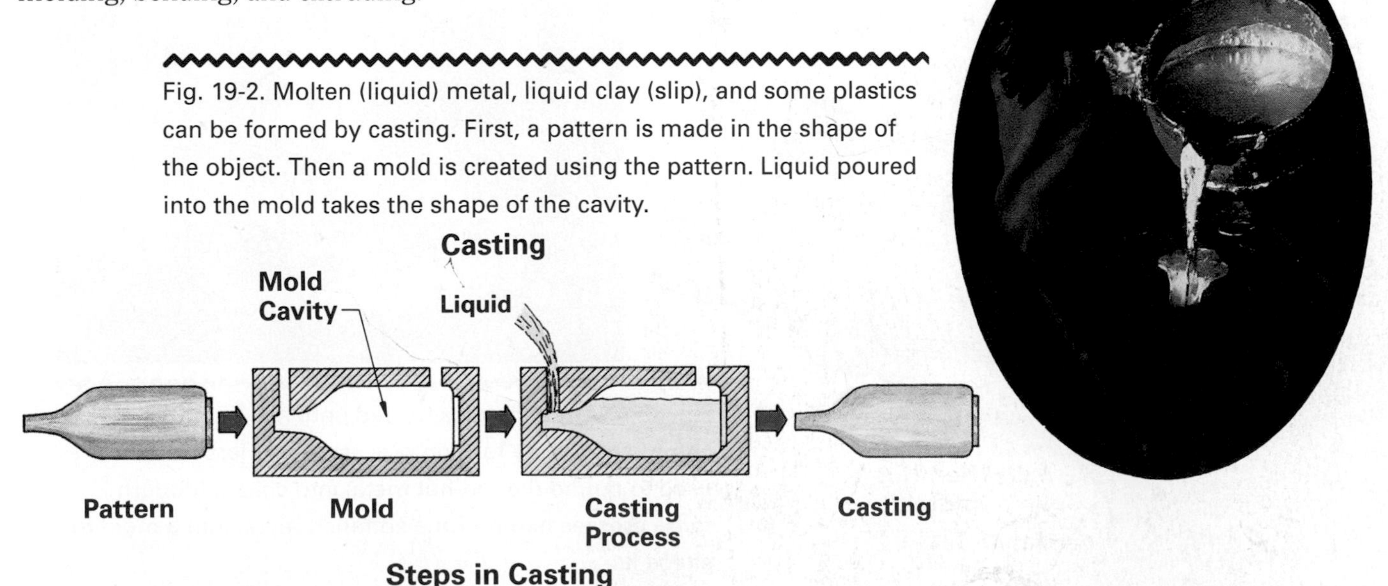

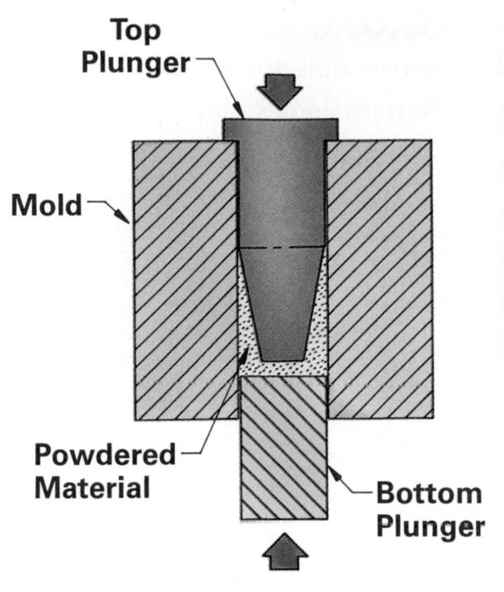

Fig. 19-3. Powdered ceramic, metal, and plastic can be formed by pressing. Using pressure, the plungers squeeze the powdered material against the walls of the mold. Heat is applied to the material during or after forming to bond the particles together.

Forging

Forging is the process of shaping metal by heating it and then hammering it into shape. You probably have seen pictures of a blacksmith forging metal on an anvil. Modern forging is done by powerful machines that use rams to hammer metal into a mold. Fig. 19–4.

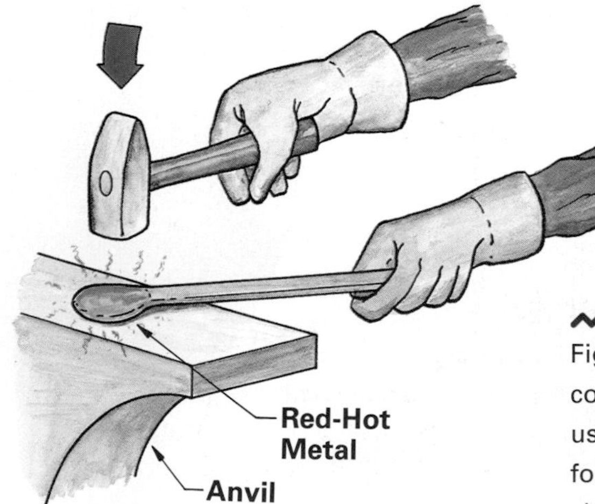

Fig. 19-4. When metal is heated until it is red-hot, it becomes soft. In the forging process, hammers or rams are used to pound the red-hot metal into shape. Modern forge presses hammer the softened metal into a mold to shape it.

Molding

Many plastic materials can be shaped by **molding**. Before plastics can be molded, they must be softened by heat. Injection molding and vacuum forming are common molding techniques. Fig. 19–5.

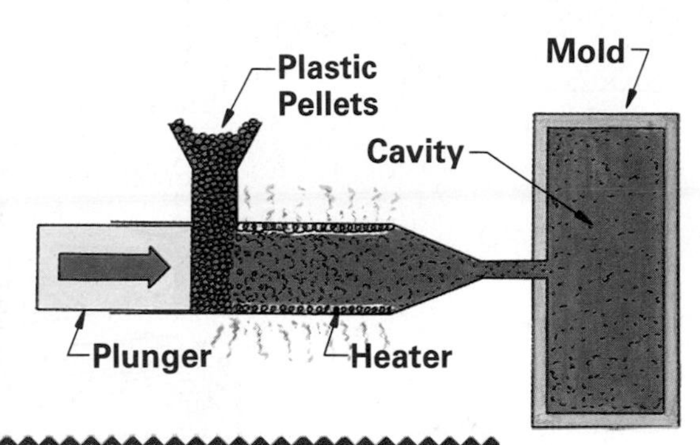

Fig. 19-5A. Injection molders soften plastic pellets by heating them. The soft plastic is pushed by the plunger into the mold.

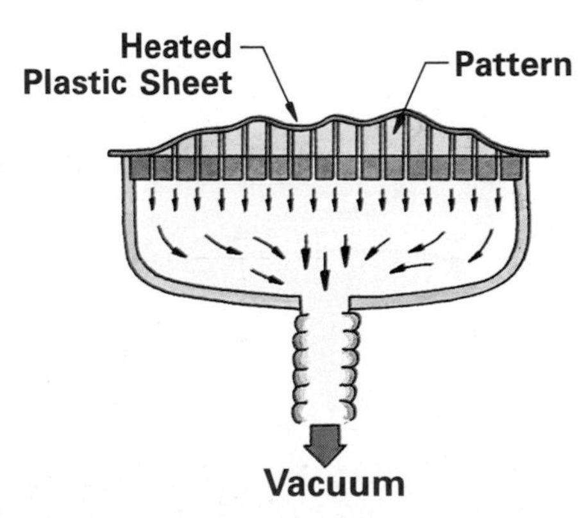

Fig. 19-5B. Vacuum forming is a process used to shape sheets of plastic. The plastic sheets are softened with heat. A vacuum pulls the plastic sheet over the pattern. When the plastic cools, it retains its formed shape.

Bending

Plastic, metal, wood, and composite materials can be formed easily by **bending**. Heat is often used to soften the material to make bending easier. Fig. 19–6.

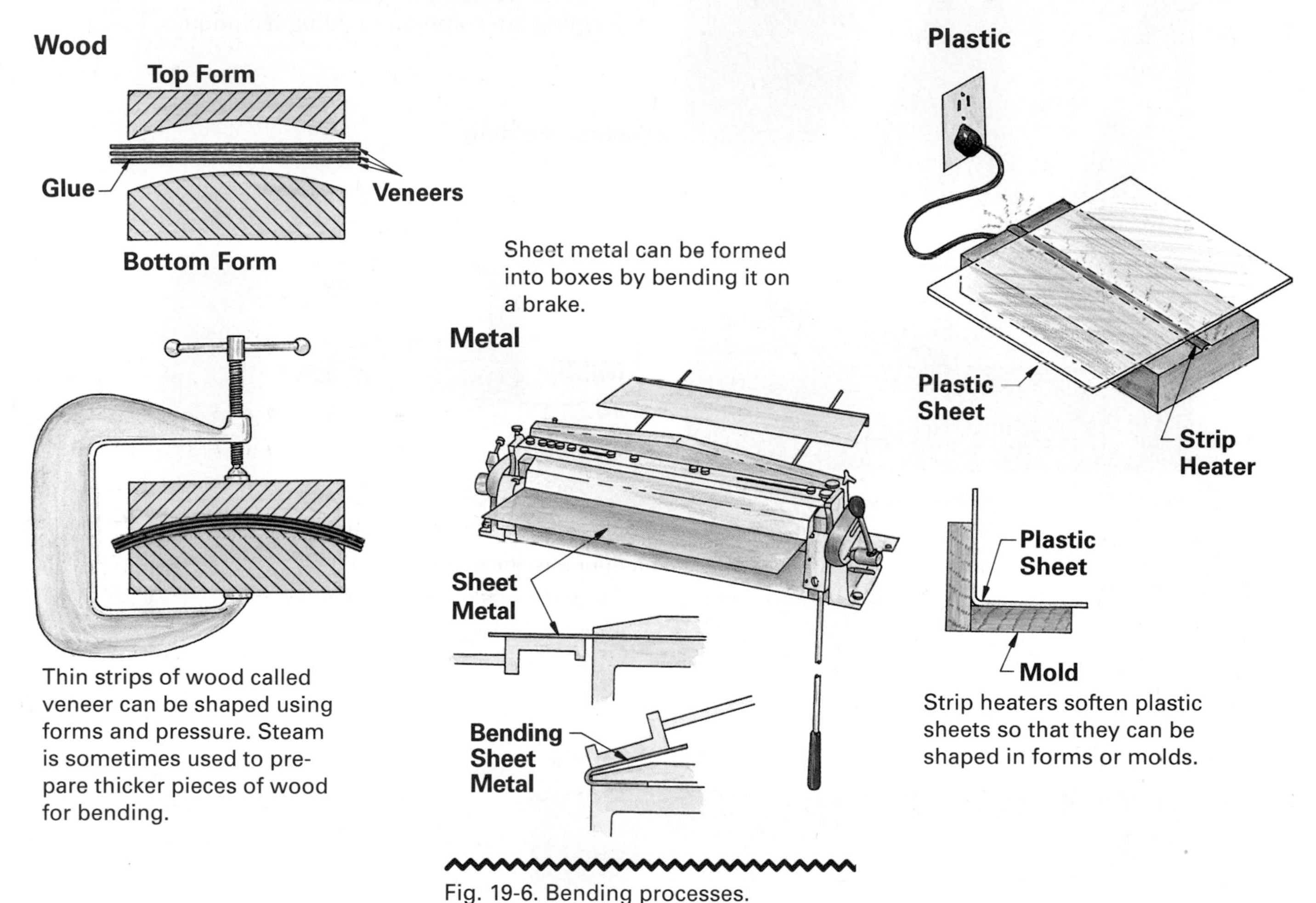
Wood
Top Form
Glue
Veneers
Bottom Form
Thin strips of wood called veneer can be shaped using forms and pressure. Steam is sometimes used to prepare thicker pieces of wood for bending.
Sheet metal can be formed into boxes by bending it on a brake.
Metal
Sheet Metal
Bending Sheet Metal
Plastic
Plastic Sheet
Strip Heater
Plastic Sheet
Mold
Strip heaters soften plastic sheets so that they can be shaped in forms or molds.

Fig. 19-6. Bending processes.

Extruding

Extrusion is a process in which a material is squeezed through an opening called a *die*. The soft material takes the shape of the die. Fig. 19–7.

▶▶▶ FOR DISCUSSION ◀◀◀

1. Describe how you could use two of the above processes to form clay bricks.
2. Which of the above processes could be used to make ice cream bars?

■ **Set up a production line and produce construction panels for the model buildings of your community.**

Separating Processes

Materials can be shaped by **separating processes**, or processes in which one piece is removed from another. People use many technological tools and machines to help separate materials. Shearing, sawing, drilling, planing, milling, turning, and shaping are some separating techniques.

When you cut paper with scissors, you are performing a shearing process. In the **shearing** process, a knife-like blade is used to slice materials. Unlike sawing, which creates sawdust, shearing does not cause the loss of any materials.

The techniques of sawing, drilling, planing, milling, grinding, turning, and shaping all remove small chips of materials as they perform separations. Fig. 19–8. For this reason, these processes are known as *chip removal* processes. The machinery used in chip removal is selected according to the materials to be cut and the type of cut to be performed. Fig. 19–9 shows some common chip removal techniques.

Fig. 19-7. The extrusion process can be used on a wide variety of materials. The machine shown here makes plastic grocery bags.

Fig. 19-8. When a material is cut using the shearing processes, no material is lost. In contrast, chip removal always removes some part of the material.

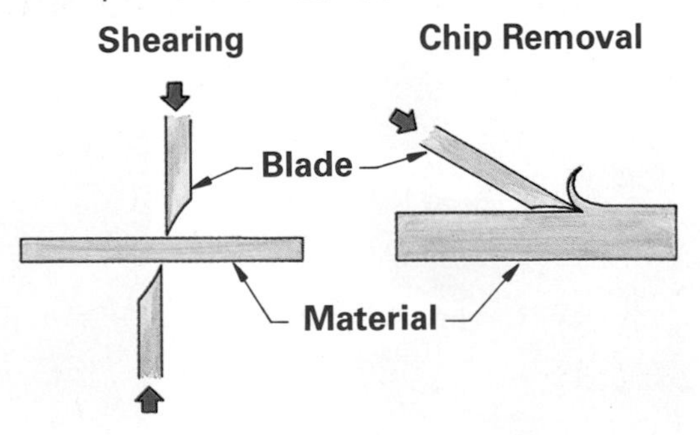

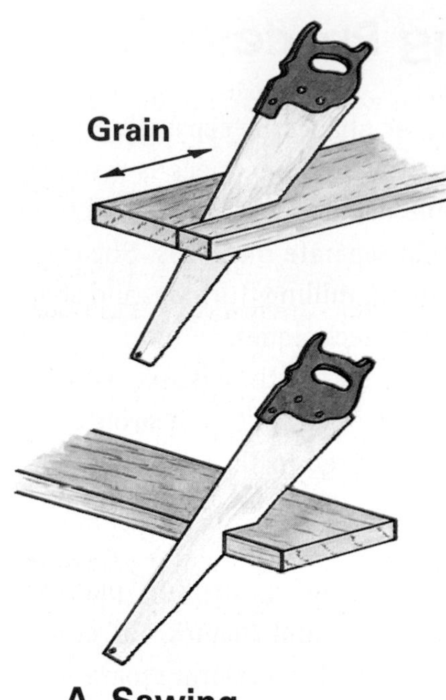

A. Sawing

Sawing is a method of cutting material with a blade that has teeth. Sawing along the wood's grain is called ripping. Sawing against the grain is crosscutting.

B. Drilling

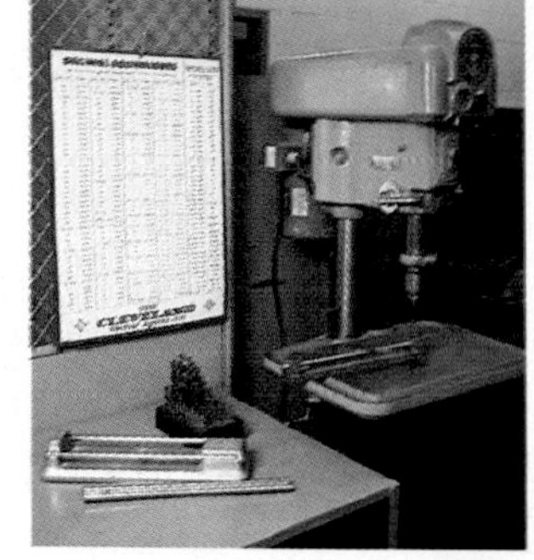

Drills are used to cut holes in material.

C. Grinding

Hard particles called abrasives are used to grind away a material. Abrasives may be glued together to make a grinding wheel or glued to a sheet to make sandpaper.

D. Milling

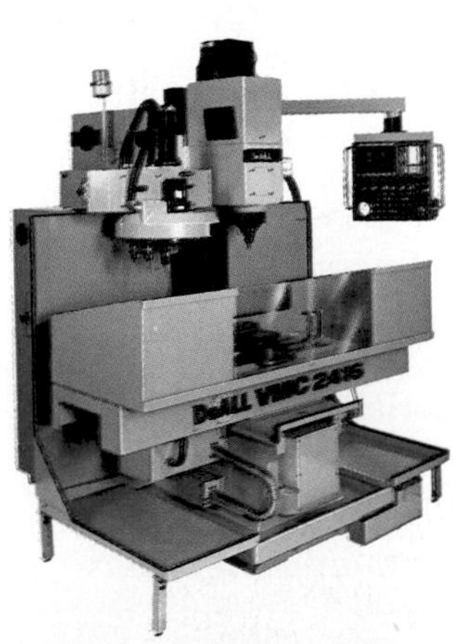

Milling machines have a spinning cutter that shapes materials.

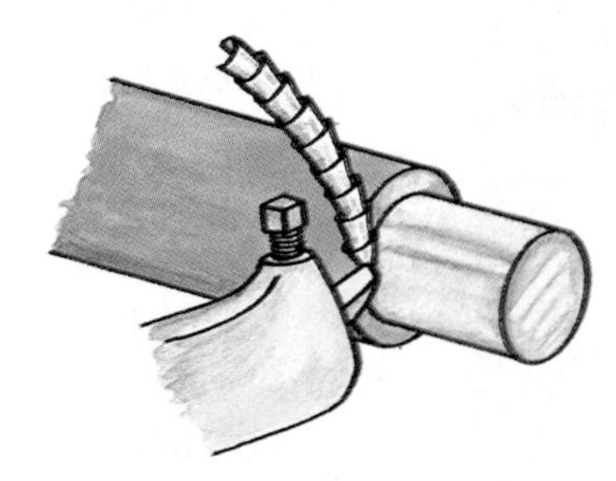

E. Turning

Turning machines rotate the material against a stationary blade.

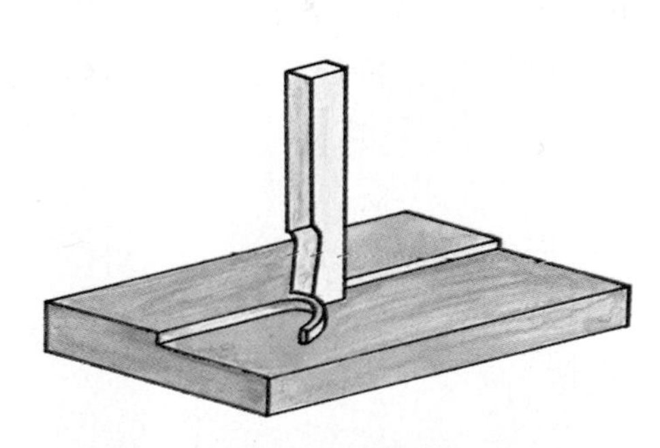

F. Planing and Shaping

Shaping tools and machines use chisel-shaped cutters to chip away at a material.

Fig. 19-9. Chip removal techniques.

▶▶▶ FOR DISCUSSION ◀◀◀

1. A cookie cutter is an example of which separation process?
2. Which separating processes might you use to cut out a template for a wall plaque?

Combining Processes

Combining processes are used to join materials together. Many different combining techniques are used in manufacturing. The technique for a given job is selected according to the materials being assembled and how strong the joint must be.

Mechanical Fasteners

Nails, screws, staples, and bolts are just a few examples of the many mechanical fasteners used in manufacturing. Some fasteners are designed to combine materials permanently, while others allow materials to be taken apart when necessary. Fig. 19–10.

Chemical Joining

Materials can be combined through chemical joining techniques. Glues (**adhesives**), solvents, and cements are used to *adhere* many materials, or make them stick together.

Adhesives form a thin film on the surface of the materials being joined. The film adheres to both surfaces, making a secure joint.

Solvents and cements are also used to combine plastic materials. Unlike adhesives, these chemicals melt or cause the two materials to flow together. This is called **cohesion**. A cohesive bond causes the molecules of materials being joined to mix together. Cohesive bonds are stronger than adhesive bonds.

Using Heat to Combine Materials

Heat joining is a process that is used mostly on metals and plastics. Welding, brazing, and soldering are heat-joining techniques that are commonly used in manufacturing. When materials are welded,

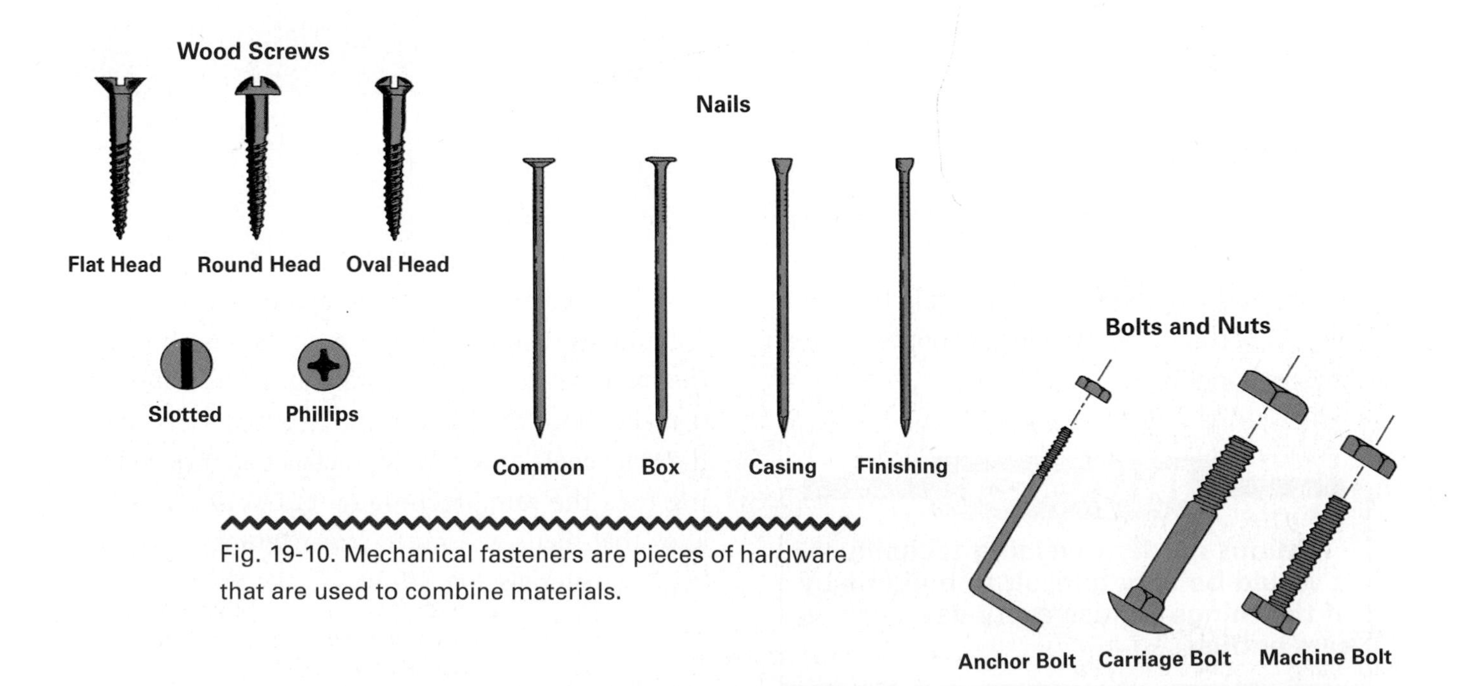

Fig. 19-10. Mechanical fasteners are pieces of hardware that are used to combine materials.

Oxyacetylene Welding

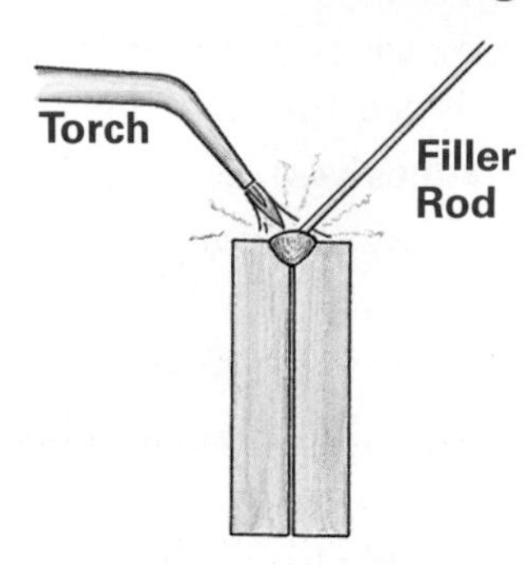

Heat produced by the torch melts the material together. The filler rod fills any gaps between the pieces.

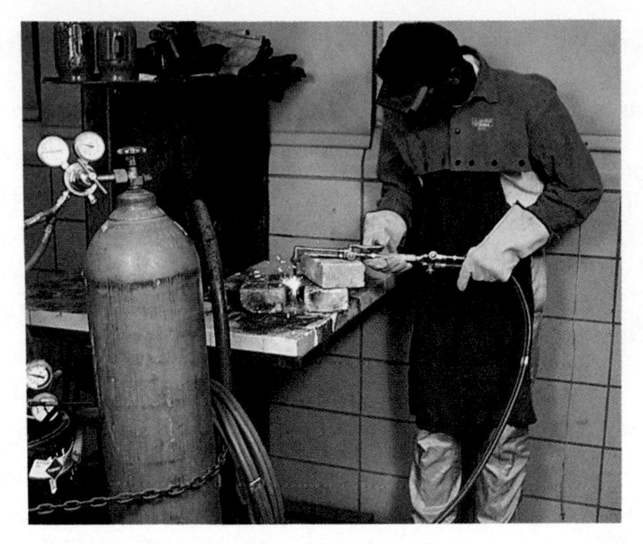

Soldering

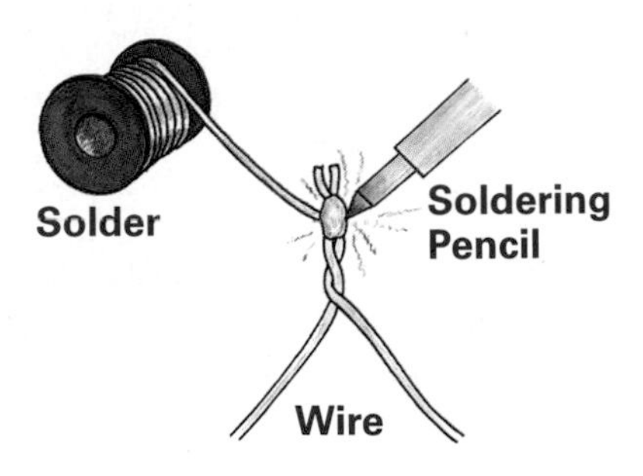

Solder melts around and between a joint to combine two materials.

Fig. 19-11. Two heat joining techniques.

high temperatures cause the materials to flow together and become one. Welding is another example of cohesive bonding.

IMPACT

Without modern welding techniques, it would be very difficult to build many of the things we use every day, such as cars or buses.

Soldering and brazing are also heat joining techniques. However, these processes do not melt the materials together. In brazing, a brass alloy rod is melted into the joint of the two materials. When the alloy cools, it bonds the metals together. Soldering uses the same technique, but solder is a lead alloy that melts at a much lower temperature than brazing materials. Fig. 19–11.

Finishing Materials

The last process performed on a material in manufacturing is usually *finishing*. Finishing can be simply polishing or smoothing the surface of a material. Some finishes coat the top of a material, providing protection and decoration. The most common surface coatings are paint, enamels, shellac, varnish, lacquer, glaze, and vinyl. Finishes are selected according to the materials being finished and the environment in which the product will be used.

▶▶▶ FOR DISCUSSION ◀◀◀

1. Where would examples of welding, adhering, fastening, and finishing be found on an automobile?
2. Explain why finishing is a combining process.

Conditioning Processes

Forming, separating and combining change the shape or size of a material. **Conditioning processes** are used to change the inner structure of a material. Athletes who work out each day condition their bodies. This means they improve the performance of their muscles and lungs through exercise.

Materials can also be conditioned to improve their performance. Manufacturing processes such as heat treating and chemical treating improve the physical properties of materials. Fig. 19–12.

▶▶▶ FOR DISCUSSION ◀◀◀

1. Give an example of a conditioned material you use every day. In what way does the conditioning help the item?
2. Compare an athlete conditioning his or her muscles to a material being conditioned. How are the two similar? How are they different?

■ Prepare a demonstration that includes showing at least one technique used in each of the following material processing activities: forming, separating, combining, and conditioning.

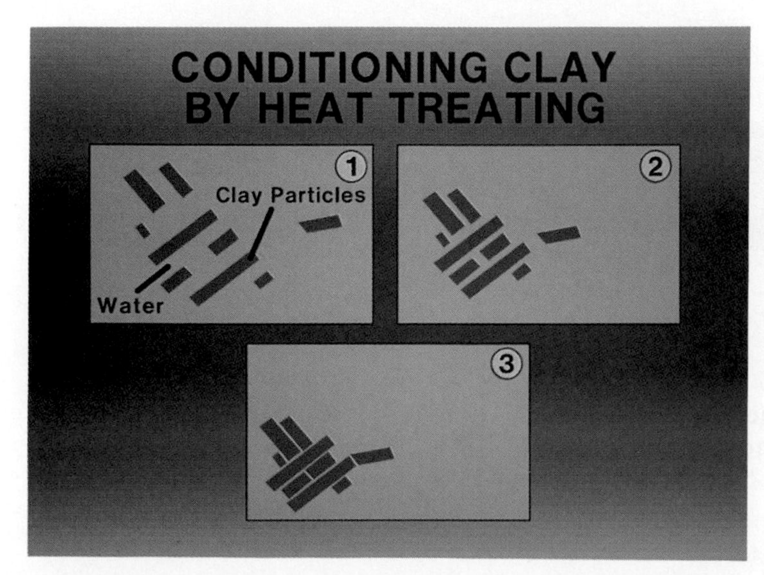

Fig. 19-12. Heat treating causes changes in the properties of materials. When clay is fired in a kiln (oven), the water between the clay particles is driven off. The particles move closer together. As the particles begin to touch, the clay becomes stiff. When all the particles compress against each other, the clay becomes hard.

Combining Processes to Make Products

In this chapter, you have read about forming, separating, combining, and conditioning processes. Various combinations of these processes are used to make products. For example, here you can see the processes used to make two familiar products: contact lenses and marbles.

Contact Lenses

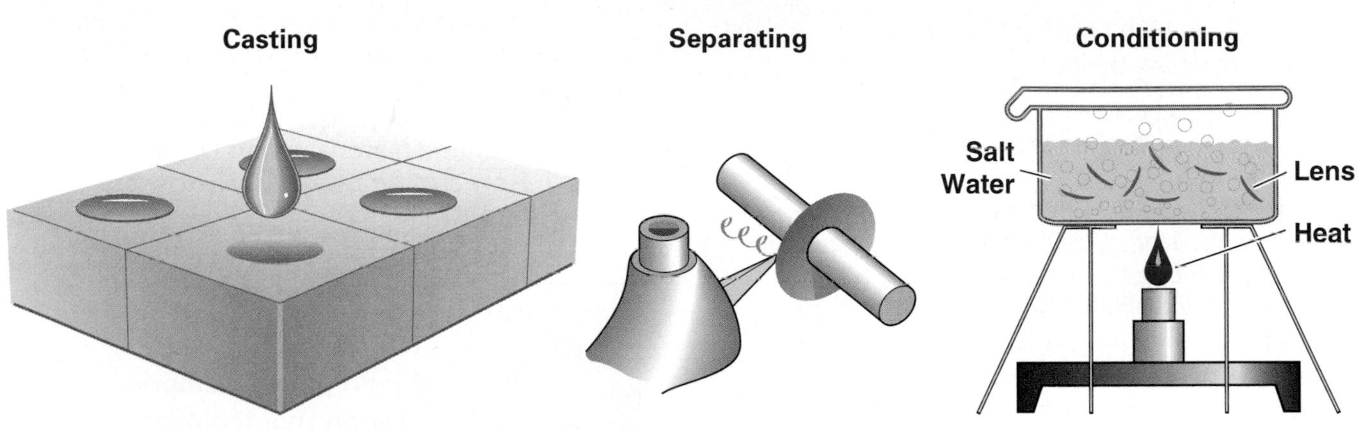

Drops of liquid plastic fill small molds. The plastic hardens to form buttons.

A turning process is used to cut the plastic buttons to the correct shape.

After being shaped, the lenses are soaked in a mixture of boiling water and salt to make them soft.

Marbles

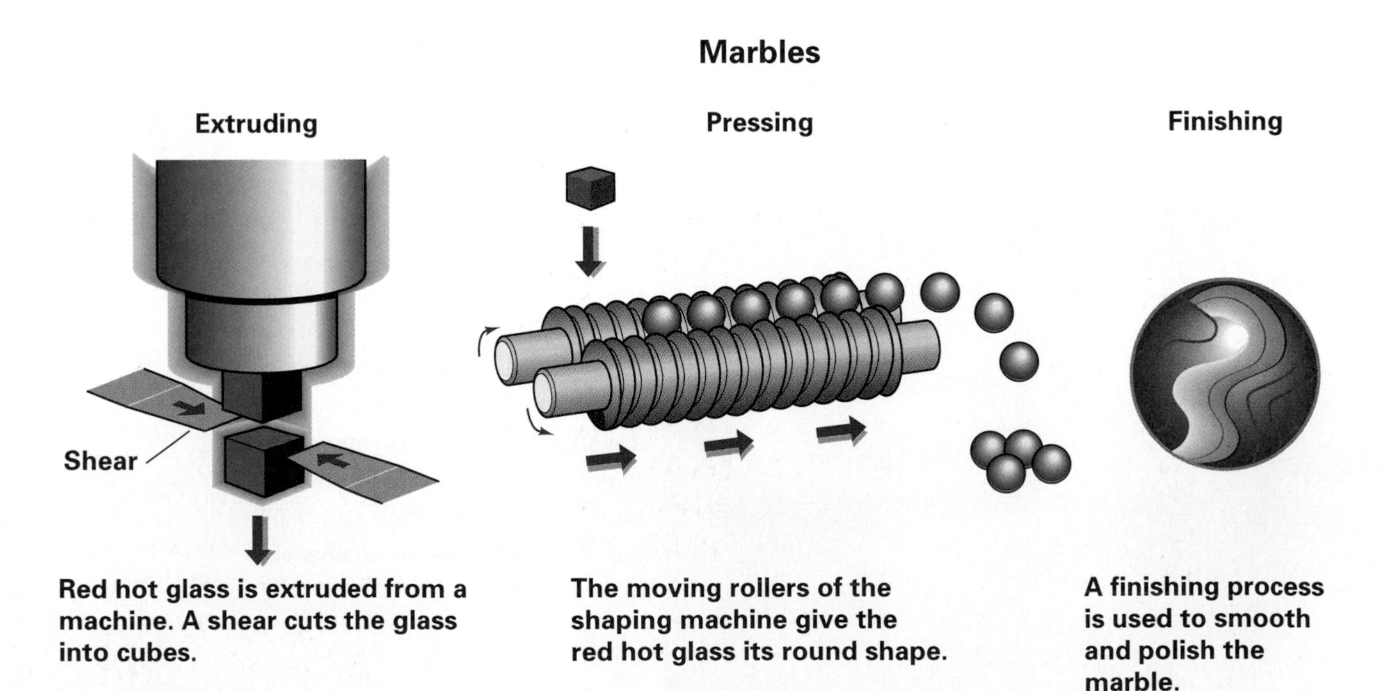

Red hot glass is extruded from a machine. A shear cuts the glass into cubes.

The moving rollers of the shaping machine give the red hot glass its round shape.

A finishing process is used to smooth and polish the marble.

CHAPTER 19

Chapter Highlights

- Secondary manufacturing changes industrial materials into finished products.
- Materials are processed into finished products through a series of manufacturing steps and operations.
- Material processing activities include forming, separating, combining, and conditioning techniques.
- Materials undergo a combination of these processes during manufacturing.

Test Your Knowledge

1. How are primary and secondary manufacturing different?
2. What four processing groups are used in manufacturing?
3. List and give a short description of three forming techniques.
4. What forming techniques can be used to make a granola bar?
5. In what way is shearing different from chip removal techniques?
6. List three methods that can be used to combine two pieces of wood.
7. How are cohesion and adhesion different?
8. Is welding a cohesion or adhesion process?
9. Why do people condition materials?
10. List and describe two conditioning techniques.

Correlations

SCIENCE

1. Metals absorb heat in different amounts, a fact that affects manufacturing processes. Try this experiment. With your teacher's help, heat metal samples of equal weight to the same temperature. Aluminum, zinc, tin, and copper are good choices. Set the heated samples on a block of wax. The samples that absorbed the most heat will sink deepest into the wax.

MATH

1. Two pieces of wood are fastened end to end. One is 5⅛ inches long, and the other is 6¾ inches long. What is their combined length?

LANGUAGE ARTS

1. Select a product, such as a piece of furniture. Describe the manufacturing processes (forming, separating, etc.) it may have undergone.

SOCIAL STUDIES

1. Make a list of the manufacturing processes described in this chapter. Next to each process, name at least one worker who uses this process. Example: heat joining—welder.

CHAPTER 20

Biotechnical Systems

Introduction

It's easy to recognize products that are produced by manufacturing systems. Automobiles, furniture, and appliances: they are all around us. These items are created using the processes described in Chapter 19.

We also use manufactured items every day that are not so visible. These items are made using **biotechnical systems**, or technologies that involve living organisms. These manufacturing processes are very different from forming, separating, combining, and conditioning techniques.

For example, biotechnical systems produce flowers sold in florist shops. They use microorganisms to produce food and to digest water pollutants. Biotechnical systems are also responsible for the development of products and techniques used in medicine and health care.

After reading this chapter, you should be able to

Discuss how biotechnical processes are different from traditional manufacturing processes.

Give examples of biotechnology processes used in manufacturing.

Explain some of the changes in agriculture brought about by biotechnical systems.

Explain the role of biotechnical systems in medicine and health care.

Words you will need

biotechnical systems
biotechnology
bioprocessing
genetic engineering
controlled environment agriculture (CEA)
hydroponics
biomedical engineering
bionics

Biotechnology

More than ever before, people are using information about living organisms to help create new technologies. Biotechnical systems are the result of combining technology and biology. The technologies that result from this mix are:

- biotechnology
- agricultural technology
- biomedical engineering technology

In **biotechnology**, living microorganisms are used to create products. Microorganisms are tiny living creatures, too small to be seen by the human eye without the aid of a microscope. Fig. 20–1. In a biotechnical system, the microorganisms do the work. This is called **bioprocessing**.

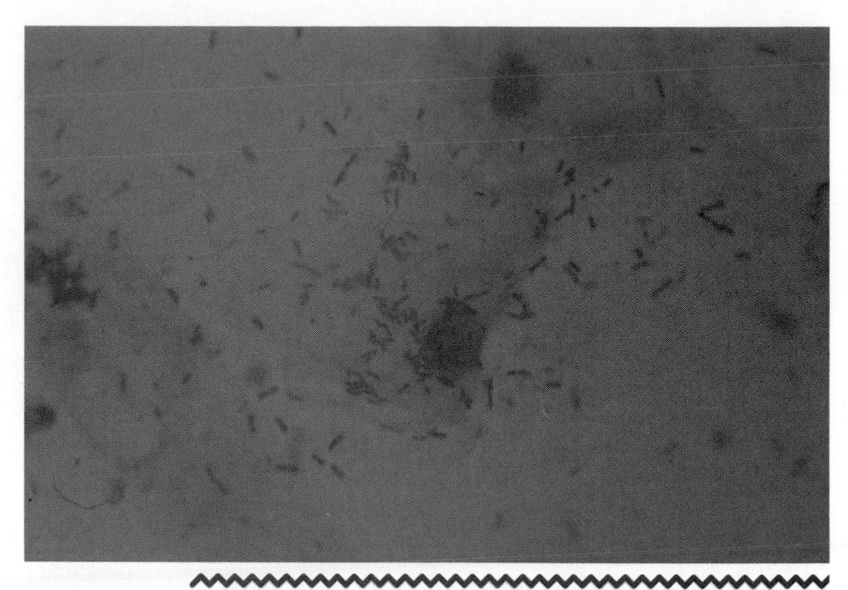

Fig. 20-1. This photograph shows one-celled organisms called *yeast*. Yeast cells are used in the manufacture of many different products.

How can tiny creatures process materials? What products do they produce? You eat one product of bioprocessing every day: bread. In the bread manufacturing process, living yeast cells are added to the dough mixture. The cells digest sugar and starch in the dough and release CO_2 (carbon dioxide). Fig. 20–2. The carbon dioxide causes the bread to rise by forming pockets of gas throughout the dough. Without yeast, the bread would be a thick, dense mass.

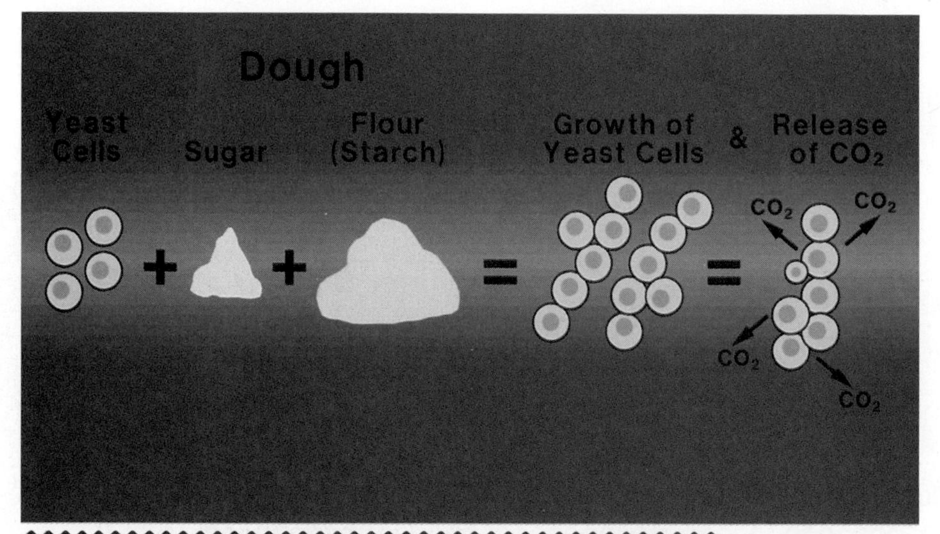

Fig. 20-2. Yeast cells multiply quickly as they consume sugar and starch in the dough. The cells release carbon dioxide gas (CO_2), which causes the dough to rise. The dough becomes the "factory" for bioprocessing.

Other microorganisms are also used in biotechnology. For example, bacteria are routinely used in the production of cheese, yogurt, and sour cream.

The manufacture of food products is not the only use of bioprocessing. Microorganisms are also used in sewage treatment plants to digest human and other waste products.

Genetic Engineering

Every cell of every living organism contains *genes*. Genes give living organisms their traits. In humans, genes determine what we look like, our hair and eye color, and even our health.

Genetic engineering is the study of genes and how they can be changed to create desired traits in living things. The science of genetics is another example of a biotechnology.

Have you ever wondered why plants of a single variety in a store all look exactly alike? It is a good possibility that the plants were a product of gene splicing or cloning.

TECHNOLOGY TRIVIA

The principles of genetics were discovered during the 1860s by Gregor Mendel, an Austrian monk, when he observed that garden peas inherited various traits in a predictable manner. He showed that organisms inherit two forms of the gene for each trait, one from each parent. Mendel's work remained unnoticed until the early 1900s when genetics became a recognized science.

Fig. 20-3. Genetic engineering processes are used to produce plants with identical characteristics.

A gene can be removed from one cell and added to another using a technique called *gene splicing*. The second cell then develops the same traits (height, fullness, color) as the first. Fig. 20–3.

Cloning can also produce hundreds of plants that look exactly alike. Clone plants are all produced from the same parent plant using a technique called *tissue culture*. Fig. 20–4.

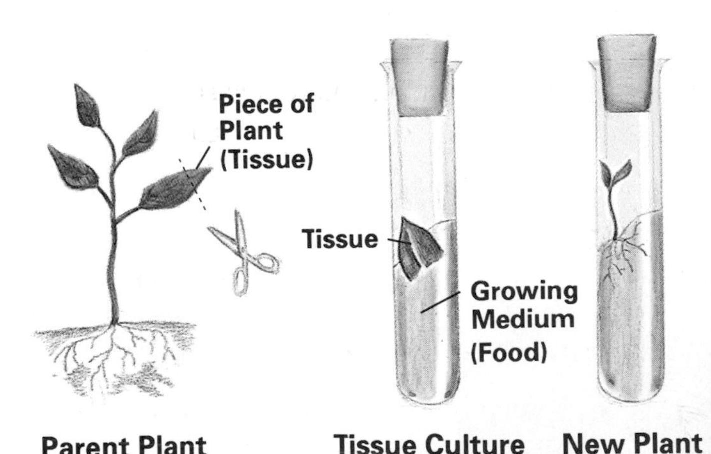

Fig. 20-4. Tissue culturing is performed by removing a piece of plant (tissue) from the parent plant. The tissue is placed into a growing medium and is allowed to grow. The new plant that results is a clone of the parent plant.

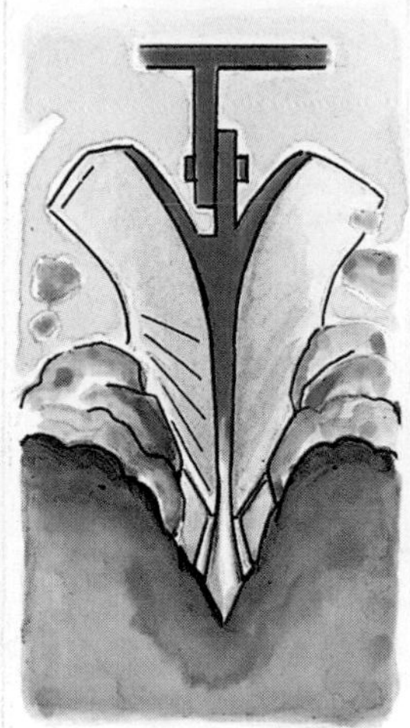

Clearing
Clearing the land gives the farmer a flat surface to plant.

Tilling
Tilling loosens the soil so that the plants' roots can take hold.

Planting
Seeds are dropped into holes and covered with soil.

Cultivating
Cultivating is the process of caring for the plant.

Harvesting
Harvesting removes the grown plants or products for processing.

Fig. 20-5. Agricultural technology involves several different types of processes.

Genetic engineering can also be used to create animals with specific traits. For example, genetically engineered sheep now produce large amounts of milk. Their milk contains an important protein that is used in medicines.

▶▶▶ FOR DISCUSSION ◀◀◀

1. **Is bread a primary or a secondary manufactured product? Explain.**
2. **What traits in a corn plant could genetic engineering change?**

Agricultural Technology

Biotechnical systems have been used for many years to make agriculture more efficient. Agriculture is the business of producing plants and animals for food, clothing, and other products. Farming is a form of manufacturing in which fields are the factories.

Farm Mechanization

Growing crops involves five activities: clearing the land, tilling the soil, planting the seeds, cultivating the crops, and harvesting the crops. Fig. 20–5. Long ago, these difficult tasks were done by hand. Later, animals were used to do the work, removing some of the burden from the farmer. Today, machines are used to accomplish most of these tasks.

Fig. 20-6. Mechanization has increased the amount that can be produced by each farmer.

Cultivators pulled by tractors till the soil, plant the seeds, and cover them with earth in one operation. In wheat fields, large harvesting combines cut the crops, separate the grain, and package the stalks. Combines can process a 30-foot path of wheat in one pass. Fig. 20–6.

TECHNOLOGY TRIVIA

Wheat is probably the earliest domesticated cereal. Wheat grown by early civilizations around the Mediterranean Sea and the Middle East was used to bake bread as early as 8,000 B.C.

IMPACT

Mechanization **(the use of machines) has had a dramatic impact on agriculture. Now only 3% of the American population is needed to supply all our food needs. At one time, 80% of the population was needed to accomplish the same goal.**

Controlled Environment Agriculture

Is it possible to grow tomatoes without soil? Can flowering plants be produced without natural sunlight? A biotechnical system called **Controlled Environment Agriculture (CEA)** makes both of these possible.

Each type of plant requires a different set of growing conditions. Humidity, temperature, and light are a few of the conditions that make up a plant's growing environment. The idea behind CEA is to produce a perfect growing environment by controlling these conditions.

A greenhouse or building is the factory for CEA. Often, computers are used to control the temperature, light, humidity, and feeding of plants. Flowering plants, herbs, and food products are produced using this technology. Fig. 20–7.

Hydroponics is also a form of CEA. This biotechnical system is used to grow plants in materials other than soil. How is this possible?

The plants' roots are fed nutrients (food and vitamins) dissolved in water. These are the same nutrients found in soil. The plants may be grown in a solution, or the nutrients may be sprayed directly onto their root systems. Fig. 20–8.

Hydroponic production has many advantages. It allows plants to be grown where soil is poor or where soil does not exist. It also allows the farmer to prepare special nutrient solutions specifically designed for each plant.

▶▶▶ FOR DISCUSSION ◀◀◀

1. Modern farm equipment is very expensive. How has this affected agriculture?
2. Why might hydroponic farming be a good method of food production in space?

■ A biosphere is an area that contains all things essential to support life. Design and construct a model of an artificial biosphere for the community.

Fig. 20-7. In this greenhouse, a computer system controls light, temperature, humidity, and the amount of food and water each plant receives.

Fig. 20-8. Plants do not need soil to grow. They only need the nutrients that are normally stored in soil. In this hydroponic greenhouse, nutrients are provided in other media, such as liquids or vermiculite.

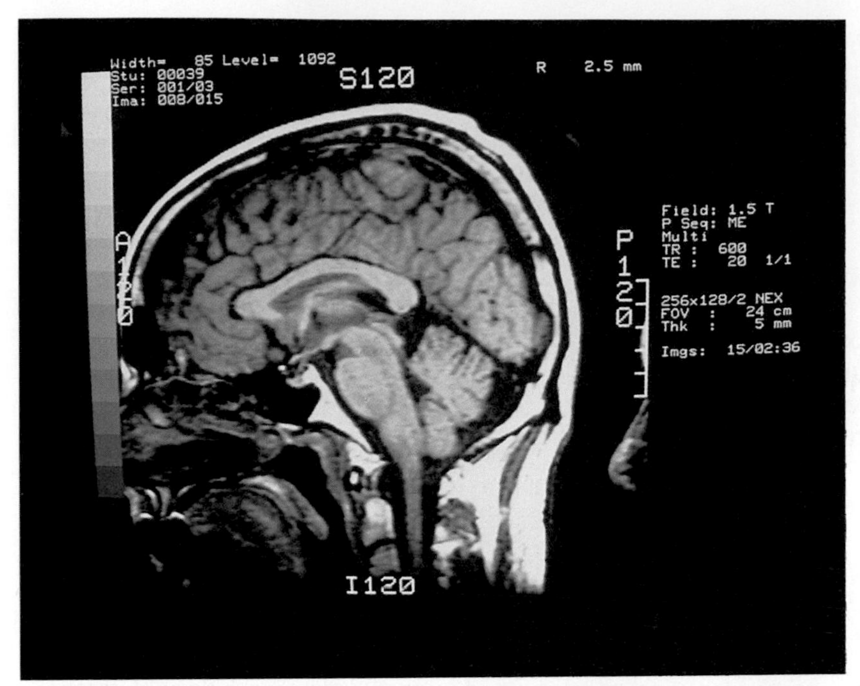

Fig. 20-9. Machines use many different techniques to take pictures of the inside of the body. This NMR image is one example of these non-invasive procedures.

Biomedical Engineering

Biomedical engineering combines technology and medicine to keep people healthy. New equipment and techniques help physicians find the causes of illnesses and treat them. Medical technologies are among the most exciting of all technologies. Science and technology together have extended the life span of humans to well above 70 years.

New Diagnostic Tools

Less than one hundred years ago, the only method available to see inside the human body was to cut it open. Surgeons relied on this exploratory technique to uncover disease. Many times, the patients did not survive the operation.

In 1895, Wilhelm Roentgen discovered a method of photographing objects that the human eye could not normally see. He called his discovery *x-rays*. X-rays changed forever how physicians diagnose illness.

Today, physicians depend on a variety of bioengineered equipment to diagnose disease. Techniques such as *computerized axial tomography (CAT) scans* and *nuclear magnetic resonance (NMR)* look deep into the body but barely touch it. Fig. 20–9. Electronic equipment can now monitor many of the body's functions and record even the slightest change in a patient's condition. Fig. 20–10.

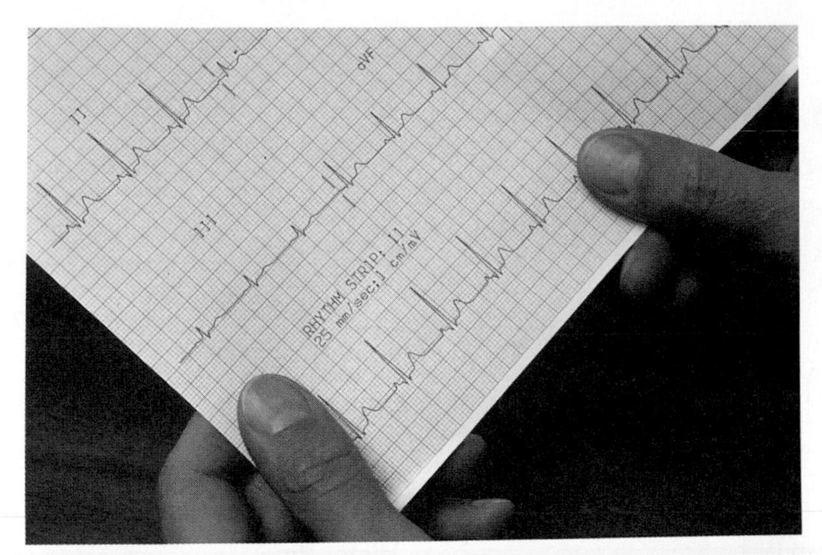

Fig. 20-10. Electronic equipment can measure even the smallest electrical charges in the body. This EKG graph was made by a machine that monitors the changes produced by the patient's heart.

Treating the Illness

When the cause of an illness is found, the physician has a wide range of new technologies to help in curing the patient. The laser beam is now replacing the scalpel in many instances, making precise cuts in living tissue. Fig. 20–11.

The *endoscope* allows surgeons to see into the body through a fiber-optic tube inserted through a small hole in the body. Using the same tool, the surgeon can insert tiny surgical instruments and perform operations that used to require large incisions. Fig. 20–12.

Biomedical technologies have even provided people with a variety of replaceable parts. When human transplants are not available or practical, manufactured parts called **bionics** can be used to replace hands, legs, bones, joints, teeth, veins, arteries, and even hearts. Refer back to Fig. 2–8 for an illustration.

Fig. 20-11. Laser technology, which is used to cut steel, can also be used for much more delicate operations, such as eye surgery.

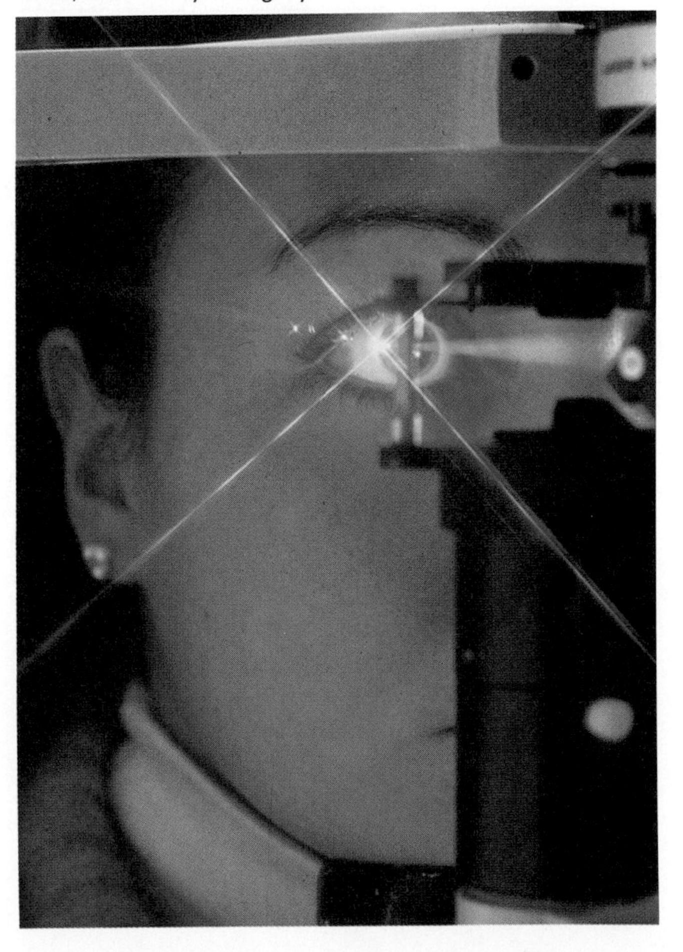

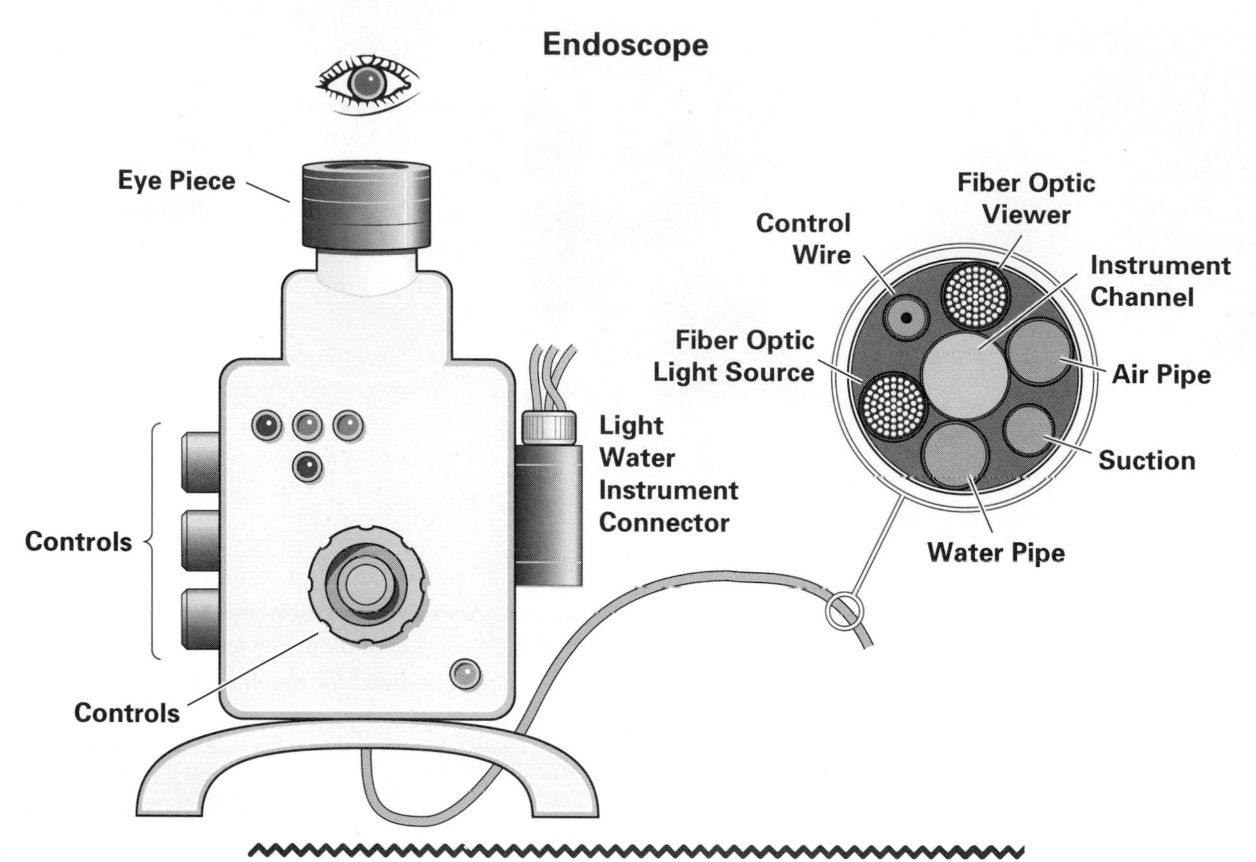

Fig. 20-12. Using an endoscope, a doctor can see what is going on in the body without making a large cut. The endoscope also allows the doctor to guide surgical instruments and perform surgery through its tube. Wires control the movement of the endoscope.

▶▶▶ FOR DISCUSSION ◀◀◀

1. How have new medical technologies affected the cost of health care?
2. Technology has provided many diagnostic techniques that did not exist 50 years ago. Name three diseases that can be caught much earlier, and therefore treated more successfully, using modern technology.

■ Find a recent article in a magazine or newspaper that tells about a new device or procedure used in the medical field. Do further research and write a report telling about the new device or procedure, the device or procedure it replaced, and the promises it holds for the future.

CHAPTER 20 Review

Chapter Highlights

- Biotechnical systems are technologies that involve living organisms.
- Biotechnical systems include biotechnology, agricultural technology, and biomedical engineering technology.
- Biotechnology uses living organisms to produce products.
- Biotechnical systems are used to make agriculture more efficient.
- Biomedical engineering combines medicine and technology to create systems that help keep people healthy.

Test Your Knowledge

1. Define *bioprocessing*.
2. List two examples of products created by bioprocessing.
3. How is gene splicing used to create plants with specific traits?
4. How is cloning used in manufacturing?
5. How has mechanization changed farming?
6. How does Controlled Environment Agriculture help create better plants?
7. How is hydroponic growing different from traditional farming?
8. List two uses for hydroponic farming.
9. How has the combination of medicine and technology increased the average person's life span?
10. Describe one new medical tool used by physicians today.

Correlations

SCIENCE

1. Clone a common houseplant such as coleus or impatiens. Cut off a small branch and wet the cut end with water. Dip the end into a rooting hormone powder and place it in potting soil. Keep the new plant's soil moist as it develops its new root system.

MATH

1. A hydroponic farmer grows 25 rows of lettuce with 38 plants in each row. How many total plants does she grow?

LANGUAGE ARTS

1. Find out what types of laser surgery are being performed in your community or in the nearest hospital. Write an essay describing the benefits of this type of surgery to patients.

SOCIAL STUDIES

1. Find out more about plant cloning. How might this process affect farming methods? How might it affect the world's food supply?

CHAPTER 21

Managing the Manufacturing Process

Introduction

Manufacturing products is a very complicated task. It can involve thousands of people, and each person has responsibilities and decisions to make. If a company is going to be successful at manufacturing, it must manage and organize every step in the production of products.

In this chapter, we will examine the management systems used to organize the complex job of manufacturing products. We will also look at the different managing methods used to organize the process of changing materials into finished products.

After reading this chapter, you should be able to

Explain the need for management.

Describe the different levels of business management.

Explain the responsibilities that go with each level.

Describe the manufacturing systems most commonly used to organize the processing of materials.

Words you will need

management
corporation
stockholder
board of directors
automated system
Computer Numerical Control (CNC)
Computer-Aided Manufacturing (CAM)
Computer Integrated Manufacturing (CIM)

Why Do We Manage?

Companies may vary a great deal in size, but they all have one need in common. They all need a system of management. **Management** provides organization for the company. The role of management is to plan, organize, and direct the activities of a business.

Your home and family are managed for many of the same reasons. Your parents or guardians are the managers. What are their responsibilities? Their first responsibility is to set goals for the family. Maintaining a nice home, planning meals, raising children, and getting ready for a new brother or sister are typical family goals.

As managers, parents plan, organize, and direct activities to achieve these goals. They may assign different tasks to each family member. You may be asked to clean the bathroom or help cook dinner. You may even be asked to get a part-time job to earn your own spending money.

▶▶▶ FOR DISCUSSION ◀◀◀

1. If you were the manager of a school store, what might some of your goals be?
2. List three activities that parents might need to manage.

Levels of Management

In many small businesses, the company owner is also the manager. The owner/manager sets goals and organizes activities to help meet those goals. In larger manufacturing companies, the system of management is more complex. This is because there are more people and activities to direct.

Stockholders

Corporations are the most common form of large business. A **corporation** is a company that has many part-owners. They are called stockholders. **Stockholders** are people who invest their money by purchasing stock in the company. Each share of stock represents a "piece" of the company they own. Large corporations can have thousands of stockholders. Fig. 21–1.

TECHNOLOGY TRIVIA

The first European stock exchange (marketplace where stocks and bonds are bought and sold) was established in Antwerp, Belgium, in 1531. The New York Stock Exchange was organized in 1792.

Fig. 21-1. Stockholders can find the value of their stock in the newspapers. The value of the stock may change, depending on how the company is doing financially.

New York bonds

Bond sales

Approx final total	$41,210,000
Prev. full day	$41,450,000
Week ago	$58,270,000
Month ago	$39,830,000
Year ago	$29,610,000
Two years ago	$30,960,000
Jan 1 to date	$155,050,000

	Yld	Sales ($1000)	Price	Net chng
WORLD BANK				
Intl4½s90	4.5	4	99.30+	.6
CORPORATION BONDS				
ANP11¾97	11.3	5	104	+ ⅛
AVS13½00	12.7	1	106	......
AVS8¼12	cv	9	140	- ⅞
AbtP6¼93	6.7	3	93	......
AbtP11s93	11.0	35	100⁵⁄₁₀	+ ⁵⁄₁₆

	Yld	Sales ($1000)	Price	Net chng
ChipM4s90	4.5	4	99.30+	.6
ChiQm11¾	7.5	5	93⅛	+1⅞
ChiQm	11.3	5	104	+ ¼
ChyS12¾	12.1	20	105¼	-1
ChyS9¼91	8.5	2	98¾	.....
CimS3½00	12.7	1	106	......
CitP8.45S	9.2	95	91¾	-1
ClaU8¼12	cv	9	140	- ⅞
CoNw¼93	6.7	3	93	......
CoPm7s94	7.7	5	91¼	.....
DiLA10⅝	cv	47	79	+ 1
DonQ19⅛	8.9	5	99¾	- ⅝
DoR11s93	11.0	35	100⁵⁄₁₆	+ ⁵⁄₁₆
ECql9s	9.3	10	97	+ ⅞
EKqa10s01	9.6	91	90⅛	+ ½
FMx9½	9.8	30	97	......
FIlm9½	9.8	21	92	+ 2¼
GnQm4s	5.2	1	76¾	.....

	Yld	Sales ($1000)	Price	Net chng
GAPx4s90	4.5	4	99.30+	.6
GAPx1¾	7.5	5	93⅛	+1⅞
GdMF3s92	11.3	5	104	+ ⅛
GraSc12¾	2.1	20	105¼	-1
GrnHs9¼	8.5	2	98¾	.....
GrwGm3½	12.7	1	106	......
GuaRm8S	9.2	95	91¾	-1
HaQm¼12	cv	9	140	- ⅞
HecVq¼93	6.7	3	93	......
HeMcos94	7.7	5	91¼	.....
HolSk10⅝	cv	47	79	+ 1
HmeSa19⅛	8.9	5	99¾	- ⅝
HouRt1s93	11.0	35	100⁵⁄₁₆	+ ⁵⁄₁₆
InCq9s95	9.3	10	97	+ ⅞
IndRc10s1	9.6	91	90⅛	+ ½
IParx9½	9.8	30	97	......
JaCo9½	9.8	21	92	+ 2¼
JoAsc4s	5.2	1	76¾	.....

Board of Directors

It is not very practical to have thousands of stockholders managing a company. Stockholders elect a **board of directors** to do this. A corporation may have many directors on its board.

The board of directors sets policy and goals for the corporation. It also elects the corporation's president and vice presidents. A vice president is usually assigned for each important operation within the company. These corporate officers, along with the board of directors, begin the chain of leadership that reaches to the individual workers. Fig. 21–2.

Let's follow the chain and see how management plans for a new policy. The board of directors for XYZ Corporation wants to increase productivity (the amount of goods produced). They issue the following policy: "XYZ will gather information to see if automated robots would increase factory productivity."

Corporate Officers

The president of the corporation carries out the policies of the board of directors. The president of XYZ assigns the vice president of production to look into new robotic technologies. The president might also direct the vice president of personnel to determine the impact this change will have on the company's work force.

Plant Managers

Plant managers direct the work in the factory. They are responsible for managing factory activities and carrying out the tasks assigned to them by the company vice presidents. The plant managers for XYZ direct their department heads to determine how robotics might be used in their departments. Fig. 21–3.

The department heads direct the activities performed by workers. After meeting with workers, the department heads report back to the plant managers. In most large companies, information travels up from the workers and down from company officers through the management system.

When the president of XYZ receives the message, he or she reports back to the board of directors. The president makes recommendations based on the information gathered through the different levels of management.

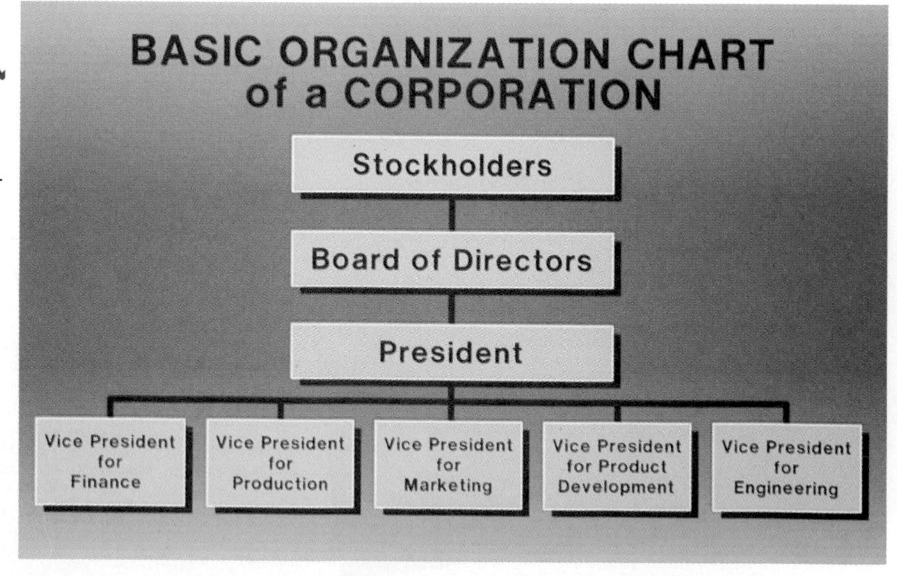

Fig. 21-2. The stockholders elect a board of directors to run the company. The board of directors elects the corporate officers, including the president and vice presidents.

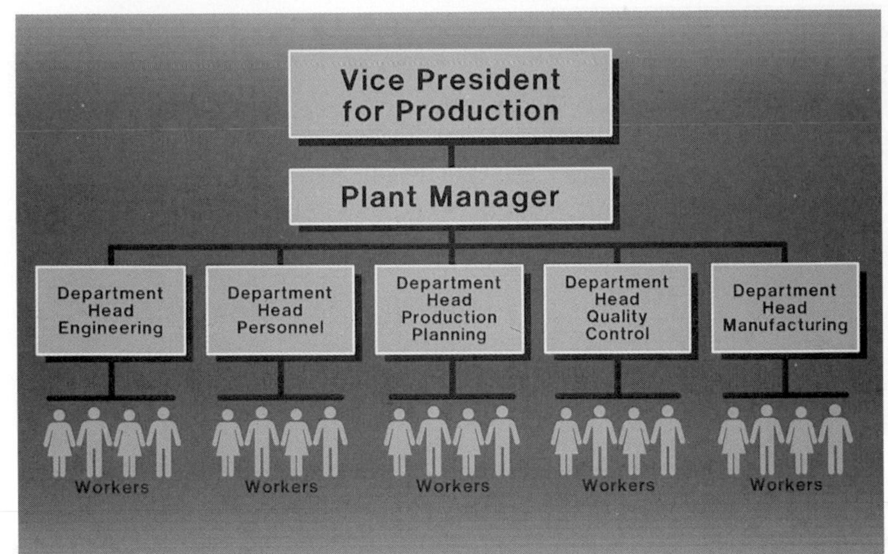

Fig. 21-3. When the vice president for production asks the plant manager for information, the plant manager requests the information from department heads. The department heads gather the information from the production employees and inform the plant manager, who gives the results to the vice president.

■ **Working with other members of your class, create an organizational chart for MegaIndustries.**

▶▶▶ FOR DISCUSSION ◀◀◀

1. If a worker has a great idea to increase productivity, how might the worker get that information to the vice president for Production?

2. Give two examples of topics that a plant manager and department heads might discuss during a meeting.

Managing Material Processing

If you owned a company that manufactured skateboards, how would you organize the production? Would you use an assembly line and mass produce thousands of look-alike boards? Would you hire a few craftspeople and have them custom-make each skateboard? Should the board be painted before or after the wheels go on?

The processing of materials also requires management. In modern manufacturing, many different systems are used to change materials into finished products. Fig. 21–4. These systems control each step of manufacturing and determine the techniques to be used in forming, separating, combining, and conditioning operations.

Custom Manufacturing

Custom-made products are specially made to meet the design and needs of individual customers.

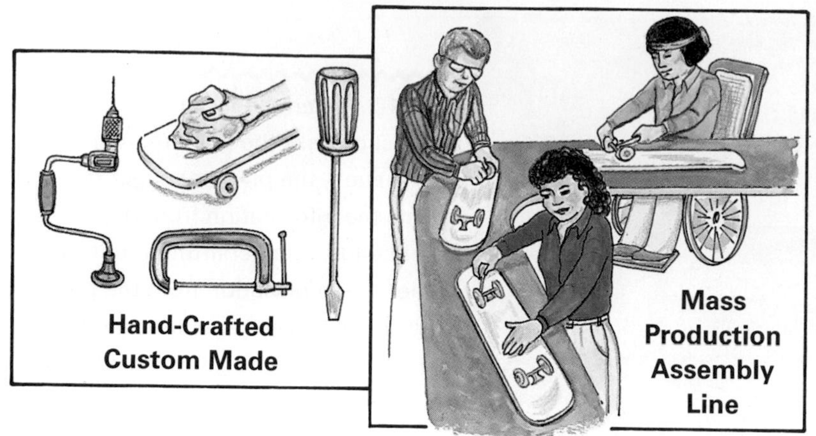

Fig. 21-4. Many systems can be used to manage the change of industrial materials into finished products. The system selected depends on the quantity and types of products to be produced.

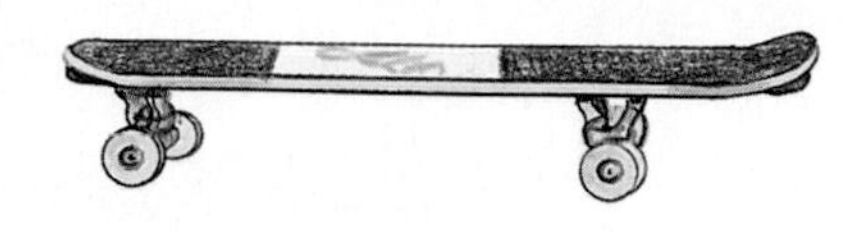

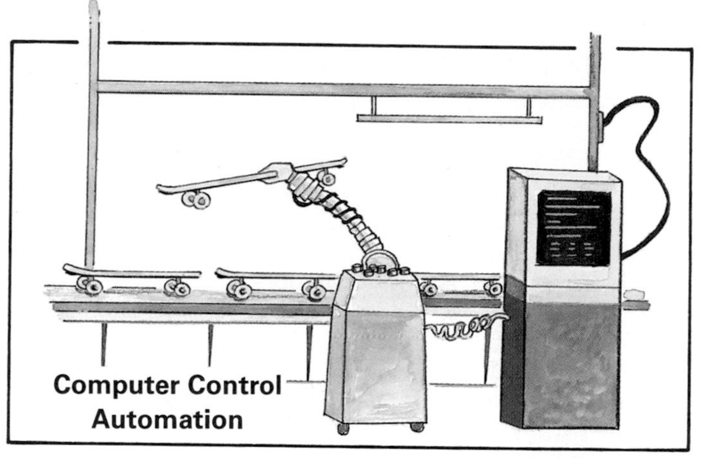

They are made in small numbers and are very expensive to produce. Boats, musical instruments, and spacecraft are examples of products that are often custom made. Fig. 21–5.

If you were to custom-make a skateboard, you would be responsible for most of the manufacturing processes. The customer would explain to you what the board should look like, what materials it should be built from, and other details. You would have to be a very skilled craftsperson to hand-make the skateboard to meet these specifications. The skateboard would be a one-of-a-kind creation.

Mass Production

Mass-produced products are manufactured in large numbers. The product moves down an assembly line from station to station. At each station, workers perform a different operation. Fig. 21–6.

Henry Ford used mass production in 1913 to manage the manufacture of automobiles. The success of Henry Ford's technique was due to his use of interchangeable parts.

Interchangeable parts are pieces made exactly the same. For example, the tires Ford used fit any automobile he produced. There was no need for

Fig. 21-5. The spacecraft used by the United States are examples of custom-made products.

Fig. 21-6. Mass-produced products are usually identical. Consumers have little input as to the design and specifications of individual pieces.

Henry Ford to custom-manufacture tires for each automobile. Mass production increased the speed with which automobiles could be produced and greatly reduced the selling price of the product.

The techniques of mass production are still used today to produce products efficiently. Modern technology, though, has provided new techniques that manage production even better.

TECHNOLOGY TRIVIA

The word automation was coined in the 1940s at the Ford Motor Company and was first applied to the automatic handling of parts in metalworking processes.

■ Set up a manufacturing company within your class. Select and design a product that your company could mass produce.

▶▶▶ FOR DISCUSSION ◀◀◀

1. Why are custom-made products more expensive than mass-produced products?
2. Give three examples of interchangeable parts found on a bicycle.

Age of Automation

More and more, manufacturing is being managed by machine. Computers make this possible. Any system in which machines control machines is an **automated system.**

Automation is usually more efficient than other systems because machines don't need coffee breaks, comfortable working conditions, or salaries. People are still needed, though, to set up, operate, and repair the automated equipment. Fig. 21–7.

IMPACT

As machines take over the work that used to be done by people, those people will need to find new jobs. Many people will change careers, not just once but several times during their working lives.

Fig. 21-7. Automation may wipe out some positions in a manufacturing plant, but it also creates new positions. Operators, programmers, and repair technicians are needed to use and maintain the equipment.

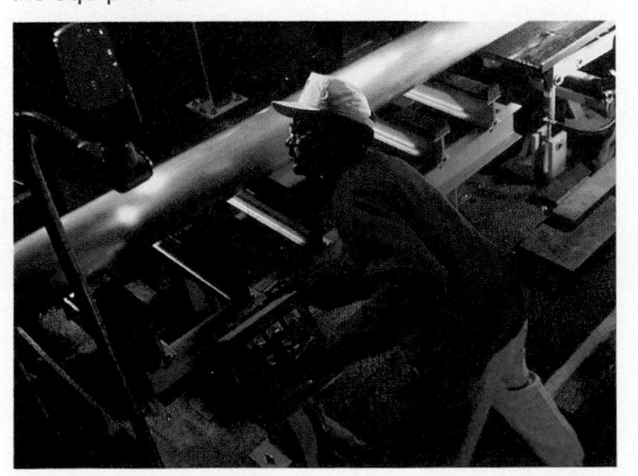

Fig. 21-8. A CNC machine can be programmed to perform a series of operations. When the job is finished, the machine can be reprogrammed to do something else.

Computer Numerical Control

Computer numerical control (CNC) is a system in which computers are used to operate machines. Cutting, drilling, milling, and turning are a few of the separation processes that can be organized by CNC. Fig. 21–8. **Computer-aided manufacturing (CAM)** is an example of a system that organizes manufacturing using computer numerical control.

CAD/CAM

CAD/CAM combines the technology of computer-aided design (CAD) and computer-aided manufacturing. Using this system, products designed on a computer can be sent directly to a machine to be produced. The same computer system that aided the engineer in designing now instructs the machines in manufacturing.

Computer Integrated Manufacturing

Computer integrated manufacturing (CIM) is the ultimate in modern manufacturing management. It combines CAD and CAM to design and manufacture, but it also provides computer software to manage finances, employee records, inventory, and other business functions. Fig. 21–9.

▶▶▶ FOR DISCUSSION ◀◀◀

1. In your opinion, what are some of the negative impacts of automation?
2. Describe how a donut shop could be automated.

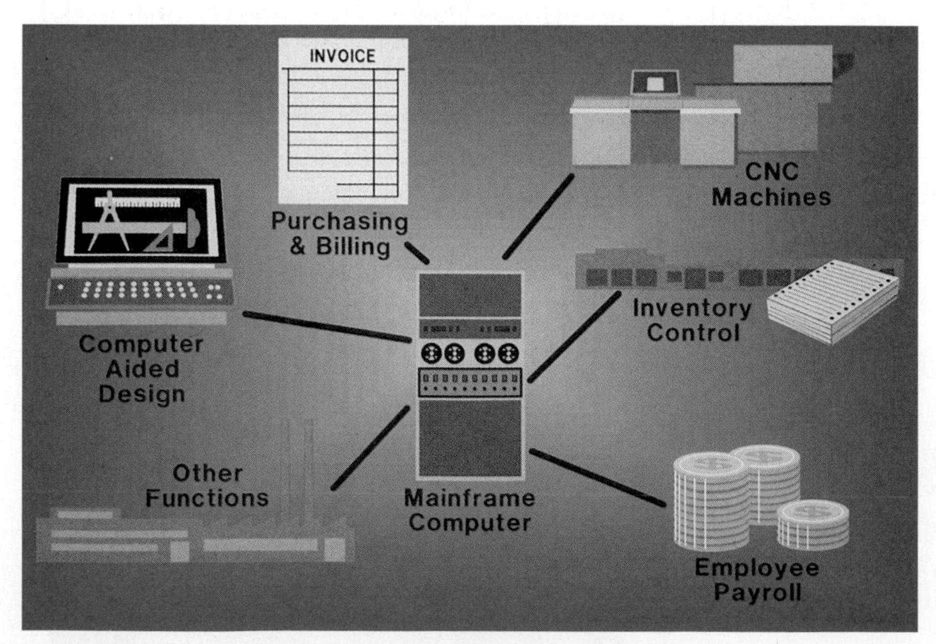

Fig. 21-9. Computer integrated manufacturing uses computer systems to control all aspects of manufacturing.

CHAPTER 21 Review

Chapter Highlights

- The role of management in manufacturing is to organize, plan, and direct the activities of the company.
- Different levels of management are used to transfer information from the owners of the company to the workers and back.
- Several different techniques can be used to manage the production of products.
- Manufacturing now relies heavily on automation to produce products.
- Computer systems now manage and control many manufacturing activities.

Test Your Knowledge

1. Who represents the stockholders in the management of a corporation?
2. What is the main responsibility of the president of a corporation?
3. What level of management brings the concerns of the workers to the plant manager?
4. Are the tour buses used by many rock groups an example of custom or mass production? Explain.
5. List two advantages and two disadvantages of mass production.
6. Why is automation a very efficient form of manufacturing?
7. What is CAD/CAM?
8. Explain how CNC works.
9. What is CIM?
10. List three activities in a manufacturing corporation that could be managed by CIM.

Correlations

SCIENCE

1. The computers used in CAM and CIM need to be networked in order to communicate with each other. Find out more about how computers "talk" to each other and report to the class.

MATH

1. Good money management is important in personal life as well as in business. Matt is saving for a custom skateboard. He needs $105 more. If he earns $15 per lawn for mowing grass, how many lawns will he need to mow in order to have enough money for the skateboard?

LANGUAGE ARTS

1. Create a diagram to illustrate the levels of management and describe the responsibilities at each level.

SOCIAL STUDIES

1. Find out how a person can become a stockholder in a company. How does the person know if his or her investment was a good one? How are the stockholders paid their share of the profits?

CHAPTER 22

Impacts of Manufacturing Systems

Introduction

If you stayed up past midnight to watch television, what effect would this action have on you? You probably would not be able to concentrate the following day. You might even fall asleep in class. This behavior would probably cause your grades to suffer.

There is a connection between staying up late and receiving poor grades. Staying up late has a negative **impact**, or effect, on your grades.

The impacts of any event can be good or bad. In some cases, the impacts of an event are not even predictable.

In this chapter, you will discover the impacts of manufacturing systems on people and the environment. You will look at the many positive and negative events that result from manufacturing technologies.

After reading this chapter, you should be able to

Discuss the impacts of manufacturing on people as consumers and producers of products.

Discuss the impacts of manufacturing on the economy.

Explain the impacts of uncontrolled manufacturing on the environment.

Suggest some ways to control pollution.

Words you will need

impact
purchase power
acid rain
sanitary landfill
toxic
conserve
recycle

Impacts of Manufacturing on Society

Manufacturing affects society at all levels. Local communities, nations, and the world community are all affected by manufacturing processes.

The products of manufacturing help make day-to-day living easier and more comfortable. This is true no matter where you live. Today, people can select from a large variety of products produced by many different manufacturers. Fig. 22–1. Mass production has made many of these products affordable to most people.

The products and processes of manufacturing influence many aspects of society. They play a major role in the economics of our communities and affect the environment in which we live.

Fig. 22-2. Manufactured products affect our daily routines. Our choices of clothing, sports equipment, and other products are often influenced by manufacturers.

Manufacturing Influences Our Daily Lives

Can products actually influence your daily routines? Products of manufacturing have become as common as technology itself. For example, people today are very concerned about good health. As a result of this concern, people are eating better and exercising more often. Manufacturers have created new product lines to satisfy this consumer demand. Food and drink manufacturers have filled the market shelves with caffeine-free soft drinks, bottled water, low-fat foods, and hundreds of other health-related items.

Fig. 22-1. Never before have consumers had such a large selection of manufactured products from which to choose. Manufacturers develop products to make our everyday lives easier.

Manufacturers have also had a major impact on how people exercise. Sports clothing, footwear, and sporting equipment are popular items at many department stores. Looking good while exercising has become almost as important as the activity itself for many people. Fig. 22–2.

Occasionally, people hurt themselves while exercising. Physicians rely on new products to treat sprains and muscle damage due to exercise. Braces, supports, and soothing medications are manufactured just for this purpose. Fig. 22–3.

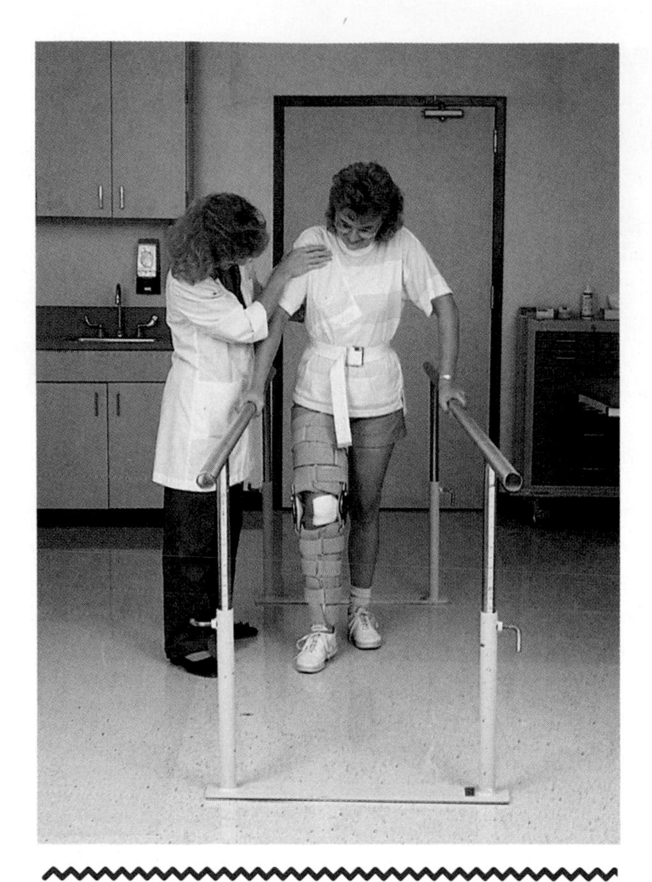

Fig. 22-3. Custom-manufactured products can be created to satisfy most needs.

Manufacturing Influences the Quality of Life

People today enjoy a better quality of life than ever before. This is partially a result of manufacturing systems. How can we measure the quality of life we enjoy? To begin, people live longer today than ever before. Health-care products and medical tools help physicians diagnose illness and treat it earlier than they could before, keeping people healthier longer.

People today also have more time to enjoy the many things that interest them. New techniques, equipment, and materials have made production faster. People can spend less time at work and still be extremely productive. During the 1700s, the average work week was 72 hours long. This dropped to 65 during the early 1900s. Today the standard work week is only 40 hours long.

We live longer and have more time to ourselves. Another measure of the quality of life is what we do with that time. People today can use that time to satisfy their need for shelter, communication, transportation, recreation, and protection by selecting from a vast number of manufactured products.

▶▶▶ FOR DISCUSSION ◀◀◀

1. Make a list of manufactured products that influence your leisure time.
2. How has manufacturing changed the way leisure time is spent compared to 20 years ago?

The Economic Impacts of Manufacturing

Manufacturers often employ a large number of people in a community. Jobs for community residents strengthen the local economy in many ways. The wages earned by employees are put back into the community when the employees make purchases at local stores. Fig. 22–4. Manufacturers also contribute to the strength of the community by paying taxes and fees to the local government. This reduces the burden placed on families to pay for schools, highways, and community development.

In Chapter 3, you learned that the strength of our nation's economy is dependent on our ability to produce (Gross National Product). When people lose their jobs, a terrible cycle of events begins. This cycle influences our nation's economy.

Fig. 22-4. How might this mall be affected if a large manufacturing company started a new factory in the same community?

When people lose their jobs, they also lose their **purchase power** (ability to buy goods). When they buy less, factories produce less. When a factory is producing fewer products, it needs fewer workers, so more people lose their jobs. The cycle repeats itself, and the GNP falls. This is one sign of a weakened national economy. Fig. 22–5.

Economy and the World Community

Many countries throughout the world have large manufacturing systems. Japan, Korea, and Brazil are only a few of the countries that produce high-quality products at reasonable costs. American manufacturers compete with these foreign manufacturers for world-wide customers.

Does price determine what you buy? Many times foreign competitors can produce products that cost less than American-made products. There are a variety of reasons for this: lower labor costs, modern factories, and government assistance are just a few.

Foreign competition is forcing American industry to rebuild. Trends toward more automation, better engineering, and a better-educated work force are changing how Americans produce products.

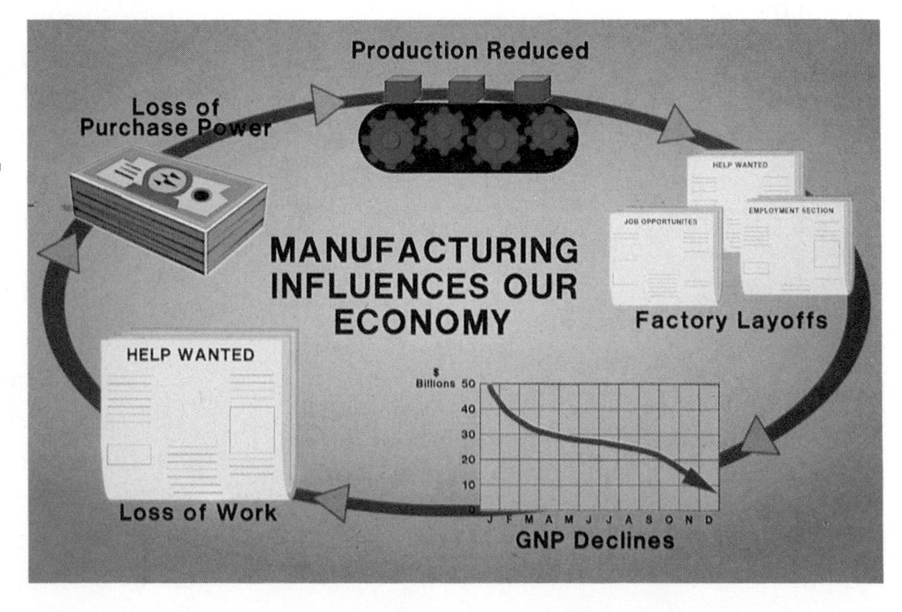

Fig. 22-5. The amount of money people make influences the amount of money in circulation. This, in turn, affects the number of products purchased, which influences our nation's economy.

Fig. 22-6. This keyboard, made by an American company, has Chinese characters as well as a Western alphabet. It is one example of the international nature of business today.

Manufacturing has encouraged many countries to become business partners. It's very common for large companies to own factories in many locations around the world. Fig. 22–6. The success of a company can have international impacts. If Mazda decided to stop production, what impact would that have in the United States?

■ **Plan an assembly line for producing a specific product in class. (See Extension Activity, Chapter 21.) Figure the cost of manufacturing your products and decide how much you would have to charge for each product. Prepare a marketing plan for selling the products.**

▶▶▶ FOR DISCUSSION ◀◀◀

1. What might be an economic impact of automating a local factory?

2. Look at the manufacturer's labels on products in your home. Make a list of the different countries they come from.

Environmental Impacts of Manufacturing

Manufacturers produce more than just products. Pollution and waste materials are often additional outputs of manufacturing systems. When left uncontrolled, these outputs dirty our air, foul our waterways, and contaminate our soil.

What causes pollution? Heating and burning materials to produce products or energy is a major cause of air pollution. Improper disposal of **toxic** (poisonous) waste products may allow them to drain through the soil into water supplies. People even pump chemicals and human waste into rivers and lakes. Many of these causes of pollution result directly from manufacturing systems.

TECHNOLOGY TRIVIA

As much as 25 percent of all toxic waste originates in individual households.

Acid Rain

One common form of pollution is acid rain. **Acid rain** is created when sulfur dioxide and nitrogen oxide mix with rainwater. These two chemicals are produced in large amounts by automobile exhaust and manufacturing processes in which materials are burned.

Fig. 22-7. Undesired outputs from manufacturing include sulfur dioxide and nitrogen oxide, which combine with rain to form acid rain.

When the three ingredients combine, they form sulfuric and nitric acid. Fig. 22–7. Acid rain changes the chemical makeup of lakes and rivers. This results in the death of fish and plant life. Acid rain also causes buildings, cars, and monuments to decay.

Water Pollution

Will the water coming from your kitchen sink always be drinkable? Water pollution occurs when groundwater supplies are contaminated with waste materials.

One of the most common causes of groundwater pollution is agricultural chemicals. Fig. 22–8.

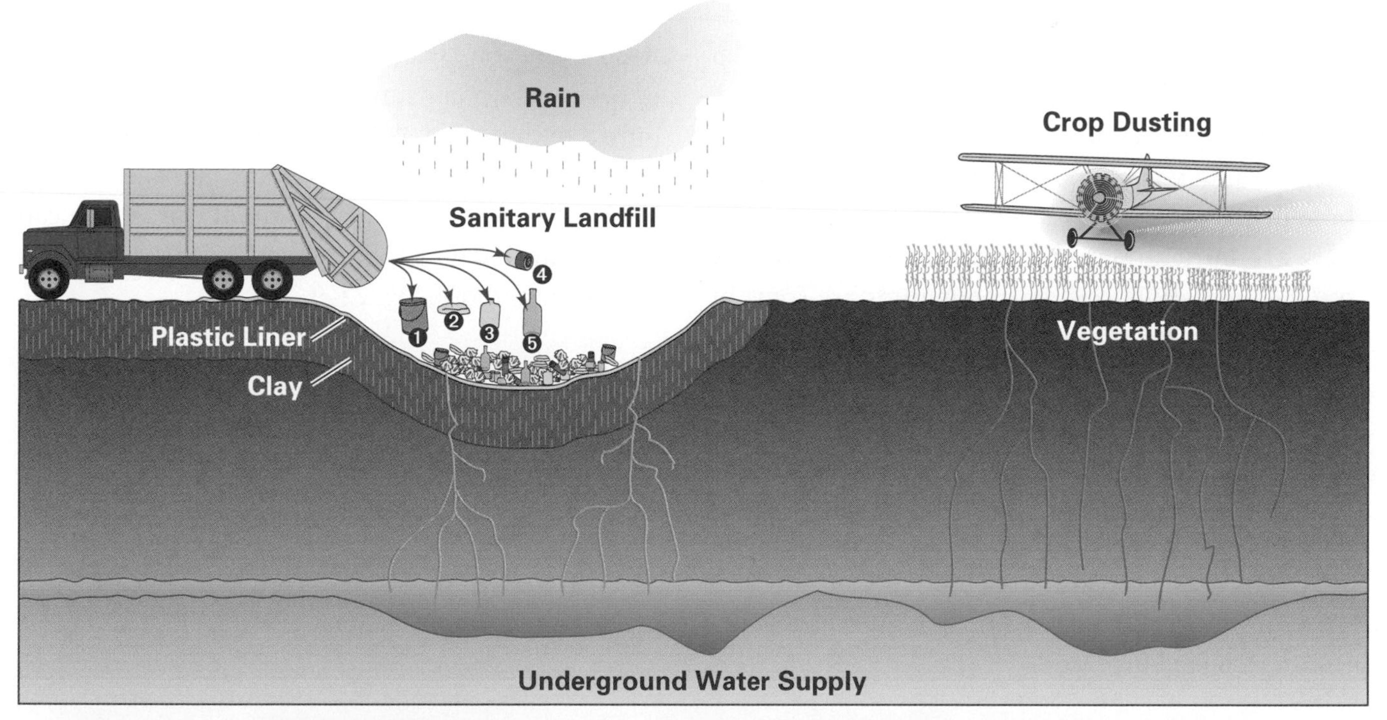

Fig. 22-8. Many communities obtain their drinking water from underground springs and wells. Leaking landfills and the heavy use of chemicals pollute these natural resources.

Solution

Problem	Clean Pollutants from Smokestack	Cleaner Burning Engines	Recycle	Conserve	Solar Energy	Proper Disposal	Sewage Treatment Plant	New Technologies
Acid Rain	✔	✔			✔			✔
Contaminated Ground Water						✔	✔	✔
Polluted Lakes & Rivers	✔	✔	✔	✔		✔	✔	✔
Leaking Landfills			✔	✔		✔		✔
Contaminated Soil			✔			✔		✔
Too Much Garbage			✔	✔				✔

Fig. 22-9. Many of the negative environmental impacts of manufacturing can be controlled.

Chemical fertilizers and pesticides used in agriculture find their way into groundwater by seeping deep into the earth.

Another major cause of groundwater pollution is sanitary landfills. A **sanitary landfill** is a large hole in the ground where garbage and waste products are buried. Materials such as food products, package materials, and lawn clippings are some of the materials that take up a great deal of room in the landfills.

By law, sanitary landfills must now have a base layer of clay, and heavy plastic linings are sandwiched between layers to stop leaks. However, toxic materials such as batteries, paints, motor oil, bug spray, and even human waste still find their way into underground water supplies through cracks and defects in the landfill.

Solid Waste Pollution

Manufacturers produce large amounts of products, and people consume them at an amazing rate. What do we do with a billion used tires each year? How do we deal with polystyrene (plastic) containers that will not break down or decay? Solid waste pollution is caused by the incredible amount of products we consume and throw away. We have only a few choices. We can bury them in landfills, dotting the land with mountains of garbage. We can burn the waste and further pollute the air, or we can change our habits and **conserve** (waste less) and **recycle** (reuse) more. Fig. 22–9.

■ **Design and construct a waste management plant for your model community.**

▶▶▶ FOR DISCUSSION ◀◀◀

1. List three ways the average person can help solve some of our pollution problems.
2. How has your local community tried to solve some of your pollution problems?

CHAPTER 22

Chapter Highlights

- The products of manufacturing influence our daily routines.
- The products of manufacturing influence the quality of our lives.
- Local, national, and international economics are influenced by manufacturing systems.
- Waste and pollution are also outputs of manufacturing systems.
- Uncontrolled pollution is damaging our air, land, and water resources.
- Conservation, recycling, and new technologies can help solve our pollution problems.

Test Your Knowledge

1. Define the word *impact*.
2. Give three examples of positive impacts of manufacturing on society.
3. Give three examples of negative impacts of manufacturing on society.
4. Describe the effects the telephone has had on society and tell whether the effects were predictable or unpredictable.
5. Give two examples of communication products that have a positive impact on the quality of our lives.
6. Describe one way American manufacturers could become more competitive in the world market.
7. List two economic influences a factory can have on a community.
8. How is acid rain produced?
9. Describe how groundwater can be contaminated from a landfill.
10. Describe two methods of reducing solid waste.

Correlations

SCIENCE

1. Find out more about acid rain. How does it harm plants and animals? Prepare a diagram that shows how the burning of fossil fuels makes rainwater more acidic.

MATH

1. The average work week in the early 1900s was 65 hours long. If a person worked five days per week, how many hours per day did she or he work?

LANGUAGE ARTS

1. Think of a way to reduce waste in your school. Put your idea in a friendly letter to your principal or student council.

SOCIAL STUDIES

1. Select a foreign country and find out what impact its manufacturing systems have on the United States. What impact do manufacturing systems in the United States have on the foreign country?

CHAPTER 23

Manufacturing in the Future

Introduction

Have you ever heard the expression "time flies"? Does it seem like just yesterday you entered first grade? Today, college and the world of work is just a few short years away. Time passes by quickly.

Time flies in the world of technology also. Changes are rapid, and new inventions become part of our everyday lives in a very short time.

In 1936, the world was introduced to public television for the first time. Today, we can flip through many different channels fed by cable and satellite systems around the world.

In 1903, the Wright brothers made their first flight in a powered aircraft. Only 66 years later, Neil Armstrong set foot on the moon.

Futurists (people who study the future) predict that the "time flies" trend will continue. New inventions will become part of our daily lives even faster than before.

What kind of changes will the future bring to manufacturing technologies? This chapter will attempt to answer this important question.

After reading this chapter, you should be able to

Discuss how manufacturing in the future might influence the way people live and work.

Gve examples of changes that may take place in how products are produced.

Discuss the industrialization of space.

Words you will need

automation
polymer chemists
machine vision
solar voltaic energy
microgravity

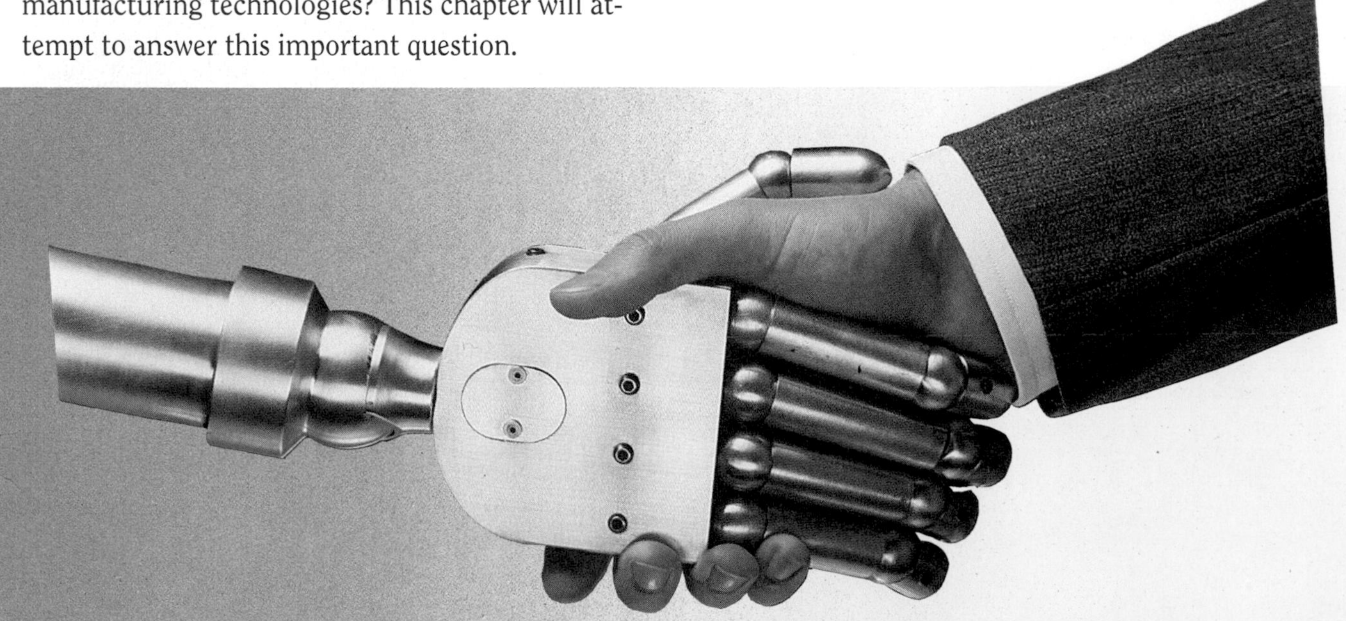

Fig. 23-1. In the past, manufacturing was labor intensive. The factories of the future will rely on robots and machines to do much of the work people used to do. (Henry Ford Museum & Greenfield Village—A)

The Workplace of the Future

In Chapter 17, you learned that futurists rely on trends to predict the future. Trends represent choices that people make. If today's trends in manufacturing continue into the future, production of products will change dramatically. For example, factories are relying more and more on **automation** (computer control) to produce products. With continued advances in computer technology and robotics, fewer people will be needed in the factory by the year 2000. Fig. 23–1. Some factories may be totally automated; they may employ only a handful of people.

Finding manufacturing jobs in the future may be a problem for some people. Complicated automated machines will require workers with new skills. People will have to be well-educated and have a good understanding of computers and automation. As manufacturing becomes less labor intensive (requires less human muscle), unskilled workers will find it more difficult to find work. Fig. 23–2.

Fig. 23-2. In the future, machines will do most factory labor, but skilled people will be needed to maintain and repair the machines.

Computer control will make the factories of the future more efficient; each task will take less worktime and create less waste. People will be able to produce more products in less time. If this trend continues, people might work a shorter work week and be able to devote more time to family and leisure activities. Fig. 23-3.

Working from home will become common in the future. Advances in telecommunication will allow more people to conduct business from home workstations. Product design, production control, and company management may be achieved without setting foot inside the factory.

IMPACT

Today, many students come home to empty houses because their parents are at work. How might society change if parents worked at home?

Increased Global Competition

Which country will lead the world community in manufacturing in the year 2000? As countries become more interdependent (dependent on each other), manufacturing may become a multi-national activity. Today, for example, many American companies rely on foreign countries to manufacture their products. Labor costs are cheaper in countries where the standard of living is lower than ours. Cheaper labor costs result in lower product prices. If this trend continues, American products may be more competitive in the world market, but the cost to American workers will be great.

▶▶▶ FOR DISCUSSION ◀◀◀

1. In the year 2000, what skills will a factory worker probably need to have? Be specific.
2. List two reasons why the automated factories of the future will be more efficient.

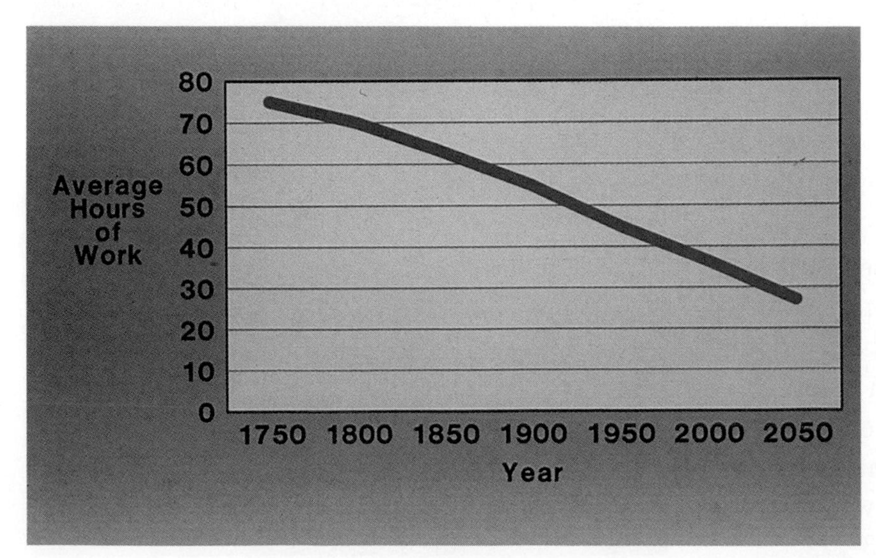

Fig. 23-3. As manufacturing becomes more efficient, the amount of work and the length of the work week will be shorter. Today, people work about 40 hours a week. By the year 2050, this may drop to 27 hours a week.

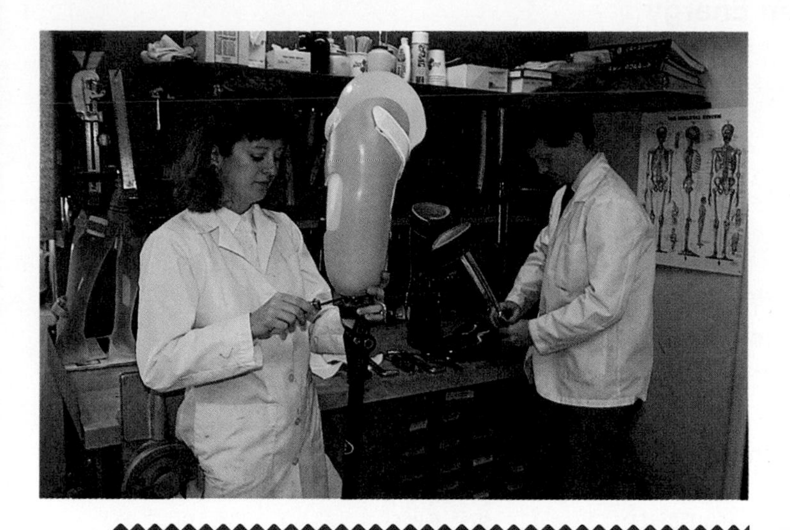

Fig. 23-4. As traditional manufacturing materials become depleted, new materials will be developed to replace them. Use of plastics will continue to grow in the future as new polymers are invented to meet our manufacturing needs.

Manufacturing Materials in the Future

Manufacturing consumes a great amount of material. Some materials are used as fuel to provide the energy needed to run machinery. Other materials are used to create the actual products. Manufacturing will continue to use up our natural resources at a fast rate. In the future, manufacturing will require new materials to replace old ones.

Plastic materials will become even more common in the manufacture of products. **Polymer chemists** (those who work with plastics) will continue to invent new plastics with special properties to help replace nonrenewable materials. Fig. 23–4. Fibers made of plastic (Kevlar™) are already in use. These fibers are five times stronger than steel fibers.

Researchers are now conducting experiments on using plastic materials to conduct electricity. In the future, these materials may replace metals such as copper.

Automobile manufacturers have already produced engines made of plastic and ceramic materials. A plastic car engine is lighter, more fuel-efficient, and longer-lasting than a metal engine.

Recycling and Conservation

Recycling and conservation of materials will continue to grow in importance in the future. Discarded materials such as plastic, metal, glass, and papers are already being collected from communities and sent to factories. New technologies will allow manufacturers to make new products from old ones. Fig. 23–5. Plastic lumber made from recycled plastic containers is an example of a remanufactured product available today.

Recycling materials also conserves energy. Changing scrap aluminum cans into new aluminum products uses less energy than changing raw materials into aluminum. As new uses for recycled materials are developed, the cost of manufacturing products from them will continue to drop.

Fig. 23-5. Products will continue to be made from recycled materials in the future. This photo shows aluminum scrap, ready for recycling.

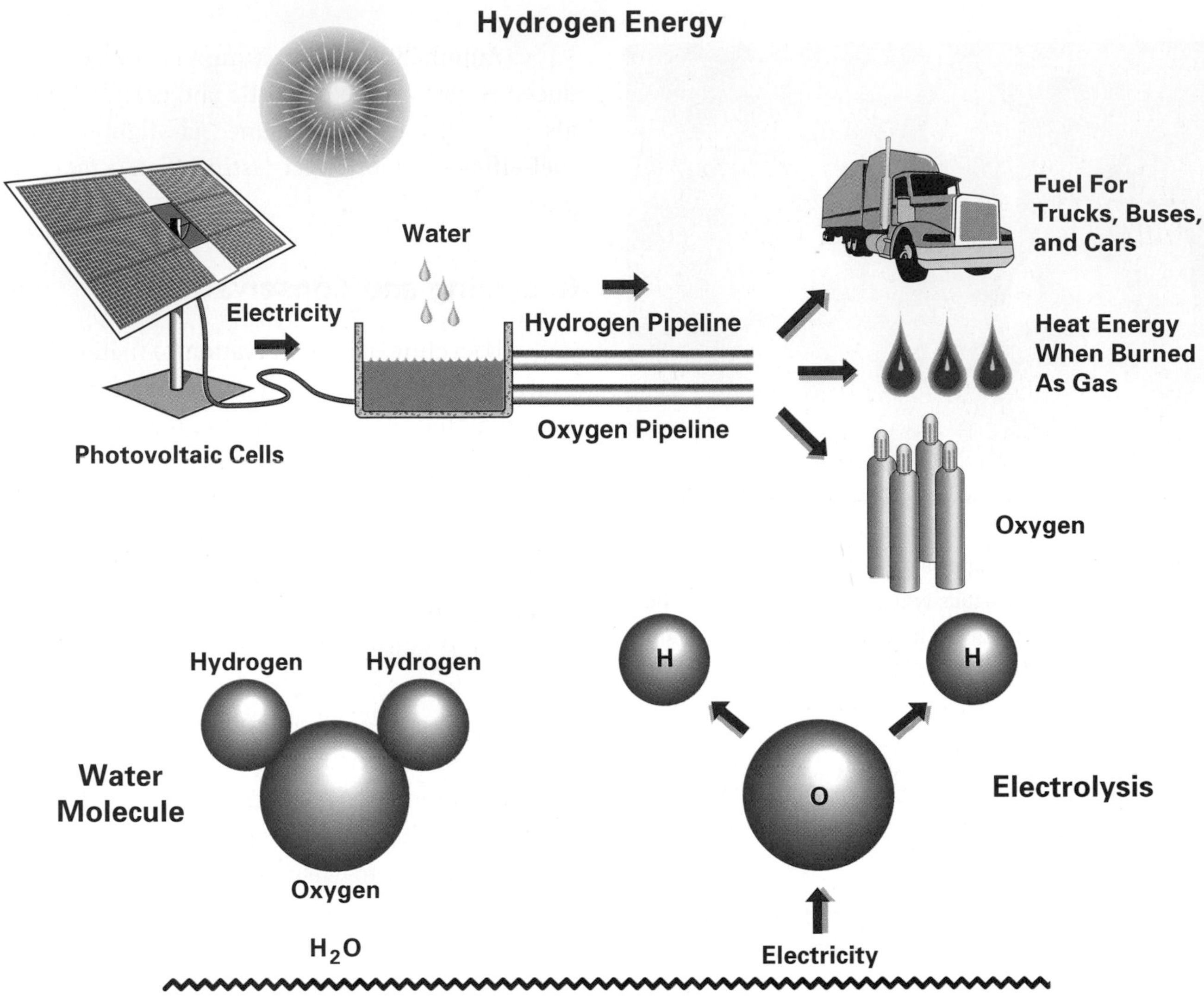

Fig. 23-6. Hydrogen, one of the elements present in water, is an almost unlimited source of energy on the earth. When hydrogen is removed from water, it can be burned in place of fossil fuels. To obtain hydrogen, an electric current is passed through the water molecule in a process called *electrolysis.*

Future Energy Sources for Manufacturing

Manufacturing in the future will still consume large amounts of energy. Today, most manufacturers rely on fossil fuels to generate the electrical energy needed to process materials and to fuel vehicles. Some scientists predict that our supply of fossil fuels may only last another 100 years. Alternative energy sources and conservation will be necessary for future manufacturing. Fig. 23–6.

TECHNOLOGY TRIVIA

The city of San Francisco, California, already gets half its energy needs from natural steam created by geothermal energy.

■ Design a new use for a product (such as a plastic container) that is presently considered to be a "throwaway."

▶▶▶ FOR DISCUSSION ◀◀◀

1. Do you think clothing manufacturers in the future will rely more on synthetics (polyester) or natural fibers (cotton) in manufacturing? Explain.
2. Why should we care about the amount of fossil fuel we use today? After all, we still have a hundred-year supply. Explain.

Fig. 23-7. In the future, light energy and even chemical energy will be used to separate materials. Here, a beam of laser light is being used to cut materials.

The Manufacturing Process in the Future

The factories of the future will produce more products in less time. How will this be achieved?

Automation, combined with new machines, will make production soar. Machines that speed up separating, forming, assembly, and conditioning processes will replace slower, less efficient machines. The cutters on new machines will run at lightning speeds. Fig. 23–7.

New super-strong adhesives will replace metal fasteners. Your entire automobile may someday be glued together.

"**Machine vision**" will scan parts as they are made. Like human eyes, sensors will scan the material being processed, searching for mistakes. When a mistake occurs, the machine will automatically identify it. Fig. 23–8.

Automatically guided vehicles will move materials and products throughout the plant. Computer sensors might even allow robots to pick one particular type of part from a container of many different parts.

▶▶▶ FOR DISCUSSION ◀◀◀

1. If you worked in an automated factory, what might your job be?
2. How will new synthetic materials affect the development of new machines?

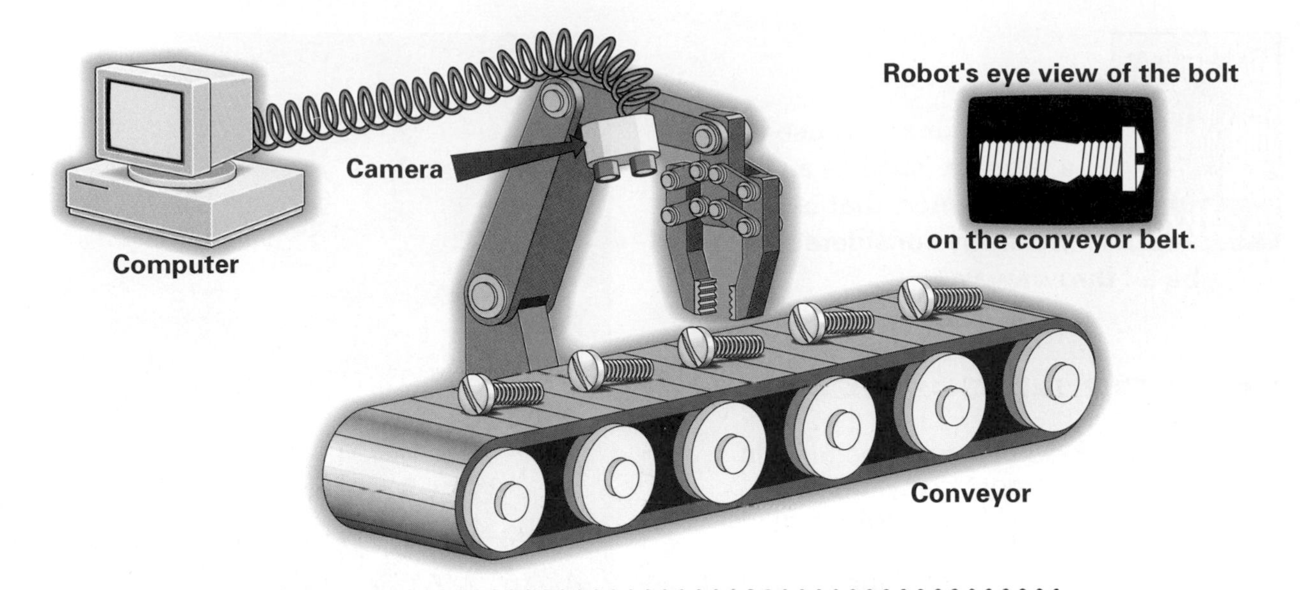

Fig. 23-8. Robots of the future will be fitted with cameras. Computers will analyze the pictures from the robotic vision. If a part has an imperfection, the computer will instruct the robotic grippers to remove the part.

Manufacturing in Space

Someday you may be using products manufactured in space. You might even work in space. Experiments conducted by the National Aeronautics and Space Administration (NASA) have already proved that products can be manufactured successfully in the far reaches of space. Fig. 23–9.

What advantages can manufacturing in space offer? Planets may contain huge supplies of natural resources that can be changed into products. Scientists have already created "space concrete" from moon soil.

Deep below a planet's surface we may find rich supplies of new fuels. These fuels could supply the energy needed for the manufacture and transportation of space products. Fig. 23–10. Solar panels anchored to orbiting platforms could supply enough **solar voltaic energy** (electrical energy) to power an entire space factory.

TECHNOLOGY TRIVIA

The first solar heating panel was designed in the 1700s by Swiss scientist Horace B. de Saussure.

Fig. 23-9. The crystals above were made on the earth. The ones to the right are the same material, but the crystals were made in space.

Fig. 23-10. Planets such as Mars may provide new sources of energy and manufacturing materials.

Processing in Space

Experiments conducted on the space shuttles have shown that some materials actually process better in a space environment. Space has almost no gravity (**microgravity**). Things float in space without the force of gravity to hold them down.

Manufacturers can take advantage of this environment. Materials that do not normally mix on the earth might combine in a microgravity environment. For example, if you tried to create an alloy of lead and aluminum on earth, the lead would sink to the bottom of the melting pot because of its weight. In microgravity, both lead and aluminum would weigh about the same, so they could combine easily. Scientists hypothesize that many new metal alloys could be created using this process.

Imagine melting glass without a container. This is also possible in a microgravity environment. The raw materials used to make glass can be suspended in microgravity using sound waves. Fig. 23-11. The sound waves form a container. While they are suspended, the raw materials can be melted into glass. Glass and other materials melted in this way are extremely pure. On the earth, materials melted in containers take on some of the impurities of the container. Lenses, fiber optics, electronic chips and other products that require purity can be made in space.

■ **Design and construct a model of a space-based MegaIndustries factory.**

▶▶▶ FOR DISCUSSION ◀◀◀

1. What things will probably contribute to the cost of manufactured space products?
2. How will working in microgravity be different from working on earth?

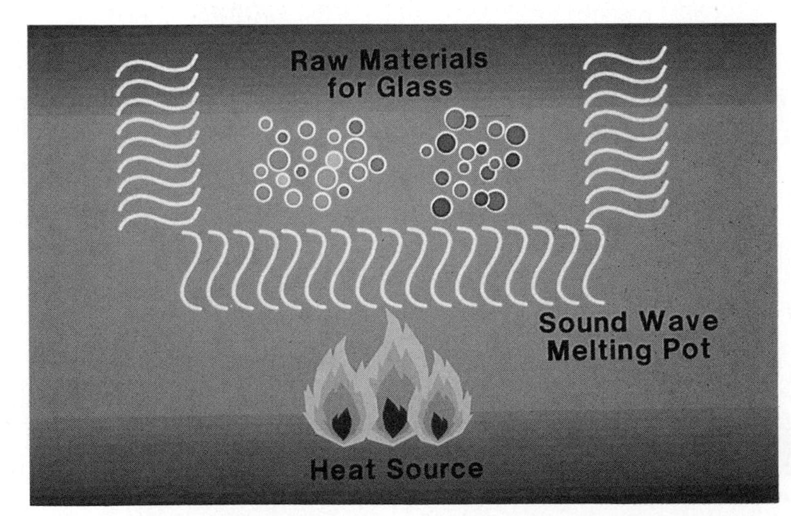

Fig. 23-11. Glass made on the earth has impurities and imperfections. The molten glass picks up impurities from the crucible (pot in which it is melted). In microgravity, sound waves can be used to suspend the material while it is being melted. The result is extremely pure glass.

CHAPTER 23 Review

Chapter Highlights

- The factories of the future will use automation and computer control.
- People will need advanced skills to work in the factories of the future.
- New synthetic materials will replace depleted resources.
- New techniques will separate, form, combine, and condition materials at high speeds.
- Space may provide new resources to replace depleted ones on the earth.
- The microgravity of space will allow materials to be processed in new ways.

Test Your Knowledge

1. How will automation increase production in the future?
2. Why will the number of people working in factories decline in the future?
3. Why will employees have to have advanced skills to work in the factories of the future?
4. What manufacturing material will increase in importance in the future?
5. Describe machine vision.
6. What are some of the advantages of manufacturing from recycled materials?
7. What advantages can space manufacturing offer?
8. Describe microgravity.
9. Explain how microgravity can help in material processing.
10. Give an example of a product that might be produced in space.

Correlations

SCIENCE

1. Consult a science book or encyclopedia to find out more about plastics. What elements are combined to make plastics? How are the molecules formed? What are some typical properties of plastics?

MATH

1. Contact a local recycling company and find out how much the company pays for scrap aluminum. How much money would you make if you turned in 63 pounds of aluminum cans?

LANGUAGE ARTS

1. Advances in manufacturing will greatly affect the skills needed by factory workers. Describe the kinds of skills that you think will be necessary for factory workers in the next century.

SOCIAL STUDIES

1. To learn how technology can change a household, ask your parents to list the items in your home that did not exist when they were your age. Now ask them to list items that *were* in their home when they were your age. Show that second list to your grandparents (or someone else in that age group). Which items on the second list were *not* in your grandparents' homes when they were children?

Make a Difference

StarServe

All across the country, young people are demonstrating that they can be winners—that they can be stars—by sharing their talents and gifts with others in their communities.

In 1990, StarServe—a national nonprofit organization funded by the Kraft General Foods Foundation—launched a nationwide educational project to help make community service a part of classroom and schoolwide activities.

From Texas to Maryland and from Florida to California, students began using StarServe resources to look at national and local issues and to match their interests and skills with community needs.

StarServe provides free materials, resources, and ongoing assistance to plan and carry out service projects. Kits are distributed to every school in the nation, and a letter from celebrities encourages students to become involved and "Be a Star."

At Lauderdale Lakes Middle School, seventh-grade students involved in the Drop-Out Prevention Program found that they could make a difference through a range of service projects. Students visited the elderly in nursing homes, worked with children in day care, collected canned goods for the hungry, and started a newspaper-recycling program.

In another StarServe project, young people at Dooley Elementary held a school carnival to teach small children about drugs. Through a puppet show and skits—plus the assistance of local social service agencies—they were able to reach others and felt a part of the community.

For community service classes at Surratsville High School, students launched a variety of projects—from tutoring middle and elementary school students to working in a hospital supply room to helping at a nature center. Students completed 2,425 hours of service. In doing so, they also learned about careers they might want to follow.

Elsewhere, students have helped beautify a park, they've helped their teachers with special projects, and they've sorted canned food at a community agency. They've made greeting cards for convalescent home residents, and they've recorded stories for visually impaired children.

After completing a project, students have looked back at their experiences to see what they've gained—and what they've given. Through the projects, students have also developed a sense of civic responsibility.

StarServe operates in partnership with the Kraft General Foods Foundation, The Love Foundation for American Music, Entertainment and Art, and the United Way of America. StarServe is also one of the first independent initiatives of the Points of Light Foundation, a nonprofit, nonpartisan foundation set up to encourage all Americans to become involved in serving others.

SECTION SIX

PRODUCTION SYSTEMS: CONSTRUCTION

Activity Brief

Technology in Construction

PART 1: Here's the Situation ....

Construction is a production system that exists to satisfy people's needs and wants. The need for shelter is a basic need. People learned very early in time how to fashion natural materials such as wood or stone into temporary dwellings as they traveled around in search of food. When they learned to raise their own food, people formed settlements. They then needed permanent shelters and other constructed products such as wells. Trade developed between settlements, creating a need for roads and bridges. As construction tools and techniques gradually became more complex, people created new ways to shape the natural world to better suit their needs.

Construction technology continues to increase in scope, variety, and complexity today. As you do this activity and as you read Chapters 24 through 27, you'll learn more about construction and the impacts it has on our lives.

PART 2: Your Challenge...........

A major challenge to early people in their travels was finding ways to cross obstacles such as rivers. Perhaps the first bridge was a fallen tree that extended from one shore to the other. Once people grasped the idea, they found ways to build better bridges.

In today's world, there is a great variety of bridges. Still, these have many features in common. For example, all bridges are either *deck* bridges or *through* bridges. In a deck bridge, the roadway runs on top of its structural members. In a through bridge, the roadway runs between them. Fig. VI-1.

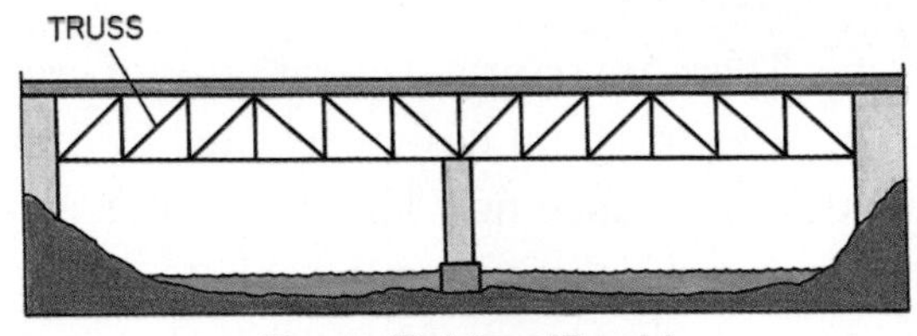

Truss Bridge (Deck)

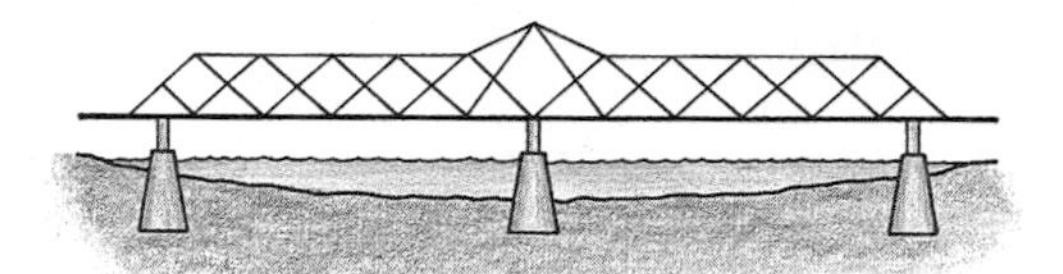

Truss Bridge (Through)

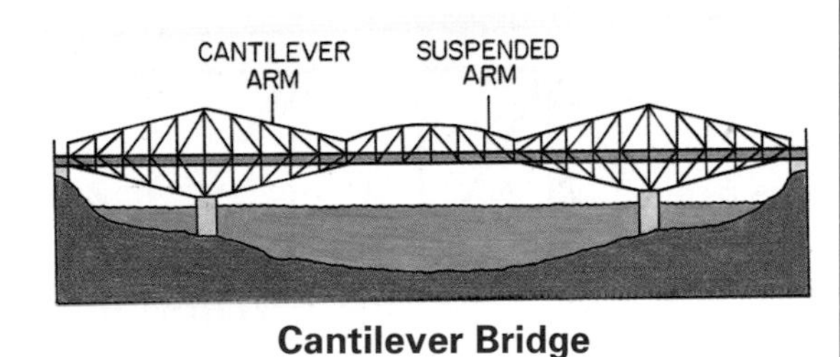

Cantilever Bridge

Fig. VI-1. Types of bridges.

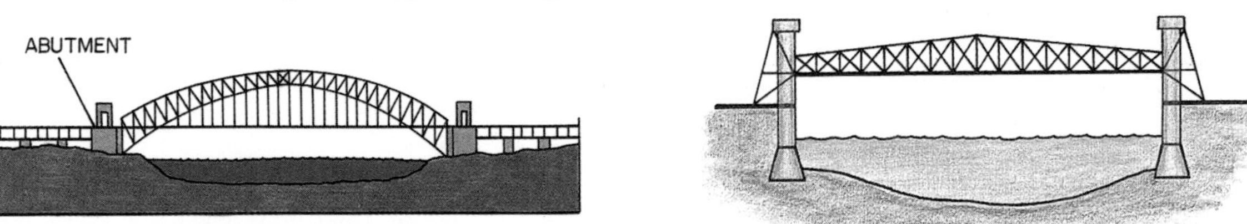

Arch Bridge (Through)

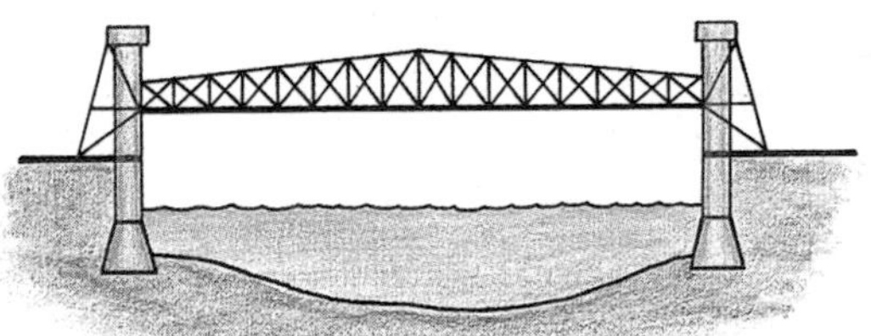

Movable Lift Bridge

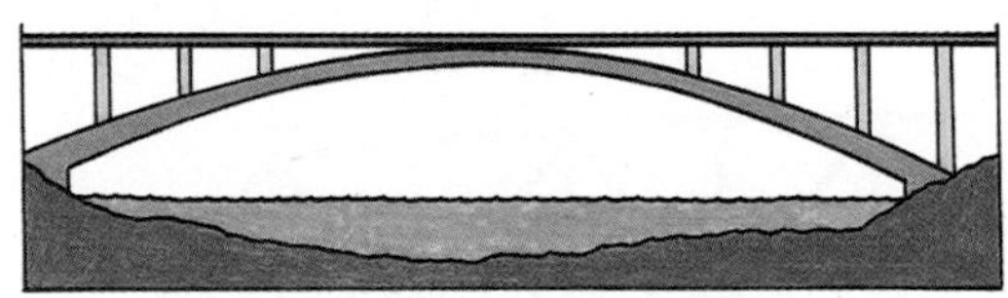

Arch Bridge (Deck)

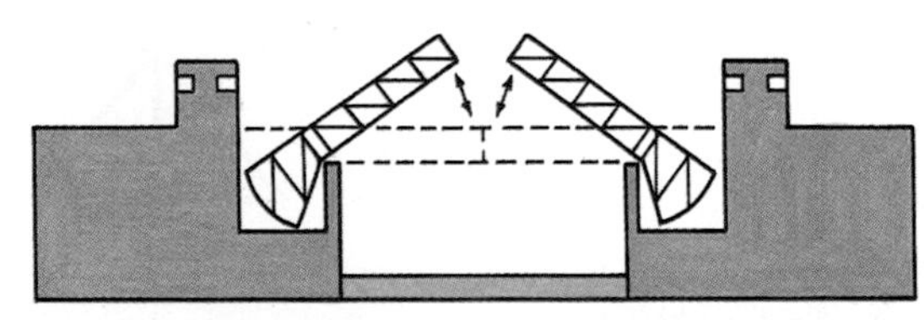

Movable Bascule Bridge

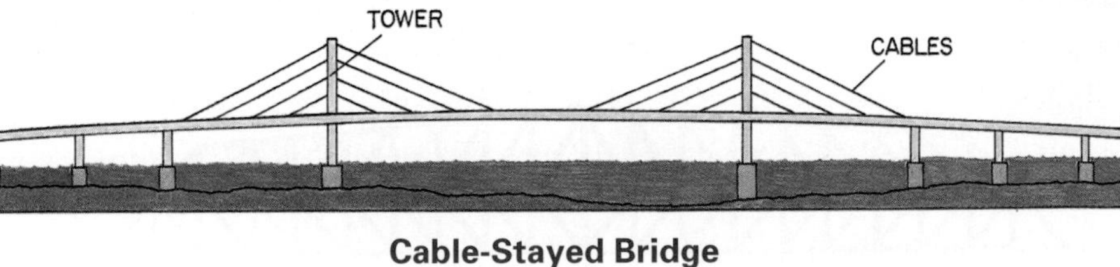

Cable-Stayed Bridge

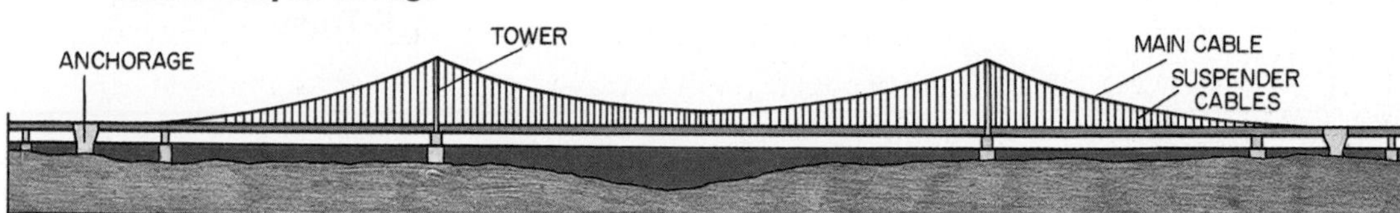

Suspension Bridge

There are several general types of bridges. Fig. VI-1. The structural design selected by builders depends mostly upon how long the bridge is to be and how much weight it must be able to support. Truss bridges or other types of bridges that include trusses in their structures are often used because the triangle is a very strong structural shape. Fig. VI-2.

In this activity, you will study the design and construction techniques used in various types of bridges. Then you will design and construct a model bridge.

Fig. VI-2. Types of trusses.

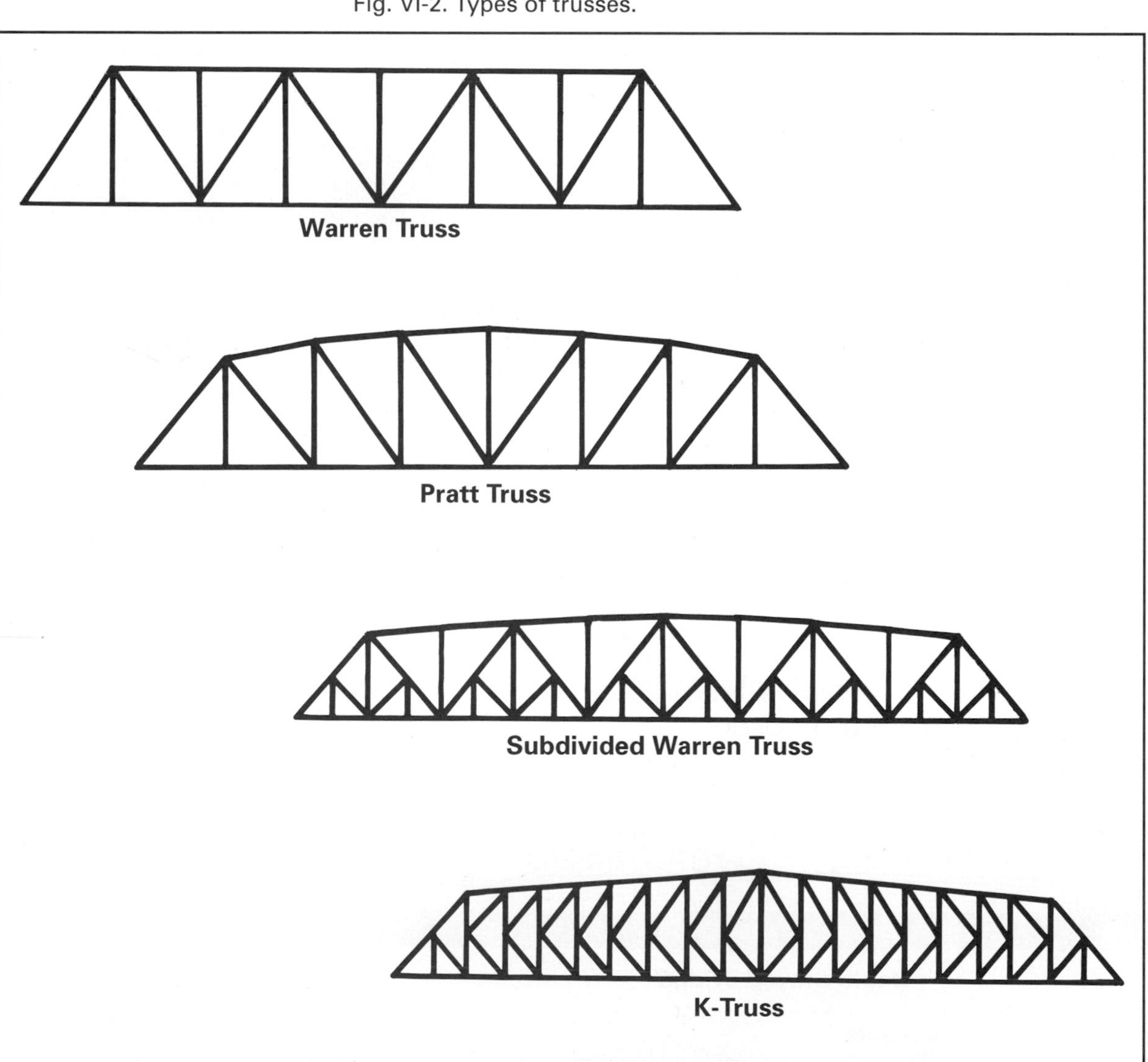

PART 3: Specifications and Limits

Your model bridge will need to meet certain standards. Read the following specifications and limits carefully before you begin.

1. Work in teams of 2 or 3 students.
2. A *span* is the distance between supports. If you construct a single-span bridge, make it 12 inches long. If you wish to construct a bridge with more than one span, your plans must be approved by your teacher and your bridge should not exceed 20 inches in length.
3. Design and construct your bridge to achieve the greatest strength possible using the least amount of materials.
4. Single-span bridges should have no vertical supports besides those used to support the ends.
5. You must hand in the following:
   - sketches
   - working drawings
   - bill of materials
   - written report

Safety Notes

- As you do this activity, remember to follow all the safety guidelines your teacher has explained to you.
- Use all tools properly. Use special care with tools that are sharp.
- If you should use any power tools, be sure you understand how to operate them and always get your teacher's permission.

PART 4: Materials, Tools

There are many materials you might use to construct your bridge. Following is a list of possibilities. Also listed are some tools you might use.

Materials

- model wood (balsa, basswood, or pine), 1/8" × 1/4"
- glue and/or other adhesives
- paper
- cardboard
- poster board
- string
- fasteners

Tools

- drafting instruments
- utility knives
- modeling knives
- glue gun
- hand tools
- scales

PART 5: Procedure..........

The bridge you choose and how you construct it will be up to you. Still, there are certain steps to follow that will make your work easier.

1. Working as a team, discuss the various types of bridges and select the type you intend to construct. Refer again to Fig. VI-1.
2. Brainstorm ideas for designs. Because trusses are commonly used in many bridges, you may also brainstorm ideas for truss designs. Look over the examples that are shown in Fig. VI-2.
3. Make sketches of your ideas. Remember, your bridge must be strong, but require a relatively small amount of materials in its construction.
4. Choose your best design and prepare working drawings. You must include a front view and a top view.
5. Prepare a bill of materials for your bridge.
6. Gather all needed materials and construct your model bridge.

PART 6: For Additional Help...........

For help with this activity, do more research on the type of bridge you are building. Also look up the following terms. You'll find some of them in this book. (Check the index.) You'll find others in dictionaries, encyclopedias, and other resource materials.

abutments
beams
bill of materials
brainstorming
bridge span
cables
piers
substructure
superstructure
trusses
working drawings

PART 7: How Well Did You Meet the Challenge?..........

When you've finished building your bridge, evaluate what you have done.

1. Prepare and give a presentation for the class. Identify the type of bridge you built and point out its design features. Describe how you designed and built your bridge. Relate any problems you encountered and tell how you solved them. If you built a movable bridge, demonstrate how it works.
2. Weigh your bridge and then test its strength. Place your bridge on two end supports. Set weights on the roadway until the bridge begins to fail. How much weight did it hold before it began to fail?
3. Discuss the test results in class. Consider the following:
   - Does the length of the span make a difference?
   - Did it matter where you placed the weights on your bridge? (Weight distribution)
   - Which design achieved the greatest strength possible using the smallest amount of materials? What features does it have that give it strength?
4. Write a report comparing your bridge and the results you achieved with those of your classmates. How could your design be improved? If you had it all to do over again, what would you do differently?

PART 8: Extending Your Experience..........

This activity helps you learn about bridge construction and structural design and analysis. Think about the following questions and discuss them in class. You'll find more about construction in Chapters 24 through 27 in this section, "Production Systems: Construction."

1. Many bridges today are made of concrete. How could you build a model bridge using concrete?
2. Some bridges are important because they are great achievements in design or construction. Others are important for historical or other reasons. Find out why each of the bridges listed below is important. Add two other notable bridges to this list. Share your findings with the class.
   - Golden Gate Bridge
   - Brooklyn Bridge
   - London Bridge
   - Bridge over the Royal Gorge
3. Bridges may go over highways, railways, and valleys as well as over bodies of water. Look for bridges in your area or as you travel. Identify which types they are and look for any interesting design and structural features they might have.

CHAPTER 24

Structures

Introduction

Imagine what the world might be like if there were no buildings. Where would you live? This was a big problem for the first humans. They attempted to solve their problem by using natural shelters, such as caves and cliff overhangs. For traveling, however, they had to invent portable shelters. People learned to make tents, teepees, and other types of temporary structures.

Eventually, people developed tools and discovered various materials that allowed structures to be designed for many different purposes. This trend has never let up. For shelter, for pleasure, for commerce—the list goes on and on—people build structures.

After reading this chapter, you should be able to

Define structure.

Give examples of public works, residential, commercial, and industrial structures.

Discuss the types of workers involved in designing a structure.

Describe the four forces that affect a structure.

Discuss the reasons why builders use various structural shapes.

Words you will need

structure
public works
live load
dead load
compression
tension
torque
shear
column
beam
truss

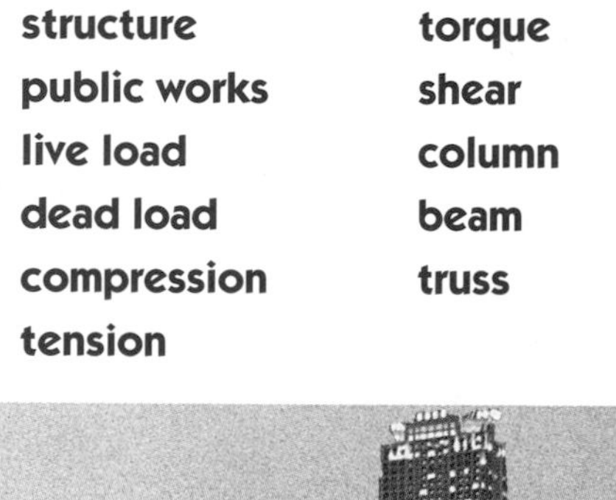

Structures

What is a structure? Is it a building? A carton? A skeleton? In fact, a structure could be any of these. Any arrangement of parts can be considered a **structure**. Although we tend to think of structures as rigid, they can also be flexible. Some structures have both rigid and flexible parts. Fig. 24–1.

Fig. 24-1. Some structures are both rigid and flexible. This structure has a rigid entryway, but the rest is flexible.

How are structures classified? People generally categorize structures by their use or function. Of course, some structures have several purposes, and others do not seem to fit neatly into any one category. These structures must be categorized by other methods.

Public Works

If a structure is built by a governmental body (local, state, or national), it is called a civil or **public work**. Public works are structures built for the "public good." In the United States, many such projects have been constructed. These include schools, roads, bridges, tunnels, parks, dams, airports, libraries, government buildings, public housing, sewer systems, bus and train stations, towers, and monuments.

TECHNOLOGY TRIVIA

Two irrigation dams built by Roman engineers 1800 years ago, in the 2nd century A.D., in Spain, are still in use today. The only major maintenance work they have needed has been the renewal of their stone facings, done in the 1930s.

Residential

Residential structures are those designed for people to live in. Single and multi-family homes, apartment buildings, hotels, trailers, and other units are included in this category. These places can be rented or privately owned. Other arrangements, such as condominiums and cooperatives, allow people to own a structure without owning the land on which it sits. The land may belong to someone else.

Commercial

Commercial structures are those whose primary purpose is commerce. Grouped here are wholesale and retail stores, shopping malls, warehouses, shipping containers and vessels, aircraft, restaurants, banks, and so on. Commercial structures vary in size and operation from small, family-owned candy stores to large corporations owned by stockholders.

Industrial

Industrial structures are primarily manufacturing facilities, factories of one type or another. Power plants, industrial "parks" (groups of such structures), and large warehouses are also included in this category. The materials produced in industrial structures are used in other factories or delivered to various commercial outlets for sale to the public.

▶▶▶ FOR DISCUSSION ◀◀◀

1. Why could a skeleton be considered a structure?
2. Describe ways to categorize structures, other than by use or function.

■ Design the MegaIndustries processing plant. Be sure to consider the product being made. The equipment that will be used in the plant and the procedure for making the product will also influence the design of the building.

Designing Structures

Who designs structures? Ultimately, there is input from many sources, but the primary designer often depends on the type of structure. For most structures, the primary designer is the architect. Architects are trained to design safe, efficient structures. They consider the appearance, cost, material, location, and function of each structure they design.

Structural engineers are responsible for the strength and stability of the design. Civil engineers are in charge of public works projects such as water supplies and transportation systems. Electrical engineers, interior designers, surveyors, and others all work together to make the best structure possible.

How does anyone know that the structure in question is strong enough, that the building will not fall? Experience and research allow people to determine the minimum standards required to prevent structural failure.

For example, the load a building is expected to bear is usually analyzed and planned for. **Live load** is weight that may vary, such as traffic or snow. **Dead load**, or static load, is the weight of the structure itself. By calculating the dead load and carefully estimating the possible live load, designers can be reasonably sure that the structure is stable. Fig. 24–2. Specifically, these loads exert four basic types of forces on the structure.

Fig. 24-2. The estimated live load for a skyscraper is 80 pounds per square foot, and the dead load is about 100 pounds per square foot. For a typical 50-floor skyscraper, the combined load may equal 101,250 tons!

Compression

Compression is the force that tends to press a material together. Fig. 24–3. Gravity is usually the cause of this force. Since compression increases as height (or depth, in the case of undersea structures) increases, the parts at the base of the structure have to be made stronger than those near the top.

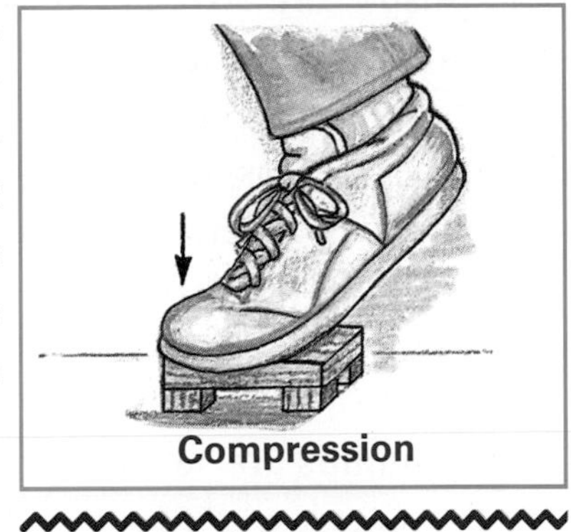

Fig. 24-3. Compression, forcing one portion of material against another, can crush an object.

Tension

Tension is the force that exerts a pull from the center toward the ends of an object. Fig. 24–4. When a rubber band is stretched, it is said to be under tension. Building materials such as rope and cable work very well under tension.

Some structural members may be influenced by two or more forces at once. For example, horizontal members may be subject to both compression and tension at the same time. Fig. 24–5.

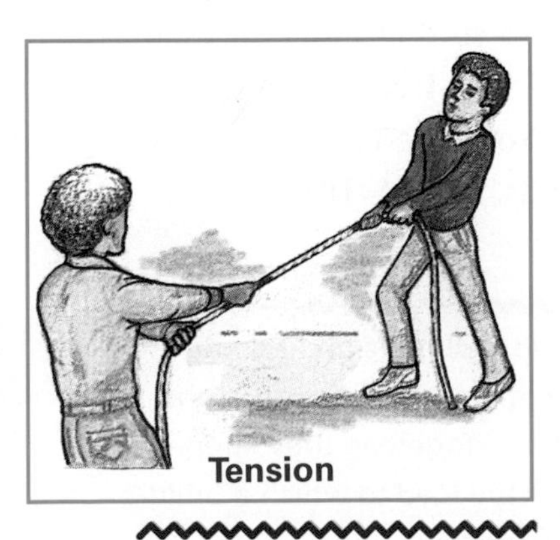

Fig. 24-4. Tension is the force that tends to pull things apart.

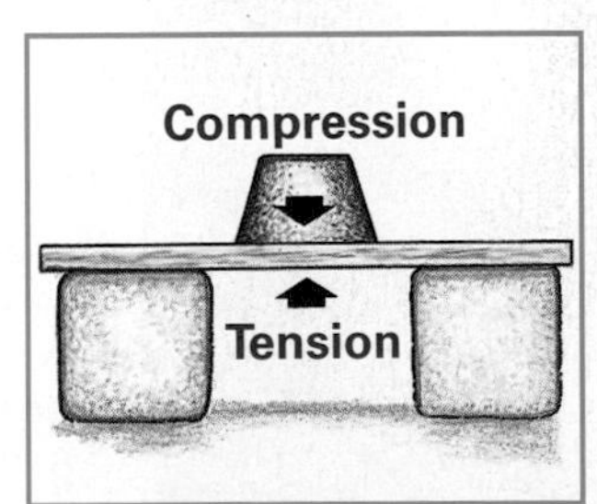

Fig. 24-5. A beam's top surface is in compression, but its lower surface is stretched under tension.

Torque

The twisting force that may remind you of a barber shop pole, or perhaps a twisted licorice stick, is called torsion or **torque**. Winds and earthquakes are examples of natural forces that may exert torque on structures. Fig. 24–6.

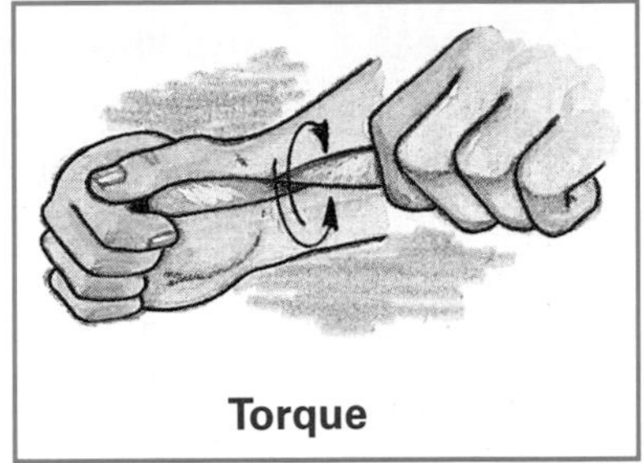

Fig. 24-6. Torque is the force that causes an object to twist along its axis.

Shear

When you cut a piece of paper with scissors, the blades cut the paper by forcing the material in opposite directions. This force is called **shear**. (This explains why scissors are sometimes called *shears*.) Fig. 24–7.

Fig. 24-7. Shearing forces push material in opposite directions.

▶▶▶ FOR DISCUSSION ◀◀◀

1. Who has primary responsibility for designing structures?
2. Why should structural designers be interested in things such as live and dead load?

■ Prepare a presentation that illustrates compression, tension, torque, and shear forces. Your presentation could be a poster, a cartoon, models, or a skit.

Structural Shapes

Over time, people have gathered much information about structural shapes. As it turns out, some shapes are stronger than others, and some shapes resist certain forces better than others. To design a structure properly, the designers must take these properties, as well as the characteristics of different building materials, into account.

Columns and Beams

Columns are vertical supports. To resist the force of compression, columns are made larger as their height increases. In other words, the taller a column is, the more weight it is designed to support.

Beams are horizontal supports. Fig. 24–8. As with columns, their shape and size may vary, and they can be made of many kinds of materials. Fig. 24–9. Experimentation has shown that the middle of a supporting member (called the *neutral axis*) will support the least stress. Therefore, most common structural shapes are designed so that much of their weight and strength are shifted away from the center, toward the ends. Fig. 24–10.

Fig. 24-8. Columns and beams are the most common structural members. Together, they support most of the load to which a building is exposed.

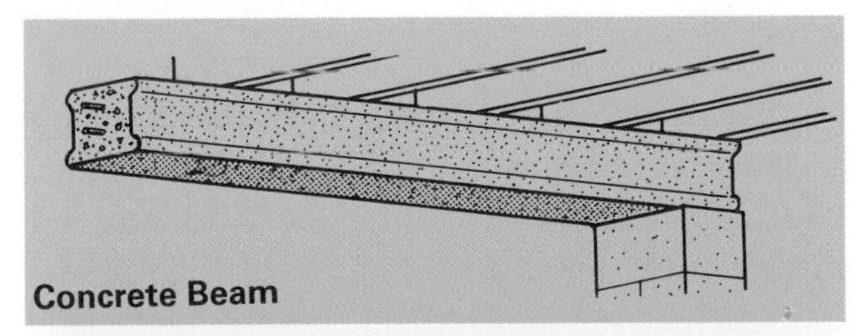

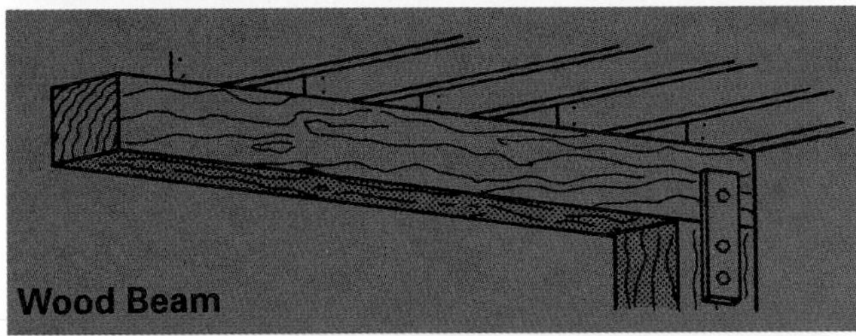

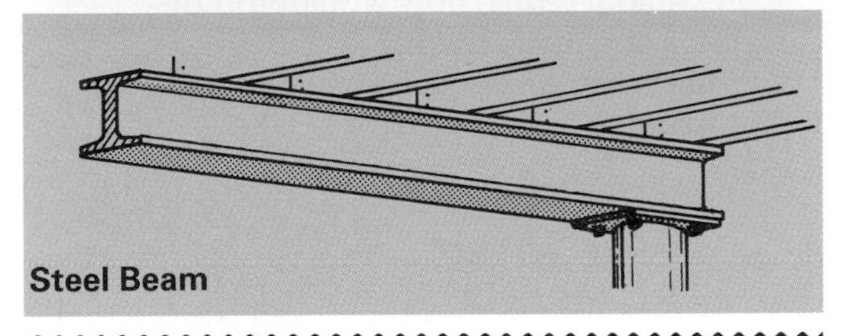

Fig. 24-9. Many materials are suitable for construction purposes. In a well-designed structure, various materials are used to their best advantage.

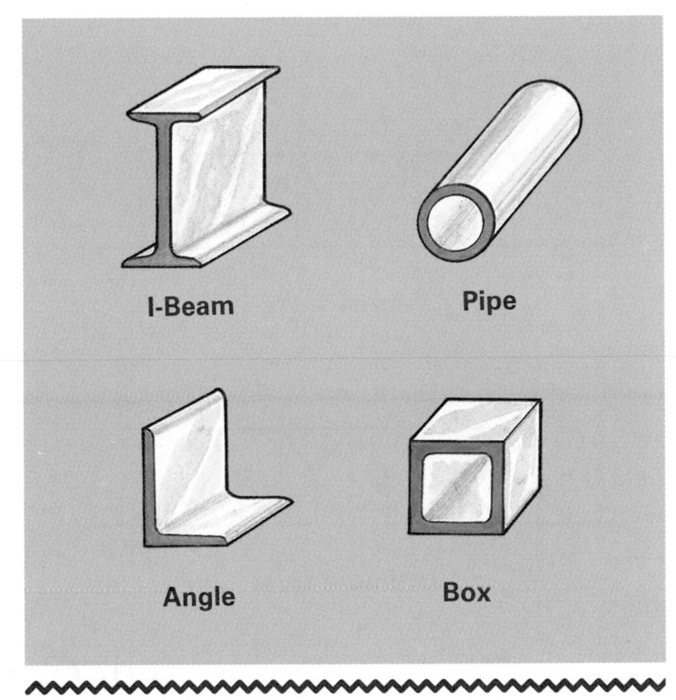

Fig. 24-10. On a pound-for-pound basis, solid structural members are not as strong as those designed with less weight toward the center, such as the ones shown here.

Arches

The arch is a structural shape that has been used for thousands of years. Arches shift their load to the ground. Since all of the force in an arch is compressive, arches are usually built of material that works well under compression, such as masonry. Fig. 24–11.

TECHNOLOGY TRIVIA

A well-known use of arch construction was a 19-arch stone bridge across the Thames River, London Bridge, built in 1209 A.D. by an engineer called Peter of Colechurch. This bridge remained in use for more than 600 years until it was replaced in 1831.

Fig. 24-11. Arches, best known for their widespread use in Roman times, work by transferring the load to the earth below.

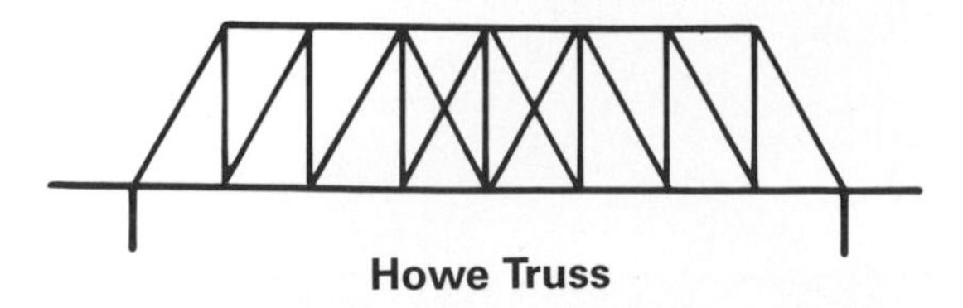

Howe Truss

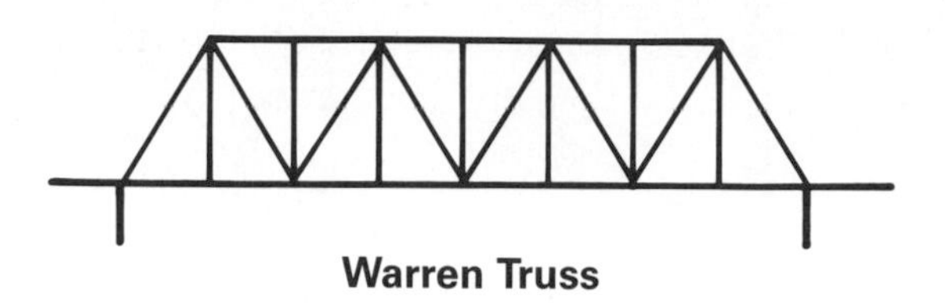

Warren Truss

Baltimore Truss

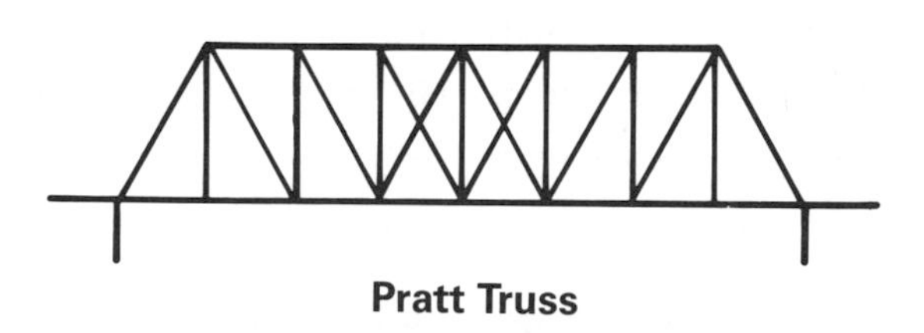

Pratt Truss

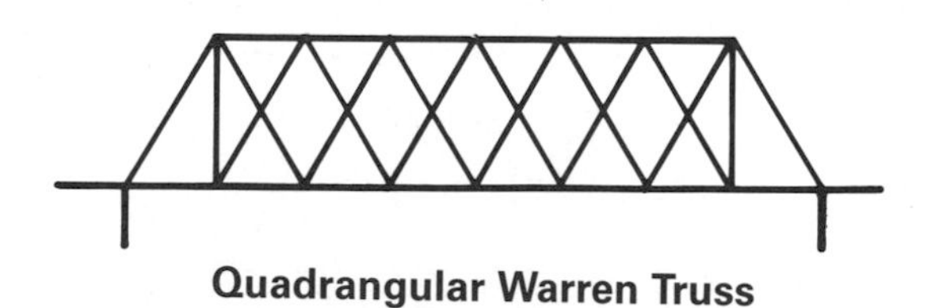

Quadrangular Warren Truss

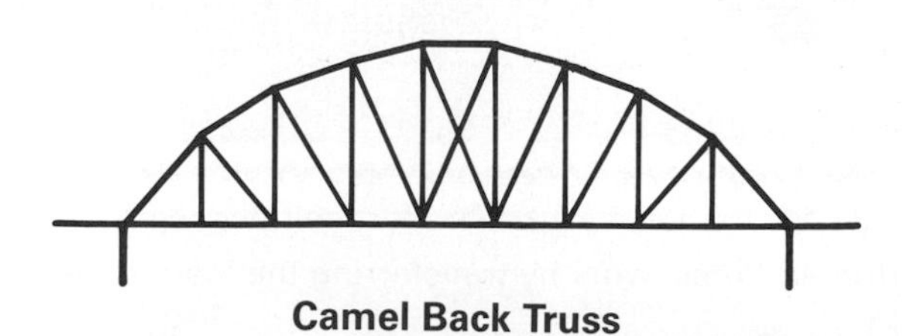

Camel Back Truss

Fig. 24-12. Trusses are composed of many open triangles. They are used as particularly stable beam structures.

Trusses

A **truss** is a structural shape that provides great weight-to-strength ratio and excellent stability. Trusses are commonly used for roofs, bridges, and any structures for which an open expanse is desirable, such as warehouses and gymnasiums. Many trusses incorporate the triangle shape, which is exceptionally stable. Fig. 24–12.

▶▶▶ FOR DISCUSSION ◀◀◀

1. In general terms, how would you describe a structural shape?
2. What is the distinction between columns and beams?

IMPACT

We shape our structures to suit our needs. In a way, our structures also shape *us*. They can affect our health, our safety, and even our moods. Would you rather be in a bright, cheerful room or in a damp, dark cellar? To most of us, our surroundings are important. We feel better when we are in a place that looks friendly and safe.

CHAPTER 24 Review

Chapter Highlights

- A structure is an arrangement of parts—structures can be as small as single atoms or as large as the entire universe.
- Designers must consider appearance, cost, material, location, and function when they design structures.
- Structures must be able to resist both live and dead loads.
- The four basic forces that act on a structure are compression, tension, torque, and shear.
- Columns, beams, arches, and trusses are the basic structural shapes used in building.

Test Your Knowledge

1. How are structures usually classified?
2. What is a public works project?
3. Give some examples of residential structures.
4. Why have industrial "parks" flourished during the past 10 years?
5. Explain the role of the architect in construction.
6. How is live (dynamic) load different from dead (static) load?
7. Why does compression increase at the base of a structure as its height increases?
8. Give an example of a structural member that can be subjected to tension and compression at the same time.
9. Why are arches often made from stone or other masonry products?
10. What is the most common shape found in trusses? Why?

Correlations

SCIENCE

1. Torque is a force that causes an object to rotate around a center point. One example of torque is a downward force applied to one end of a seesaw. That end of the seesaw will rotate downward (until it hits the ground). The other end will move upward. Torque is calculated by multiplying a force times its distance from a pivot point. To see torque in action, hang a yardstick or meter stick from its center so that it is balanced. Use lead sinkers for forces and hang one on one half of the stick. Hang two sinkers on the other half to balance the stick. Where do the sinkers have to hang in relation to the center point?

MATH

1. Architects and engineers use many geometric shapes, such as triangles and squares, in their designs. Make a list of the geometric shapes you see in your school, a bridge, an office building, and a gymnasium. Which shape do you see most often?

LANGUAGE ARTS

1. Look through old newspapers or magazines and find a photograph of an unusual structure. In a brief essay describe the structure and its function. Photos and their descriptions may be displayed on a bulletin board in the class.

SOCIAL STUDIES

1. Edison's addition of an electric motor in the elevator made the development of skyscrapers possible. What designer was responsible for the first skyscraper? Where was it built and how tall was it? Compare that skyscraper with a modern skyscraper built within the last decade.

CHAPTER 25

Construction Processes

Introduction

Did you ever wonder how things get built? The knowledge required to build all but the smallest structures has evolved over many years. It usually takes a number of people, filling different roles, to plan and build a structure. Because larger structures are usually more complex, they require more people. They must be planned by people who have specialized training. Almost all such projects, however, are built using similar processes.

After reading this chapter, you should be able to

Define construction.

Explain the role of the architect in planning a new structure.

Describe the process of site selection and preparation.

Explain the purpose of a foundation.

Illustrate the concept of framing.

Give examples of work done to enclose or finish a structure.

Words you will need

construction
architect
zoning
site
survey
foundation
soil sample
contractor
framing
insulate

Preparing for Construction

Construction is the process of building structures. Like all systems, construction has input, process, and output elements. Fig. 25–1. We often think of construction simply as a means of producing a house or building, but many other types of structures may also be constructed. Tunnels, dams, roads, bridges—these are all construction projects, even though none of them are considered buildings.

From the earliest times to the present day, construction has been one of civilization's major activities. Almost anyone may provide the idea for a new building: a business executive looking for a new store, a landowner hoping to attract investors, perhaps someone seeking a new home.

Why do people want to build new structures? Why not look for something already built? Once again, there are many possible reasons. It could be that there are no suitable structures available, or that no suitable structures exist at the right location. In short, structures are conceived by any number of people for all sorts of reasons.

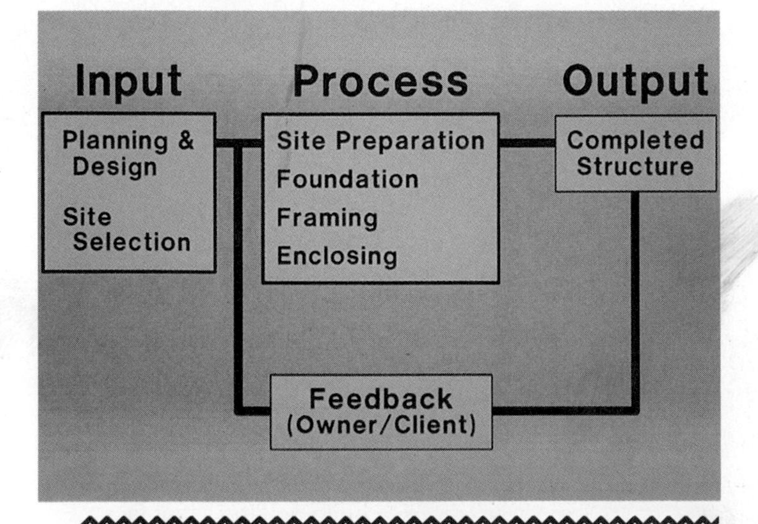

Fig. 25-1. All construction has input, process, and output components. The process is different, however, for different types of construction. This is a typical system diagram for the construction of a framed structure.

Fig. 25-2. This architect is showing clients several popular styles for houses. The clients must make many decisions before the house can be designed.

Planning and Design

For the moment, let's assume that the structure in question is a private home. How is the house's appearance determined? Some people might select their home from an inventory of previously prepared plans that are sold as a complete set of drawings. This is much less expensive than having an architect draw up custom-made plans. An **architect** is a person trained to design structures.

Buying previously prepared plans does not allow for input from the client as the plans are being prepared or supervision by the architect as the house is being built. Therefore, many people who want to build a new home consult an architect. It is common for architects to specialize in certain areas, such as residences, restaurants, or office buildings. People look for an architect who is familiar with their intended type of structure.

One of the first things the architect does is ask the client about his/her wants and needs. If the people are not sure what they want or what is available, the architect assists them. Fig. 25–2. There are so many decisions to be made that the architect might even give the client a checklist to review. Checklist items may include questions about the size of the family, hobbies, how many cars are driven, whether climbing stairs is a problem, and so on.

In addition to the client's wishes, there are also other factors to consider. For example, can the client afford to pay for everything requested? Is the property large enough to accommodate the structure? Can everything requested fit into a reasonably sized house? Will **zoning** (local building codes controlling the type of structures that can be built) be a problem?

TECHNOLOGY TRIVIA

Until the invention of the elevator in 1854 by Elisha Otis, the reluctance of people to climb stairs restricted most buildings to fewer than six stories. The first passenger elevator was installed in a 5-story building in 1856. From then on, tall buildings sprouted everywhere.

Site Selection and Preparation

One of the first challenges to a budding builder is to acquire a suitable **site**, or space to build on. What is meant by "suitable"? If a client insists on waterfront property, then all other places will be unacceptable. Other clients may want privacy or a beautiful view from the backyard. A common requirement is that the site be convenient to the person's job or to a means of mass transit. Other concerns include the type of neighborhood, location of shopping, reputation of nearby schools, and cost of land. Fig. 25–3. It is no exaggeration to say that locating an appropriate parcel of land may prove a difficult task.

After a site has been selected, it generally must be prepared for building. Typical requirements include removal of undesired structures (such as an abandoned shed), clearing the land (removing rocks or trees that are in the way), and *cutting and filling* (removing and adding earth to acquire a more level surface). Fig. 25–4. A surveyor must make a **survey** to determine the exact property and building lines.

Fig. 25-3. Many factors affect the choice of a building site. What factors may have influenced the choice of each of these sites?

Fig. 25–5. Together, the architect and the client must determine where on the plot of land to locate the house and in what direction the house will face.

▶▶▶ FOR DISCUSSION ◀◀◀

1. What makes up a "suitable" site for a building project? Explain.
2. Name two procedures people can use to select a design for a house.

Fig. 25-4. Preparing a site for construction often means clearing the land of unwanted debris.

Fig. 25-5. Using instruments, a surveyor determines the precise boundaries of a piece of property and the location of the proposed structure.

Building the Structure

Now that the design has been finalized and the site has been prepared, construction can begin. Construction of a building generally goes through four main phases: building the foundation, building the frame, enclosing the structure, and finishing the structure.

Building the Foundation

The **foundation** is the part of the structure that is in contact with the ground. The foundation is sometimes called the *substructure*. The prefix *sub* means "beneath" or "below." Because the entire weight of the house rests on the foundation, it must be both stable and strong. To protect against the foundation settling (sinking into the ground), the architect may request a **soil sample** to determine how soft the earth is. A soil engineer can then estimate how much weight the soil can support. Fig. 25–6. The architect designs the foundation with this information in mind.

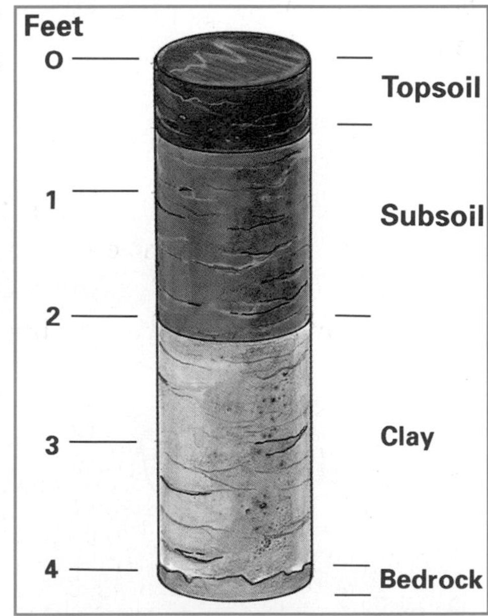

Fig. 25-6. A soil engineer can analyze a soil sample to determine how much weight the soil can support. The architect uses this information to design a stable foundation.

Foundations vary in type and material. The type of foundation and choice of material depend on local building codes (requirements), availability of materials, type of soil, and the builder's preference. A common option is to *excavate*, or dig, a trench and pour concrete. Concrete is made of a mixture of cement, sand, aggregate (small rocks or pebbles), and water. It is poured into a form, such as a trench or wooden frame, and hardens into a rigid solid. Fig. 25–7. Other builders use pilings (columns made of various materials) set into the earth. Still others use precast concrete blocks.

The person in charge of building the foundation as well as the rest of the structure is called the **contractor**. The contractor is responsible for supervising all the work according to the architect's drawings and instructions. In fact, the

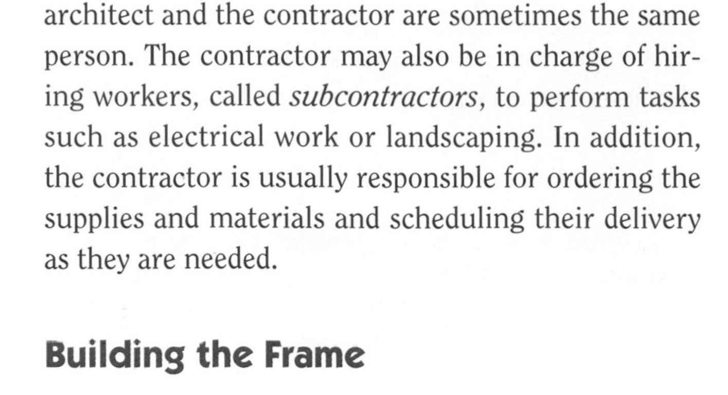

architect and the contractor are sometimes the same person. The contractor may also be in charge of hiring workers, called *subcontractors*, to perform tasks such as electrical work or landscaping. In addition, the contractor is usually responsible for ordering the supplies and materials and scheduling their delivery as they are needed.

Building the Frame

The part of the building that sits on top of the foundation is sometimes called the *superstructure*. Fig. 25–8. The frame of the superstructure provides its basic support, much as a skeleton provides the support for the human body. **Framing**, or building the frame of the structure, typically begins with the

Fig. 25-7. Foundations of concrete and other materials may take several different forms. The appropriate foundation is determined by soil type and the kind of structure to be built.

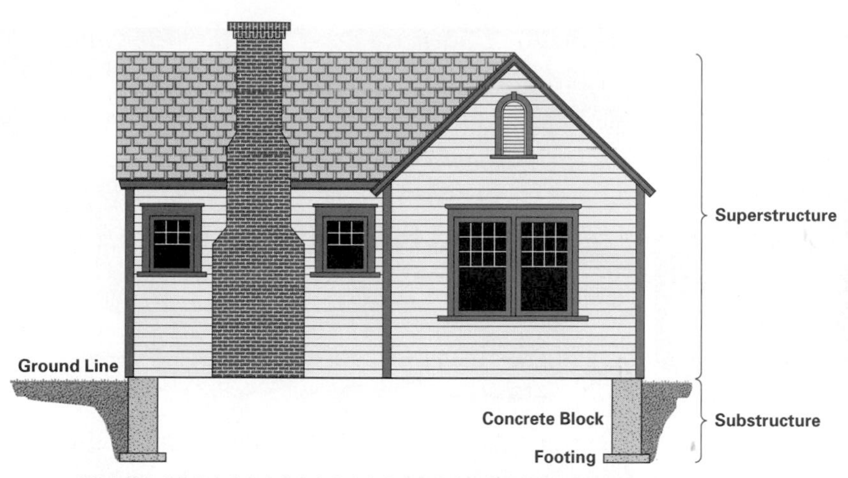

Fig. 25-8. The substructure is the foundation of the building and is usually below ground level. The superstructure includes the rest of the structure. Its weight is supported by the substructure.

Fig. 25-9. Floor trusses can be made from wood, metal, or both. They are built in a factory and installed at the building site.

floor. *Floor joists* or *trusses* support the floor, and they in turn are supported by the foundation. Fig. 25–9. Next, the walls are framed. The walls are often assembled flat on the floor and then raised as a unit. Fig. 25–10. The last part to be framed is the roof. Today, many roof rafters are made in the form of *trusses*, usually off-site, and delivered as needed. Fig. 25–11.

TECHNOLOGY TRIVIA

Pillars supported the ceiling of a structure designed in 1689 by architect Sir Christopher Wren. Officials thought the ceiling would not stay up and ordered Wren to add more pillars. Believing his ceiling needed no extra support, Wren put up four phony pillars. They served no structural purpose—they did not even reach the ceiling. The officials were fooled. The phony pillars still stand—and the ceiling Wren designed shows no sign of falling down.

Fig. 25-10. Carpenters (or "framers") build and join the walls on the site.

Fig. 25-11. Roof framing may be done by using preassembled trusses or by building separate rafters.

Enclosing the Structure

After the structure is framed, it must be enclosed. The exterior covering has a number of functions. It weatherproofs the interior, provides security and privacy, gives the structure its actual appearance, and *insulates* the structure (provides barriers against heat, cold, and sound). Many materials have been developed especially for the roofs and walls of buildings and for the windows and doors in them.

Before the interior can be finished, other work must be completed. Pipes for plumbing and wires for electricity are positioned while the insides of the walls are still accessible. Gas lines and ducts (metal sleeves) for heating and/or air conditioning are installed. Insulation must be placed inside exterior walls and above the highest ceiling to minimize heat transfer. Fig. 25–12. A well-insulated structure is energy efficient—it costs less to heat and cool.

IMPACT

In the mid-1970s, the price of oil increased sharply. This led to high prices for heating and cooling. As a result, builders began to make structures more energy-efficient.

Finishing the Structure

The interior of the building is finished in various ways. Rigid material, often in the form of 4′ × 8′ panels, is attached to the floor, walls, and ceiling. Fig. 25–13. This covers the framing and provides a surface that can be finished by painting or wallpapering. Doors, cabinets, windows, and plumbing fixtures are installed. Carpet or other floor covering is

Fig. 25-12. Insulation prevents the loss of heat in winter and helps keep heat out in the summer.

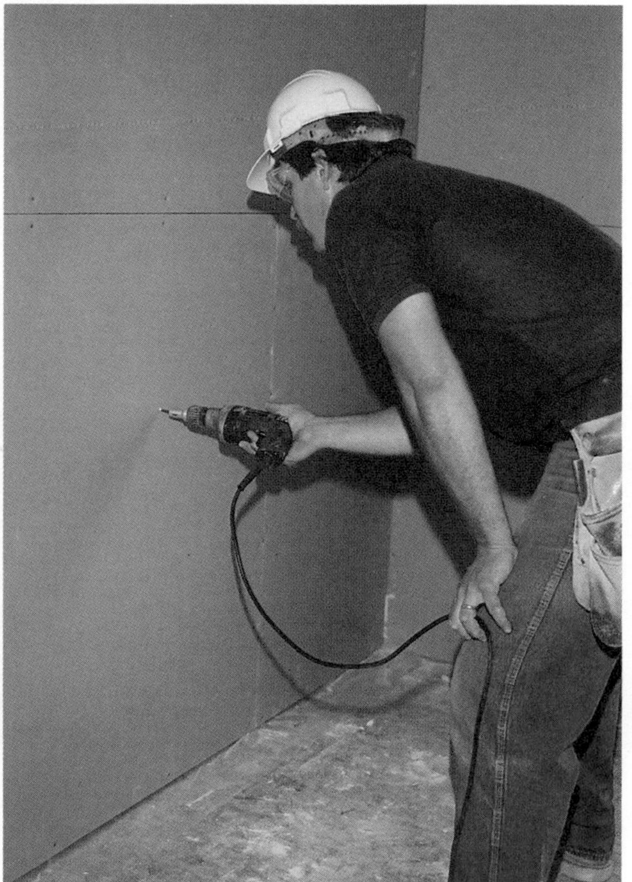

Fig. 25-13. Panels of wall and floor covering are installed over the framing to create the interior of the building.

Fig. 25-14. Molding and trim give surface intersections a finished appearance.

put down. The last interior finishing detail is usually placing trim or molding where two surfaces meet, such as the wall and the floor. Fig. 25–14.

Finally, the exterior is finished. The outside of the house is painted, and any trim details are finished. Concrete driveways and sidewalks are poured at this time, if the design calls for them. In some cases, grass seeds and attractive bushes and trees may be planted to beautify the exterior of the house.

■ **Observe a construction project underway in your community or in a nearby area. Describe what was happening on the site when you observed it. In which phase of construction was the project?**

▶▶▶ FOR DISCUSSION ◀◀◀

1. Who actually decides what a new structure will look like?
2. Why must foundations be strong and stable?

■ **Construct the MegaIndustries plant and the other structures for the community model. You may use the panels that were produced in Chapter 19's community activity.**

CHAPTER 25 Review

Chapter Highlights

- Construction is a major activity of civilization.
- Architects plan and design new structures.
- Construction sites generally require preparation before construction can begin.
- Foundations anchor and support structures.
- Framing is to a house structure what a skeleton is to the human body.
- Enclosing and finishing a structure complete the construction process.

Test Your Knowledge

1. What is meant by the term *construction*?
2. Why are architects needed for construction projects?
3. What factors play a role in site selection?
4. What is meant by "cutting and filling"?
5. What is the purpose of a foundation?
6. Why do architects sometimes request a soil sample?
7. Name the responsibilities of the contractor.
8. Which part of the framing is usually done first?
9. Why do structures need exterior covering?
10. Why is insulation installed in structures?

Correlations

SCIENCE

1. Compare the load a flat roof can support to the load a peaked roof can support. Construct roof shapes from scrap wood. Hang a bucket from the center of each roof shape. Gently pour sand into the bucket until the structure falls.

MATH

1. The living room of your new home is 15×18 feet. How many square yards of carpet do you need? If the carpet is $15.99 per square yard, how much will it cost to carpet the living room?

LANGUAGE ARTS

1. Suppose you are an architect designing a new junior high school. Make a list of questions that would need to be answered before you could begin to design the proposed building.

SOCIAL STUDIES

1. Interview an architect or builder within your community. Find out what is the most popular style of home in your area.
2. Compare a home of today with early pioneer homes. What everyday features of today's homes would be considered luxuries in the early pioneer homes?

CHAPTER 26

Impacts of Construction Systems

Introduction

Are there currently any societies that do not use construction? Not likely, since even the most primitive social groups build simple shelters. Modern industrial nations pride themselves on their architecture. Attractive, useful structures abound in the United States. We have, in the relatively short history of our nation, developed a magnificent system of roads, housing, and commercial and civil structures. It is no wonder that the impact of all this construction reaches each and every one of us.

After reading this chapter, you should be able to

Describe how construction affects people, society, and the environment.

Discuss some differences in lifestyles for people living in urban, suburban, and rural communities.

List some careers involved in construction.

Explain some of the reasons modern society is more mobile than it was 100 years ago.

Give examples of direct and indirect impacts of construction.

Words you will need

indirect employment
urban
suburban
rural
erosion

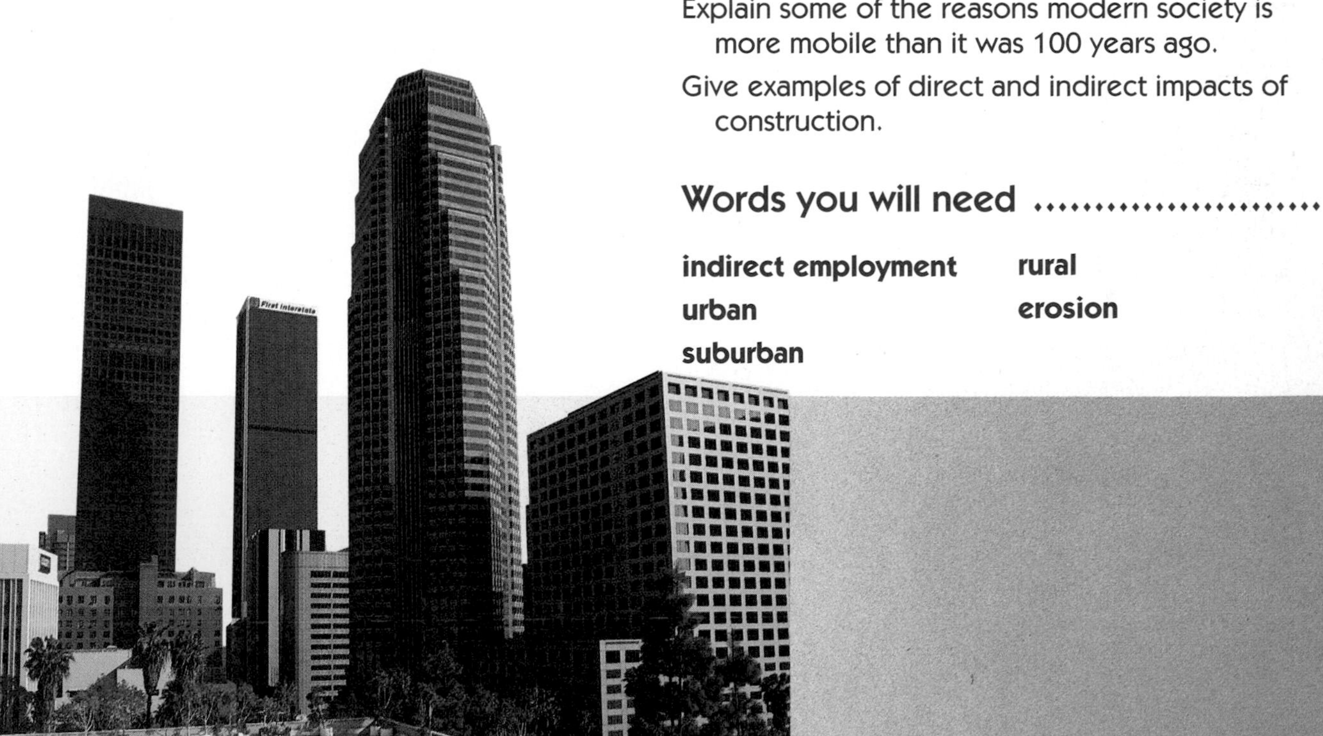

Impacts on People

In the normal course of a day, almost everyone uses structures: for shelter, for work, for transportation. Probably the most frequent use is for shelter. Whether or not a person has a job or needs to commute from place to place, the need for housing remains.

Construction Affects Lifestyles

Each person's lifestyle is molded to some extent by the type of home he/she lives in. People who live in large apartment buildings, for example, have a different routine than those who live in private homes. Fig. 26–1. The former never worry about cutting the lawn, while the latter need not worry about late evening noise disturbing the downstairs tenants.

Fig. 26-1. People's lifestyles are influenced by the type of structure in which they live. What advantages and disadvantages might each of these residences have?

A city apartment dweller might meet 20 or 30 local residents just outside his/her door, while a **rural** (country) homeowner might not have a neighbor within miles. An inner-city, or **urban**, resident might feel that owning a car is more trouble and expense than it is worth, while a **suburban** family (one that lives on the outskirts of a city) commonly owns two or more cars. Fig. 26–2.

Structures used for transportation also have far-reaching impacts on people. On land, roads,

Fig. 26-2. Many city dwellers, faced with limited parking, decide not to buy a private car. Suburban commuters, with less mass transit available, are more likely to need a personal car.

bridges, and tunnels permit easy travel between locations. To understand the importance of these structures, you have only to observe what happens when a major highway route is closed. The traffic backup and the personal and business inconvenience that result become worthy of news headlines. Fig. 26–3.

In air travel, the vital role of airports, with their runways and hangars, cannot be overestimated. With respect to water, structures such as dams and canals don't appear to have an immediate effect, but that is far from the truth. Dams provide irrigation and drinking water and power hydroelectric plants. Canals permit passenger and cargo ships to travel routes that would otherwise be inaccessible or uneconomical.

TECHNOLOGY TRIVIA

Around 500 B.C., Darius I ordered construction of a canal to connect the Mediterranean and Red Seas. Darius' Canal remained in more or less regular use for about 1300 years. Today, the major waterway connecting these seas is the Suez Canal, opened in 1869. The Suez Canal follows the same general route of Darius' Canal.

Fig. 26-3. Significant backups can occur when even one major roadway is closed or narrowed temporarily.

Construction Provides Employment

Construction is also a major source of employment for people. Many jobs are directly involved in the design, construction, and maintenance of structures. Skilled craftspeople, such as bricklayers, plasterers, plumbers, and roofers, are employed in construction projects. Construction also provides jobs for architects, drafters, and interior designers. Professionals such as structural and electrical engineers are also needed. Of course, management personnel such as contractors, project managers, and business managers are needed to control the construction process. Building inspectors, surveyors, and many other people also owe their jobs to construction. Fig. 26–4.

Construction also provides **indirect employment**, or secondary employment. It is estimated that for every job in construction, at least one other job is created or influenced by it. For example, think of all the people involved in growing, cutting, transporting, and selling the lumber that is used. Large numbers of people are similarly involved in supplying other building materials (masonry, steel, plastic, copper, etc.), hand and power tools, power machinery, earth-moving machines, and so on. In addition, food suppliers, lawyers, accountants, health care providers, real estate salespeople, and many others are indirectly involved in construction.

TECHNOLOGY TRIVIA

In 1874, steel was first used as a structural material in the U.S., in building the Eads Bridge across the Mississippi River at St. Louis.

▶▶▶ FOR DISCUSSION ◀◀◀

1. Why do so many inner-city residents decide not to own a car?
2. How might a local restaurant be affected by nearby construction?

■ Make a list of three construction-related occupations that you think might be interesting. Do research on one of these and write a report describing the skills/knowledge needed and the work tasks and responsibilities involved in that type of work.

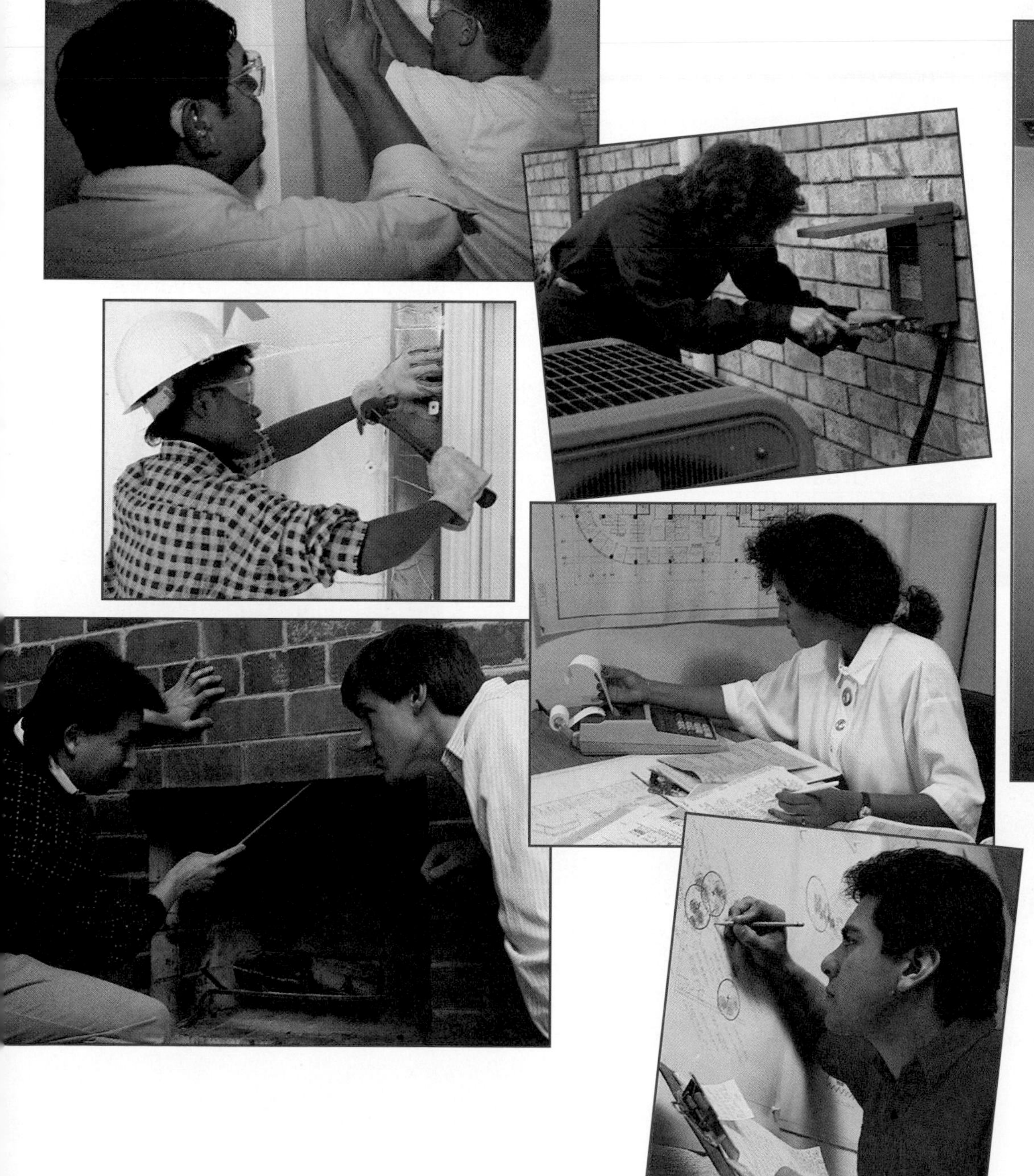

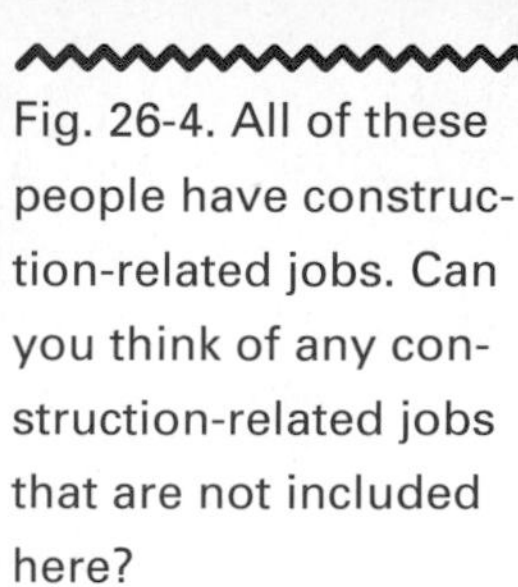

Fig. 26-4. All of these people have construction-related jobs. Can you think of any construction-related jobs that are not included here?

Impacts on Society

The word *society* means "a large group of people, bound together for the common good." In the United States, society has affected and has been affected by construction. Just as individual people are influenced by structures that give shelter and employment and allow transport, so our society is influenced by construction.

Residential Structures

Housing has developed into three general modes. Urban dwellings are typically intended for more than one family. City housing sometimes takes the form of a high-rise, and it is almost always densely packed.

Suburban communities surround the metropolitan areas. Housing is private, with some multi-family units, very few tall structures, and moderate density. Suburban communities are sometimes called "bedroom communities."

Rural housing is far from the city. It is almost always designed for single families and is rarely more than two stories high. Fig. 26–5. It is the relatively large variety of housing in this country that allows people such a wide choice in living conditions.

Urban	Suburban	Rural
High Rise	Mostly Low Rise	Low Rise
High Density	Medium Density	Low Density
Inner City	Surrounding City	Distant from Cities

Fig. 26-5. In addition to location, residential structures vary in number of stories and population density.

Transportation Facilities

Excellent transportation facilities have encouraged our society to become more mobile. It is commonplace today for people to commute twenty miles or more to work using personal or mass transportation systems. Fig. 26–6. This means that people who previously lived near their place of employment now have the option of living elsewhere without giving up their jobs. By the same token, people may change jobs more readily without moving to a new location.

Fig. 26-6. Modern transportation facilities enable Americans to travel freely for business or pleasure.

Water-Related Structures

Dams and canals also affect society. Entire communities are sometimes uprooted to make way for such construction. On the other side of that coin, it should be pointed out that new centers of commerce and housing may sprout up as a result of these projects. Dams often provide irrigation for land otherwise too dry to be used for agriculture. Fig. 26–7. Some of today's large metropolitan areas would never have gained their current size were it not for the fresh water supplied by dams.

▶▶▶ FOR DISCUSSION ◀◀◀

1. How would you describe "society" in the United States?
2. Why do excellent transportation facilities encourage a more mobile society?

Irrigation

Hydroelectric power

Flood control

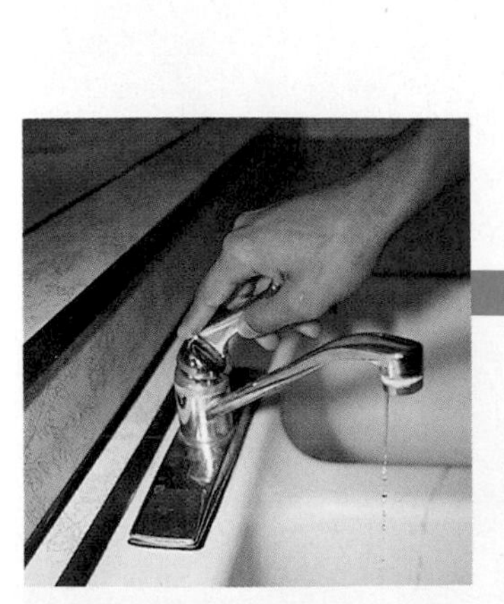

Drinking water

Water treatment

Fig. 26-7. Dams serve many purposes in a community. They help control flooding, provide drinking water, irrigate dry land, and sometimes even provide electricity via hydroelectric power plants.

Impact on the Environment

Construction has a multitude of influences on the environment. As is always the case, some of the impacts are predictable, but some are unplanned. Unintended side effects have both positive and negative results. For example, let's look at the process of cutting trees to make lumber for structures.

When trees are cut, what happens to the land that is left behind? If the forest is owned by people who are participating in a planned replanting program, new trees will soon take the place of those lost. If nothing is done, the land may revert to wild growth. In some places, the land may even be subject to **erosion**, or loss of top soil, as rains wash away the earth that was formerly anchored by the trees. Fig. 26–8. The animals and other natural wildlife that lived among the trees may no longer have homes.

Fig. 26-8. When trees are harvested without concern for replanting, top soil can be washed away, leaving the land lifeless and barren.

What about the actual construction of the structure? Were people displaced to make way for the new construction? Were other structures on the site demolished? Where does the debris get dumped? When people are attracted to the new structure as tenants, employees, or customers, what impact will they have on the surrounding community? Will the additional traffic (and the pollution that results) create a problem? Fig. 26–9. Will new electrical generating plants and other new facilities be required? If this is a large-scale, multi-structure development, what about additional hospitals, schools, and other such services? Will open spaces such as parks need to be created? As you can see, construction has quite a lengthy "chain reaction" to account for.

Fig. 26-9. Well-planned communities include open spaces, such as parks, as well as buildings and roads.

Fig. 26-10. In many areas, construction projects must be approved by civic groups or public agencies before building can proceed.

The degree of impact will, of course, depend on the type and scope of the structure in question. A small, single-family residence simply does not have the potential for change that a skyscraper, dam, or bridge might. Those responsible for planning these larger projects must consider the impact the structures will have on the community and the environment. To encourage this, most governmental bodies have agencies or public groups that review all proposals for new construction. Fig. 26–10. These agencies consider all of the known social and environmental effects of each project before they approve construction.

▶▶▶ FOR DISCUSSION ◀◀◀

1. Give an example of an unintended side effect of construction.
2. What effects might the additional traffic from new construction have on the environment?

■ **Citizens are concerned about the environmental effects of the new MegaIndustries plant. Hold a town meeting to discuss the environmental impacts and to consider ways to deal with them.**

CHAPTER 26 Review

Chapter Highlights

- Structures are vital for shelter, employment, and transportation.
- Construction provides people with employment both directly and indirectly.
- Society is very much influenced by construction and the structures that result.
- The side effects of construction may be positive or negative, expected or unintended.
- Large-scale structures have a greater impact on the community and the environment than smaller ones.

Test Your Knowledge

1. Why do modern industrial nations emphasize vigorous construction programs?
2. In what ways might a person's lifestyle be influenced by his/her type of residence?
3. List several categories of craftspeople who work in construction.
4. What is meant by "indirect employment"?
5. How do structures built for land transportation affect people?
6. In what ways do dams affect the locations in which people live?
7. What are the three modes of housing?
8. Why is it that construction has unplanned side effects?
9. What is soil erosion?
10. How does government protect against undesirable construction?

Correlations

SCIENCE

1. Investigate the impact of Florida's growing population on the Everglades.

MATH

1. Mrs. Rogers commutes 20 miles each way to her job in a nearby city. If she works five days a week, how many miles does she drive each week to and from her job?

LANGUAGE ARTS

1. Imagine a new shopping mall is to be built in the center of your community. Describe the positive and negative effects of such a proposal.

SOCIAL STUDIES

1. How did the construction of cities change people's lives? How did the presence of the people affect the structure of the city?

CHAPTER 27

Construction in the Future

Introduction

It's always exciting to guess what the future will bring. How much will change, and how much will remain the same? Although no one can say for sure, we can certainly ask the question: What will construction be like in the future? We can also think about what construction was like in the past. How has it evolved over the last hundred years—or perhaps the last thousand?

In fact, the basics of construction have remained very much the same for hundreds of years. Only with the relatively recent invention of power machinery and the development of iron and steel did the construction process change much. Even with these technical modifications, the goals of building structures—comfortable shelter, efficient transportation, and so on—have not changed very much at all.

After reading this chapter, you should be able to

Discuss future locations for construction.

Describe future construction tools and materials.

Discuss the use of electronics in design and construction.

Describe the role of the environment with respect to new construction.

Name several trends in construction.

Words you will need

project management
structural plastics
composites
pultrusion
smart houses
artificial intelligence
thermography
macroengineering
superinsulated
modular homes

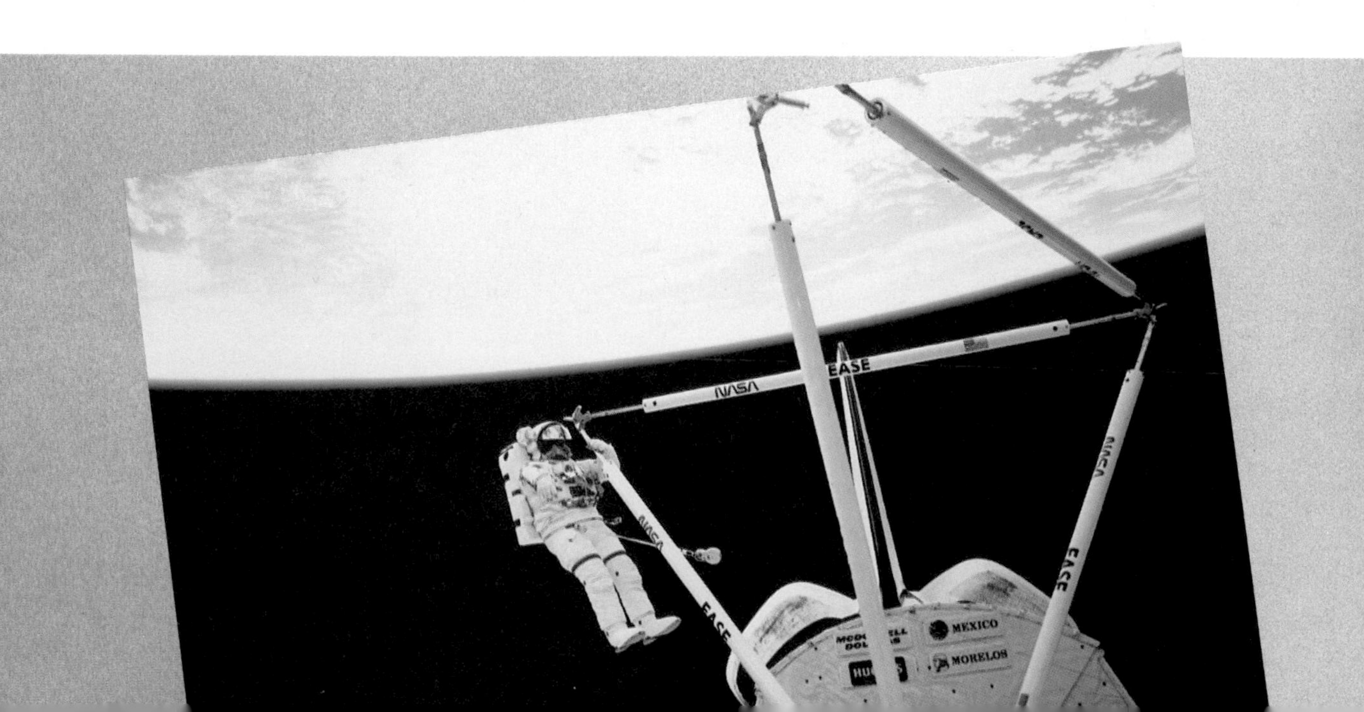

Location of Future Construction

Most structures today are built above ground, and for good reason. What about other environments? Could people live beneath the sea? Short-term experiments have shown that this is possible, but not very practical. Expense, inconvenience, and isolation combine to prevent this from being an ideal place for new construction. Similar concerns dictate that underground shelters be used mainly for storage.

What about creating an artificial environment? One experiment, called *Biosphere II*, is attempting to recreate the earth's atmosphere within a completely sealed, 3.1-acre structure. Located in the Arizona desert, *Biosphere II* is designed to determine whether such a structure is practical. Fig. 27–1. If so, it could be the prototype for others to be built elsewhere, perhaps as the first lunar community!

Fig. 27-1. *Biosphere II* is an experiment in creating an artificial environment. The experiment will last from 1991 to 1993.

Another possible building location is in space. Space structures will probably orbit the earth, much like communication satellites, but will also have living quarters. Fig. 27–2. Space will probably contain research and development facilities and allow servicing for other satellites. Among other things, this means they must have provisions for air, food, and water.

TECHNOLOGY TRIVIA

The first space station, Salyut (meaning "Salute") was launched by the U.S.S.R. in April of 1971. Salyut was manned for 23 days. The first American space station, Skylab, was launched in May of 1973. Three teams of astronauts manned Skylab for a total of 172 days.

▶▶▶ FOR DISCUSSION ◀◀◀

1. Explain why most structures are built above ground.
2. Why would a lunar community need a sealed dome environment?

IMPACT

Space stations, cities on the moon, long-term missions into deep space—all of these may become a reality in your lifetime. Perhaps the people who leave the earth to live in space will also leave behind their prejudices and their political differences. Perhaps they will take these along, and space will become another battleground. Which do you think will happen?

■ **Design and construct a space-based factory for the assembly of a spaceship.**

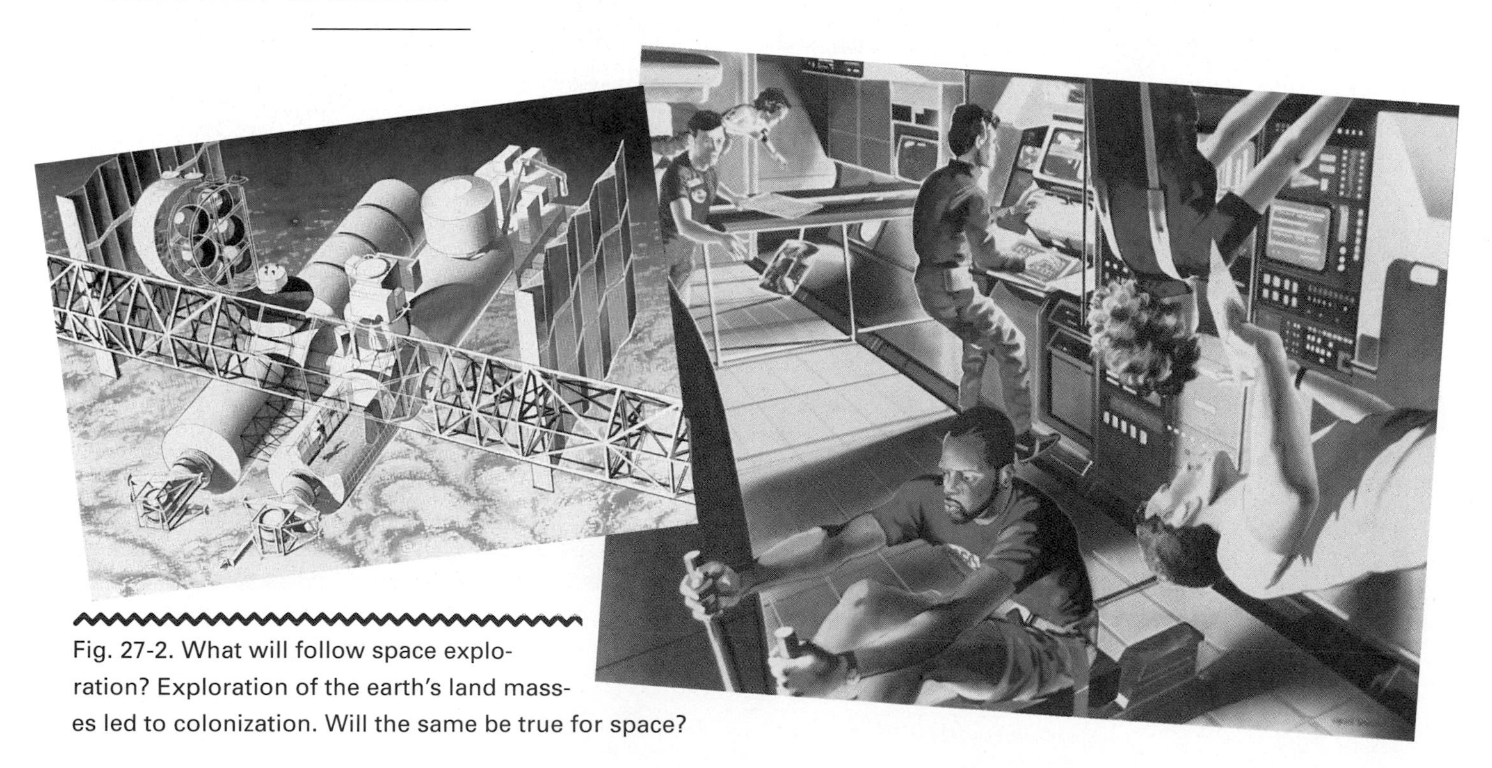

Fig. 27-2. What will follow space exploration? Exploration of the earth's land masses led to colonization. Will the same be true for space?

Tools and Materials

Will construction tools be the same in the future? They probably will, at least in the near future. Still, there has been some indication of change. Lasers, for example, have become very popular for surveying and grading. Fig. 27–3. Builders can also use lasers to determine whether a building is straight (plumb) and level.

Computers will continue to assist builders in construction. Even now, they are being used more and more for computer-assisted design, manufacturing, and engineering.

TECHNOLOGY TRIVIA

Theodore H. Maiman of the United States built the first laser, a ruby laser, and operated it for the first time in 1960.

Computers are also being used for **project management**. This allows the operator to keep track of the pace of construction, delivery of materials, number of employees, and so on for each construction project. Computerized project management will no doubt become more common because it is efficient.

What about construction materials? People are always seeking new and improved materials for construction. Although traditional materials such as masonry, architectural metals, and wood are still very popular, there is a demand for stronger, lighter, and less expensive items.

One such development is **structural plastics**. Structural plastics are immune to corrosion and electrical interference, and they are good insulators. They are also fairly inexpensive.

Another trend is the use of composites. **Composites** are made of a variety of different materials. The materials are combined to make the final product much more useful than its individual components. Composites are often used where great weight-to-strength ratios are needed.

Composites have also been developed that have other desirable traits, such as resistance to heat or to certain chemicals. One of the newer manufacturing processes, called **pultrusion**, allows compos-

Fig. 27-3. A powerful and sharply defined laser beam can be made to scan along a preset boundary, providing an accurate guide for builders and surveyors.

Fig. 27-4. In pultrusion, material is pulled out of a die. This picture shows two sucker rods being formed. The sucker rods will be used for pumping oil.

ites to be made from continuous fibers, yielding a much stronger final product. Fig. 27–4.

▶▶▶ FOR DISCUSSION ◀◀◀

1. How are computers being used in construction?
2. What characteristics are desirable for materials used in construction work?

Electronics

Electronics will almost certainly play a large role in future construction. **Smart houses** have been designed in which computers play an essential part. In these houses, computers monitor and adjust all the electrical systems, such as lighting, heating, and security systems. Fig. 27–5. The occupants can program the computer to turn lights and appliances on and off at certain times. The computer also notes any broken or defective fixtures, calculates the efficiency of heating and air conditioning, and so on.

"Intelligent" office buildings have also been planned. In these structures, built-in cabling systems can be used for voice, picture, and data transmission. Fiber-optic systems are typically included, and microwave antennas may adorn the roof. Electronic fire, security, and surveillance systems are built into the building's wiring. Computer-operated elevators figure out the best route to minimize waiting. They can even allow for parallel unit operation (so that both elevators don't stop at the same floor unnecessarily).

Computers programmed with **artificial intelligence** may also be part of the contractor's tools in the future. Artificial intelligence programs allow the computer to evaluate data and respond to a builder's questions. This technology saves research time and helps prevent needless errors and omissions.

Thermography, or computerized mapping of infrared photography, is a method by which heat loss from a building can be analyzed. It is clear that computers are going to become increasingly important in many aspects of construction.

▶▶▶ FOR DISCUSSION ◀◀◀

1. Describe a "smart" home.
2. Name some ways in which electronics may be used in a building.

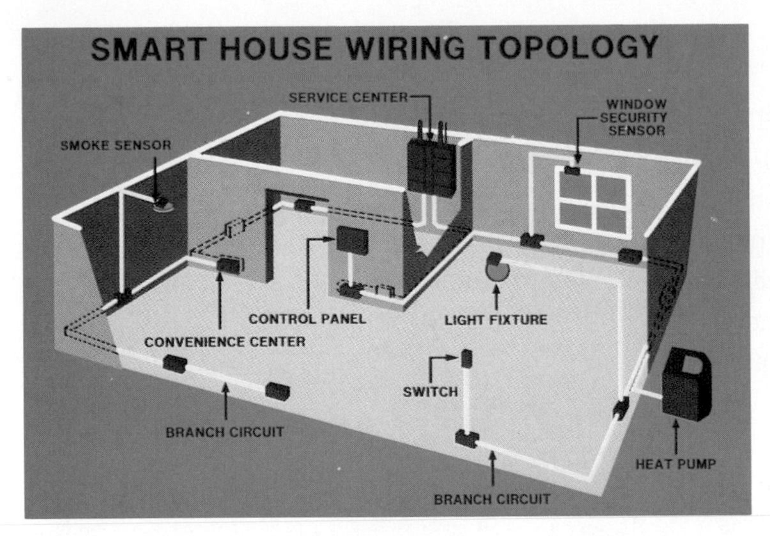

Fig. 27-5. Integrating the computer into the home can be much more than simply placing a microcomputer on a desk. "Smart" homes of the future will have a built-in computer system to help run and monitor various operations such as heating and cooling.

Construction and the Environment

It is an unavoidable fact that the earth is a place of shrinking natural resources. Fossil fuels, farmland, drinkable water, and forest products are all in short supply. Due to shrinking resources and rising costs, new construction is being tailored to use less expensive materials and smaller amounts. Decreased power consumption and energy conservation have become, and will continue to be, high-priority items on the architect's wish list.

One response to the pressures on the environment is called **macroengineering**. This term is used to describe large-scale engineering projects that affect the environment. Macroengineering may be able to relieve the pressure that our growing population is placing on the earth's resources. For example, a project that could irrigate the parched areas of the western United States without affecting other natural resources would provide more land for agriculture and housing at little cost to the environment.

Future structures will be designed to maximize solar energy, an inexpensive, "unlimited" energy resource. Fig. 27–6. **Superinsulated** buildings are being built to reduce the need for energy by conserving heat in winter and blocking out heat in summer. Architects would like to have structures that "respond" to the environment, such as tinted glass that gets darker when it is exposed to bright sunlight.

▶▶▶ FOR DISCUSSION ◀◀◀

1. Why is new construction designed to conserve energy?
2. What is the value of a superinsulated building?

Fig. 27-6. Solar energy remains the largest source of "free" energy available.

Trends in Construction

Generally speaking, new structures will be similar to current structures for the next 25–50 years. However, several recent trends in construction are now gaining acceptance. **Modular homes**, structures that are prefabricated off-site and trucked to the assembly point, will become more commonplace. Fig. 27–7. Wiring and plumbing are preinstalled in modular homes for rapid assembly. The lower initial costs and quick assembly time make modular construction ideal for low-cost housing.

Another trend is greater housing density. This density is a direct result of limited land and increasing population. More attached homes will be built, as well as more high-rise residential structures.

Homes that cater to older populations will become more necessary as the population of the United States ages. Better health and nutrition have allowed us to live longer than ever, and the post-World War II "baby boom" is now middle aged. Eventually, all of those "babies" will need housing suitable for senior citizens.

▶▶▶ FOR DISCUSSION ◀◀◀

1. **Name some advantages of modular construction.**
2. **Why will tomorrow's housing probably be of greater density?**

■ Examine a structure built many years ago and a similar structure built recently. What differences do you observe? For example, what materials were used in the newer structure that were not used in the older one?

Fig. 27-7. Assembling parts of structures off-site can result in real savings of time and money. This factory-built home is ready for delivery and final assembly.

CHAPTER 27 Review

Chapter Highlights

- Structures can be designed for use beneath the sea, underground, on the earth's surface, and in space.
- The newest, most powerful tool for future construction may be the computer.
- New materials are in demand to conserve energy and costs.
- "Smart" structures will become a larger part of future construction.
- Modular homes will increase in popularity as a result of their lower costs.
- Housing will increase in density because population is increasing but the amount of available land remains fixed.

Test Your Knowledge

1. Name two similarities between construction 100 years ago and construction today.
2. Explain the purpose of *Biosphere II*.
3. Why would people want to design structures for use in space?
4. How can computers be used for project management?
5. What is the value of a strong, light-weight structural material?
6. Describe the process of pultrusion.
7. Why might it be valuable or convenient for lights and appliances to be programmed to go on and off automatically?
8. How can computers be used in elevator operation?
9. What is thermography? Why is it used?
10. Why should architects be interested in utilizing solar energy?

Correlations

SCIENCE

1. Composites are made by combining materials. Each material keeps its own properties, but the composite also has properties that the individual materials do not have. Concrete is a composite material. Find out what the ingredients of concrete are. With your teacher's help, mix a small batch of concrete in your school lab.

MATH

1. A high-rise retirement home is being constructed in your community. How many apartments are in a twelve-story building with sixteen one-bedroom units on each of the first five floors and ten two-bedroom units on each of the remaining seven floors?

LANGUAGE ARTS

1. "Smart" structures will play a major role in future construction. In a creative essay, describe your ideal "smart" house.

SOCIAL STUDIES

1. Do you think human beings should build permanent communities on the moon? Why or why not?

Make a Difference

Habitat for Humanity

Armed with paint brushes, cleaning supplies, and grass seed, many young people today are using their skills and talents to help Habitat for Humanity build and renovate homes in partnership with families living in substandard housing.

The families work alongside Habitat volunteers on both their own homes and the homes of other families in need. Under the nonprofit program, they receive an interest-free loan from Habitat, and the money they pay back is then used to build homes for other families.

During its first 15 years, Habitat volunteers worked alongside future homeowners on more than 12,000 houses in nearly 700 cities and 32 countries. Working with adult volunteers, college and high school students have found that they too can play an important part in this worldwide effort.

While the more experienced workers tackle the actual construction projects, younger volunteers have helped with a variety of other tasks.

Confirmation classes at St. Patrick's helped paint the walls in one home, and a youth group from Eastview Christian Church graded the yard and planted grass seed. High school students also helped with last-minute cleaning—from the upstairs on out the door—to get the house ready for the family to move in.

In another project, youth and adults from a Community Presbyterian Church helped a young family finish their home by drywalling the entire inside—the walls and ceilings of seven rooms.

Six young people from South Dakota traveled to Chicago for still another project—to help clean out and renovate a house in an older section of the city.

Habitat for Humanity International was started in Americus, Georgia, by Linda and Millard Fuller. It currently has more than 200 college chapters. In 1988, the Marist School in Atlanta, Georgia, became Habitat's first official high school chapter.

The Collegiate Challenge, a special "Spring Break" program, was launched in 1990. That first year, more than 1300 college students gave up their vacations to help Habitat families with forty-five projects. The following year, nearly 2100 students from 101 colleges and universities, two high schools, and four churches worked with thirty-six Habitat affiliates.

Young people have also been successful in raising funds to help the program. Children in a Maryland church collected "A Mile of Pennies" for Habitat, and students at the University of Richmond, Virginia, raised more than $30,000 with a bicycle race.

When students at R. J. Reynolds High School worked as "extras" in a movie, they made about $30,000—and donated it to Habitat. La Plata High School students picked up recyclable aluminum cans along streets and roadways to benefit both the environment and Habitat for Humanity.

Students found that they could also help with materials for Habitat houses. One church group, for example, helped remove ceiling tiles from a bank to save for future Habitat houses!

Throughout the world—through the efforts of dedicated adults and young people—Habitat continues to turn dream homes into reality.

SECTION SEVEN

TRANSPORTATION SYSTEMS

Activity Brief

Technology in Transportation

PART 1: Here's the Situation ....

Transportation systems can move people, products, and materials to any corner of the earth. We have created a "mobile world." A maze of highways and railways connects cities, states, and even countries. Air transportation systems have made travel between continents an everyday occurrence, and travel into space seems almost routine. Our transportation systems are remarkable, but they also create problems such as traffic jams and environmental pollution. Research continues as people look for ways to improve our present transportation systems and develop new systems.

As you do this activity and as you read Chapters 28 through 31, you'll learn about transportation systems, their positive and negative impacts, and new developments that hold great promise for the future.

PART 2: Your Challenge..........

Improved mass transit systems can help solve many transportation problems that exist today. Train systems that apply the principles of magnetism are being developed and may provide one solution. These systems are called **maglev systems** because the vehicles, or trains, are **mag**netically **lev**itated. They float on a cushion of magnetic energy powered by electrical energy.

The magnetic energy, or magnetic field, on which a maglev train floats is created when like poles of magnets repel each other. Have you ever tried to stack like poles of magnets atop each other? The force you felt is the same energy force applied and controlled in maglev systems except that in a maglev system the force is thousands of times greater.

Germany and Japan have built working maglev systems. The trains are levitated and powered by the pushing-and-pulling effect created by switching like and unlike magnetic poles in electromagnets. Maglev trains can travel at about twice the speed of railed trains presently in use.

In this activity, you will design and build a prototype maglev vehicle and a one-foot guideway (track) to show how the vehicle can be levitated.

PART 3: Specifications and Limits..........

Your guideway and vehicle must stay within the following specifications and limits. Read these carefully before you begin.

1. The guideway (Fig. VII-1) should be 12 inches long and 2 ½ inches wide.
2. The polarity of the row of magnets on one side of the guideway **must be opposite** the polarity of the magnets on the other side.

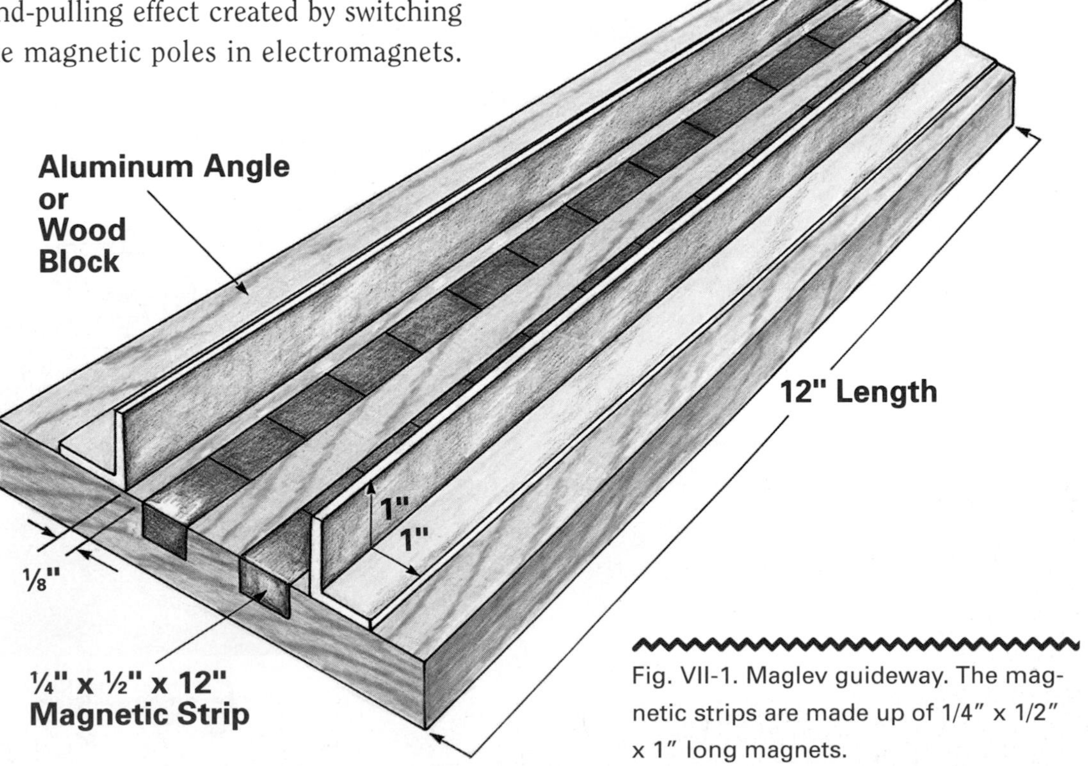

Fig. VII-1. Maglev guideway. The magnetic strips are made up of 1/4″ x 1/2″ x 1″ long magnets.

Research Tip:
Find out about aerodynamic design.

3. The vehicle (Fig. VII-2) should be 2 $\frac{15}{16}$ inches wide and between 4 and 6 inches long.
4. The vehicle shape should reflect good design qualities and apply aerodynamic principles.
5. The vehicle must levitate in the guideway and travel down it when the guideway is inclined.

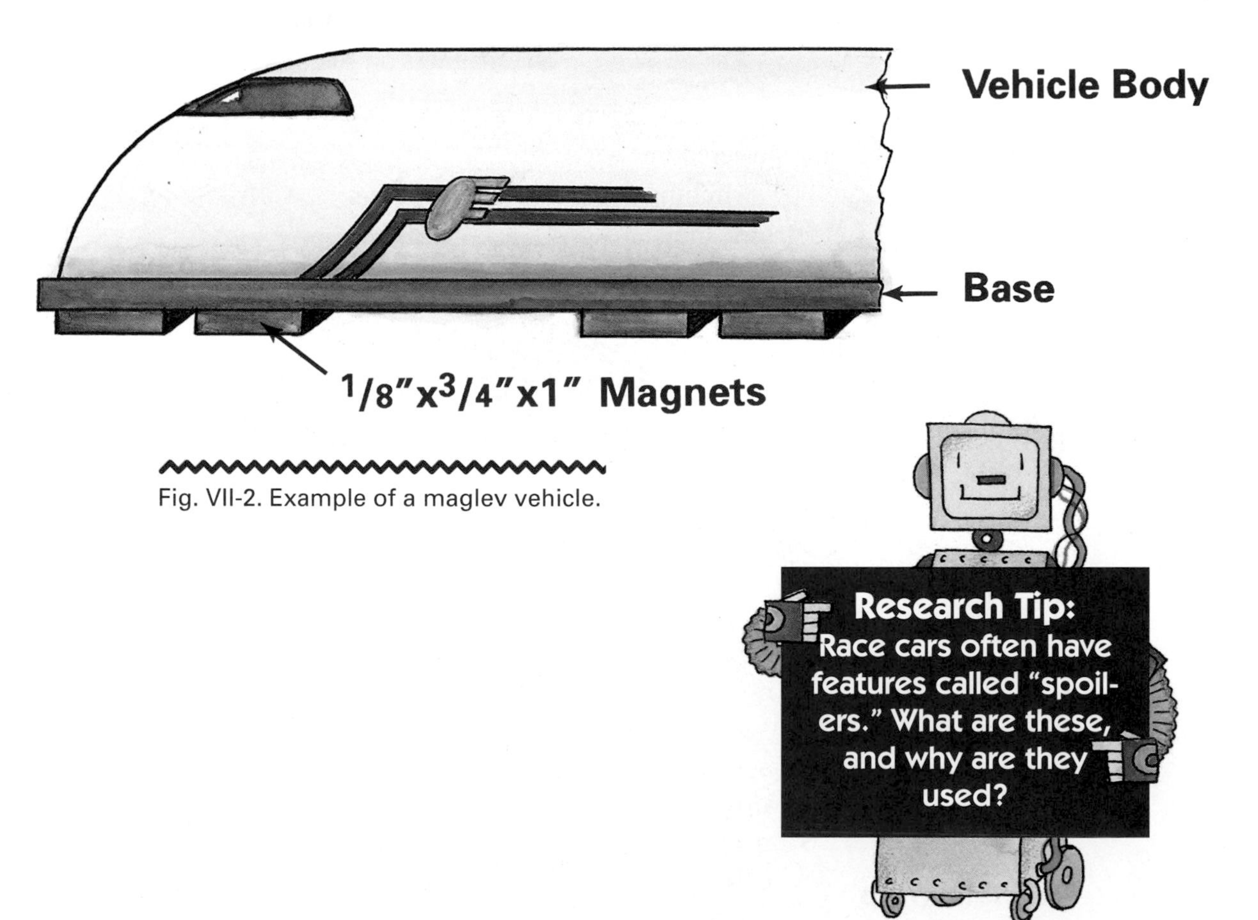
Vehicle Body
Base
1/8"x3/4"x1" Magnets
Research Tip:
Race cars often have features called "spoilers." What are these, and why are they used?

Fig. VII-2. Example of a maglev vehicle.

Guideway

magnets
fasteners
Base:
plywood with dadoes precut
Side Rails:
angle aluminum or wood blocks

Vehicle

magnets
Base:
1/8″ hardboard or plywood
Body:
cardboard
plastic foam
wood
foamboard
paper
Other Materials:
masking tape
double-faced tape
hot glue
markers
paints
adhesives

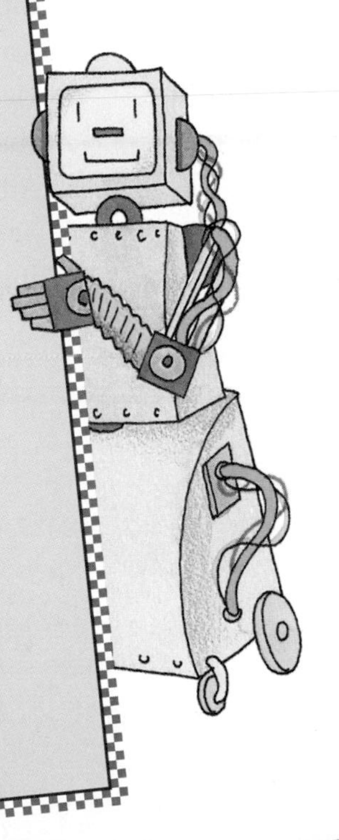

PART 4: Materials...........

Materials that you will need or could use include the list to the left.

Safety Notes

- As you do this activity, remember to follow all the safety guidelines your teacher has explained to you.
- Use all tools properly. Use special care with tools that are sharp.
- Before using any power tools, be sure you understand how to operate them and always get your teacher's permission.

PART 5: Procedure..........

The design of your maglev vehicle and the way in which you build it will be up to you. Still, there are certain steps to follow that will make your work easier.

1. Construct the guideway. Fig. VII-1.
2. Make the base for your vehicle.
3. Attach the magnets to the base. You will need to develop a system for placing the magnets to ensure that the magnetic poles are in the proper position to levitate the vehicle. When the vehicle is in the guideway, the magnets must be directly above the guideway magnetic strip of the same polarity. It's a good idea to tape each magnet in place until they are all positioned. Then glue them onto the base one at a time. (Start with the corners of the base first.)
4. Test the vehicle base to see that it levitates and moves freely down the guideway. Make adjustments if necessary.
5. Design and construct the body to fit the base. Fig. VII-2.
6. Attach the body to the base.
7. Test the vehicle. Fig. VII-3. Make adjustments if necessary.

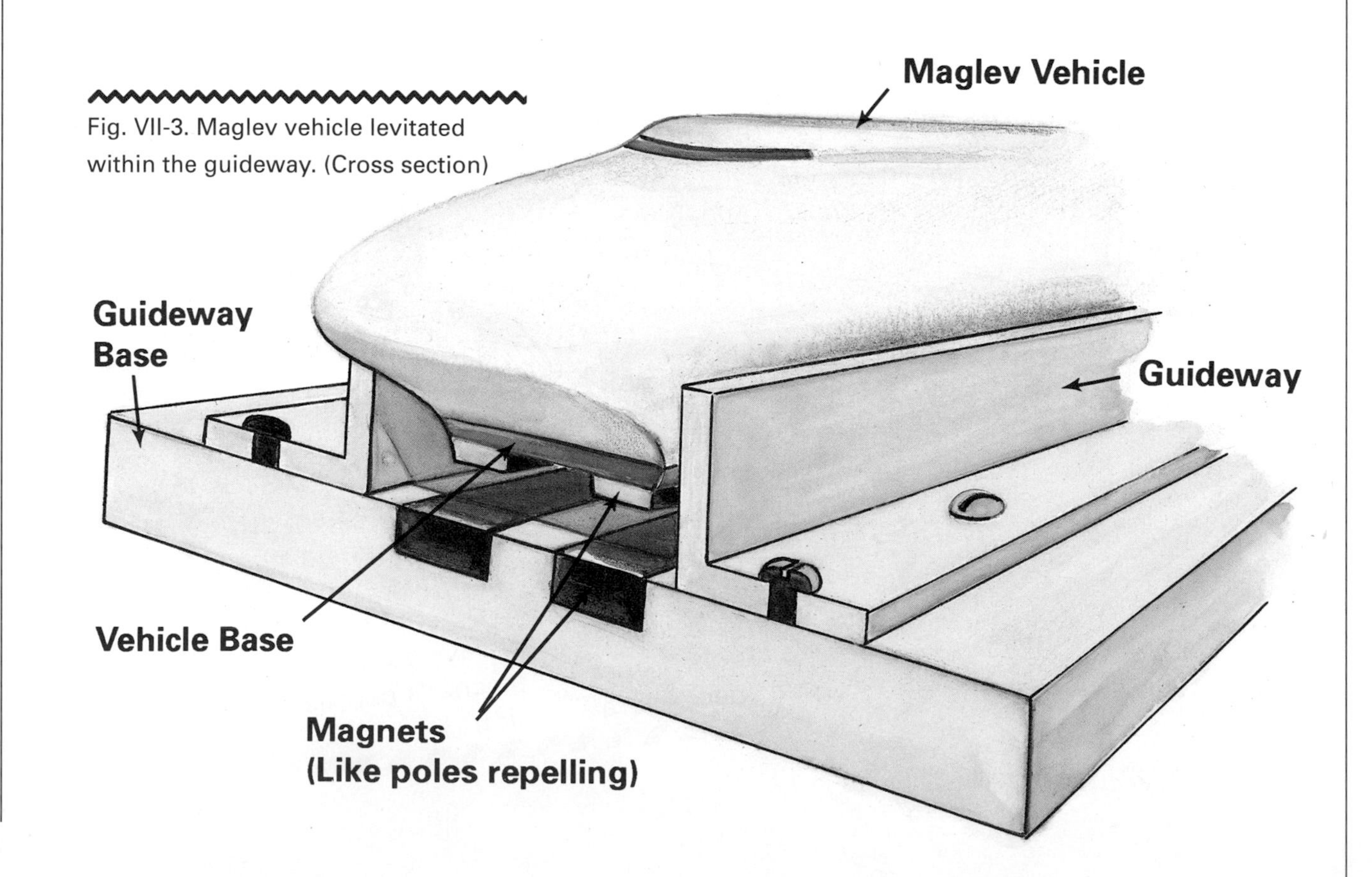

Fig. VII-3. Maglev vehicle levitated within the guideway. (Cross section)

PART 6: For Additional Help..........

For more help with this activity, look up the following terms. You'll find some of them in this book. (Check the index.) You'll find others in science books, dictionaries, encyclopedias, or other resource materials.

aerodynamics
attraction
drag
electromagnets
magnetic field
magnetic poles
magnets
mass transit
polarity
prototype
repulsion

PART 7: How Well Did You Meet the Challenge?..........

Evaluate the maglev vehicle. Does it levitate properly? Is the design streamlined and attractive? Answer the following questions about your vehicle.

1. How might you propel your vehicle across a flat track using wind or electricity as a source of power?
2. How might magnetism be used to propel the vehicle across the track?
3. How could you increase the magnetic force that levitates your vehicle so that the vehicle could carry a 100-gram (3.5 ounce) weight?

PART 8: Extending Your Experience..........

This activity has helped you learn about one of the newest and most promising methods of transportation. Think about the following questions and discuss them in class.

1. What are some of the advantages of maglev transportation technology?
2. What are some of the possible disadvantages of this technology?
3. What is the main reason that it would be expensive to replace the present steel-wheel-on-rail train systems with maglev systems?
4. Do further research. Find out more about the maglev systems in Germany and Japan. Also, find out what is being done with maglev systems in the United States.

CHAPTER 28

Types of Transportation Systems

Introduction

Transportation is one of our oldest and most basic activities. People have always felt the need to travel and to move things from one place to another. What has really changed over the years is not the need or desire to transport people and goods, but rather the technology used to do so.

As with many aspects of daily life, transportation has evolved a great deal in the last hundred years or so. This chapter will describe the development of transportation and the incredible progress people have made in transportation technology in the last 100 years.

In this chapter, transportation has been grouped into land, air, and marine (water) transportation. This is not the only way to categorize transportation, however, and you are encouraged to look at the subject from your own perspective.

After reading this chapter, you should be able to

Trace the development of early land transportation.

Discuss the rise and decline of the rail system.

Discuss the development of air and space transportation.

Describe several types of watercraft.

Explain the purpose of intermodal transportation.

Words you will need

subway **hydrofoil**
monorail **intermodal**

Land Transportation

Early land transportation was slow and crude. People simply walked and carried what they needed. Improvements came about as people began using animals to ride on and to transport goods. The first carrying containers were probably just wooden sleds dragged along the ground.

The invention of the wheel, which allowed sleds to roll with little resistance, revolutionized transportation. From that point on, efforts focused to a great degree on ways to power the rolling containers (and, as you might suspect, on roads upon which to roll them). Wagons, stagecoaches, and even bicycles, were common vehicles for many years.

TECHNOLOGY TRIVIA

The first public railway was opened in 1803 in England. It used horses to pull the cars. Locomotives weren't used for public railways until 1825.

Rail-Based Vehicles

People have used rails as guides or tracks for various containers for hundreds of years. Railroading as we know it, however, was not possible until the invention of the steam engine around 1800. In the United States, thousands of miles of track were laid between 1850 and 1900. Fig. 28–1.

The "golden age" of railroads existed from about 1920 to 1945. During that time, powerful diesel and electric locomotives pulled sleek passenger trains and long strings of freight cars. After World War II, improved highways and the development of commercial airlines provided more desirable alternatives, and the railroads declined. However, rail is still a major form of transportation for products, especially since the creation of special-purpose freight cars. Refrigerated cars, tank cars, and auto transport cars are commonly used in the United States to transport products.

Fig. 28-1. In 1869, the first transcontinental railroad link was completed in Promontory, Utah. This made it possible to travel by rail from coast to coast — from the Atlantic Ocean to the Pacific and back again.

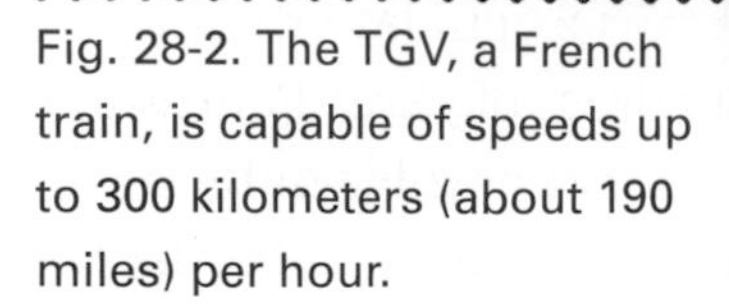

Fig. 28-2. The TGV, a French train, is capable of speeds up to 300 kilometers (about 190 miles) per hour.

Recently, people have shown a renewed interest in rail as a means of mass transit. Special high-speed trains now carry passengers in Japan and Europe. These trains are capable of speeds exceeding 125 miles per hour. Fig. 28–2. In the United States, large cities are now using more rail transportation to relieve crowded roadways.

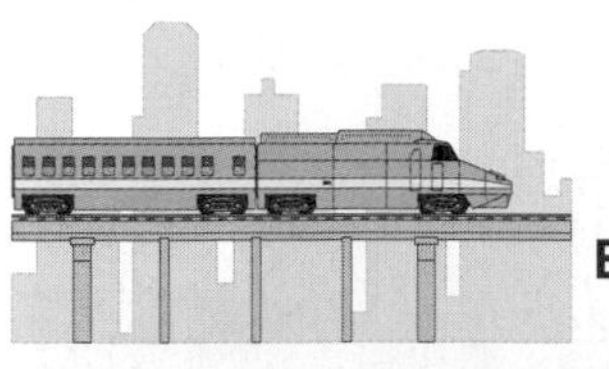

Fig. 28-3. Sometimes railroads are built underground or on elevated tracks that run above the streets. This allows rail transportation even where there is no room for it on the surface.

Railroads designed for cities are frequently built underground because buildings block the right-of-way. These subterranean systems have become known as **subways**, and they are a major form of mass transit in large cities. In some cities, elevated tracks are also used, and it is not unusual for a rail system to combine below-ground, ground level, and elevated tracks as the situation requires. Fig. 28–3.

TECHNOLOGY TRIVIA

The first subway system in the United States opened in Boston in 1898.

Monorail (single-rail) vehicles, elevated or at ground level, are used mostly for sightseeing or amusement park transportation. Other rail-based transportation includes cable cars and trolley cars. These are used mostly inside cities. Their popularity has declined during the last 50 years. However, trolleys, too, are receiving renewed interest from city governments in cities where traffic congestion is a major problem.

Motor Vehicles

The invention of the internal combustion engine sparked one of the largest, longest-lasting manufacturing programs in history. The automobile, with its ease of use and freedom of direction, has become a mainstay of modern life. Fig. 28–4. In the United States, several million cars and trucks are sold each year, and a system of interstate highways makes travel easy. Large buses serve as mass transit for people using the roads, and tractor-trailers carry freight.

Fig. 28-4. The popularity of the "horseless carriage" has never waned. Modern manufacturing techniques allow hundreds of new models to be introduced each year.

People Movers

Several inventions help move people around in special situations. Elevators and escalators allow ease of vertical movement. "Moving sidewalks" allow people to be moved horizontally without having to walk. This has proved particularly useful in airports, where people are often burdened down with luggage and find long walks difficult. Fig. 28–5.

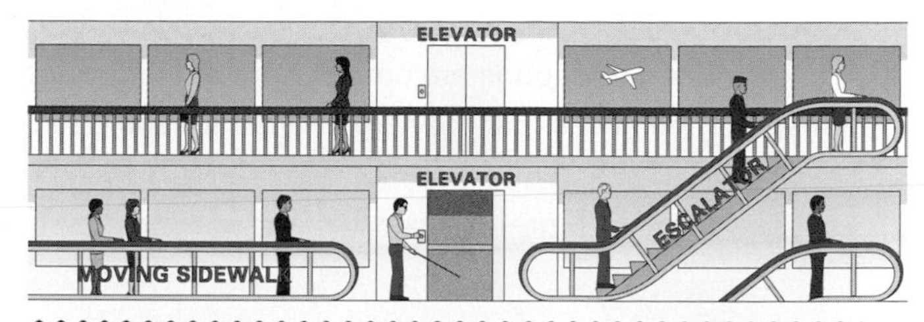

Fig. 28-5. Elevators, escalators, and moving sidewalks help move people short distances, usually within buildings.

Material Handling

Sometimes materials cannot be shipped economically by rail, car, or truck. Inside buildings, and especially for short distances, conveyor belts are used to carry these materials.

When water, oil, and gas need to be transported over long distances, the most economical route is often a pipeline system. Fig. 28–6. Although pipelines are restricted to a single route, they allow 24-hour-a-day operation and are fairly resistant to damage by weather and temperature extremes.

Fig. 28-6. Material handling devices such as this pipeline speed materials automatically to their destinations.

▸▸▸ FOR DISCUSSION ◂◂◂

1. Why did the automobile eventually cause the decline of passenger railroads?
2. What are the advantages and disadvantages of using rail transport such as railroads and trolley cars to commute in cities today?

Fig. 28-7. Hot-air balloons are not very practical for most transportation needs, but they are popular for sport and advertising.

Air and Space Transportation

People dreamed of flying through the air for hundreds of years before flying became a reality. The first successful attempts were made using hot-air balloons. Floating at the mercy of air currents, they were popular, but not very useful. Fig. 28–7.

The famous Wright brothers staged the first successful heavier-than-air flight, and aviation as we know it was born. Today, types of aircraft range from single seaters to huge jumbo jets capable of carrying hundreds of passengers.

IMPACT

Airplanes can quickly take us across several time zones. Our bodies may have trouble adjusting to such abrupt changes. We may feel fatigued, ill, and confused. This condition is called jet lag.

We also have many specialized aircraft. Helicopters, for example, can fly vertically and can hover in one place for long periods. VTOL (vertical take-off and landing) and STOL (short take-off and landing) aircraft combine the characteristics of airplanes and helicopters. Fig. 28–8.

Fig. 28-8. Short and Vertical Take-Off and Landing (STOL, VTOL) craft provide the horizontal speed of a fixed-wing airplane and the agility of a helicopter.

Fig. 28-9. The space shuttle can be flown back from its missions, refitted, and reused. This lowers flight costs and shortens the time between missions.

Space exploration was another long-time human dream. In the late 1950s, the USSR sent up the first orbiting artificial satellite, and a new era was born. In the years that followed, increasingly powerful rockets were developed, and larger payloads were sent into space. As space technology advanced, it became possible for people to fly in spacecraft, and soon it was possible to land on the moon!

Until the development of the space shuttle, each spacecraft could only be used one time. The space shuttle was designed to be flown back to the earth at the end of each mission. This made launches more economical. Fig. 28–9. The current phase of our space program is to explore other planets, starting with our own solar system.

■ **Do research on the history of transportation. Make a time line that shows when significant events in transportation took place.**

▶▶▶ FOR DISCUSSION ◀◀◀

1. Why were hot-air balloons not considered a practical means of transportation?
2. In what way does the space shuttle make space exploration more economical?

Marine Transportation

Traveling across water has long held a fascination for people. From the simplest of rafts to the awesome spectacle of an aircraft carrier (sometimes called a "floating city"), we have developed means to accommodate passengers and freight on inland waterways and on the open seas.

Ocean-going craft, or ships, are designed to be seaworthy in rough weather and over long passages. Most modern ships are designed for non-passenger use: oil tankers, ice-breakers, freighters, and so on. Cruise ships are still used for tours and vacations, but not as often as they were before air transportation became widespread.

Many specialized watercraft have been created. **Hydrofoils** skim through the water at high speeds. The hull of a hydrofoil is lifted out of the water as its speed increases. Hovercraft™ literally float on a bed of air, so they can be designed for use on both land and water. Air is pumped out below them, creating a cushion on which they ride. Fig. 28–10.

Ferries allow large numbers of people (and often vehicles) to cross small bodies of water. Tugboats and towboats help larger craft dock inside harbors, where it is difficult to maneuver. Barges allow slow movement of bulk freight. Submarines allow passage underwater.

Fig. 28-10. Pushing through the water slows most marine vessels. Since hydrofoils and Hovercraft™ can overcome this problem, they are fast in comparison.

▶▶▶ FOR DISCUSSION ◀◀◀

1. Why might people call aircraft carriers "floating cities"?
2. Why is there a need for so many different specialty watercraft?

Intermodal Transportation

Not all journeys of goods and people can take place entirely on one conveyance. Freight, especially, must be shifted between modes of transportation: from ship to train to truck. Several means of containerized shipping have been developed to make it more practical to use more than one type of transportation. Packaging of goods has been standardized in trailer-sized containers, and people have invented ways to transfer these containers from ship to rail to truck. Using this technology, freight can be shipped from one mode of transportation to the other without unpacking and repacking, saving a tremendous amount of time and labor. Fig. 28–11. This concept is described as **intermodal** transportation because goods pass from one mode to another during shipment.

▶▶▶ FOR DISCUSSION ◀◀◀

1. Why does freight, more than people, demand intermodal transportation?
2. Describe a possible path for an intermodal shipment of a product from France to Ohio.

■ Plan the layout for streets, highways, and railroads in your community. Consider in your plans that the river will also be used for transportation. Use a CAD (computer-aided design) system if one is available.

Fig. 28-11. Containerized shipping allows "loose" cargo to be transferred among modes of transportation without repacking. This method also permits bulk loading of the containers.

CHAPTER 28

Chapter Highlights

- Technology has had a major impact on transportation during the last 100 years.
- Steam, internal combustion, and rocket engines provide the power for many of today's transportation systems.
- Some transportation systems move freight, some move people, and some can move both.
- Intermodal transportation allows goods to be transferred efficiently from one mode of transport to another.

Test Your Knowledge

1. Why was the invention of the wheel so important?
2. Name at least two types of rail-based systems and explain how they are different.
3. Why did the railroads decline after the 1940s?
4. Why are some train tracks built above and below ground level?
5. What is meant by the term "moving sidewalk"?
6. What types of goods are shipped via pipeline?
7. What is unique about the design of VTOL aircraft?
8. Name three specialized types of watercraft and explain the use of each.
9. Describe the operation of a hydrofoil.
10. Why is intermodal shipment so economical?

Correlations

SCIENCE

1. Find out more about hydrofoils. How does the shape of a hydrofoil's underwater "wing" compare to the shape of an airplane's wing?

MATH

1. During the rail portion of intermodal transportation, trailers are often stacked two deep on rail cars. How many trailers could a train accommodate if it had 173 rail cars?

LANGUAGE ARTS

1. Consider the major changes in transportation during the last one hundred years. In a creative essay, describe what changes in transportation might occur during the next one hundred years.

SOCIAL STUDIES

1. How has the development of the automobile changed our lives? If the automobile had not been invented, how might this country be different?

CHAPTER 29

Power in Transportation

Introduction

Moving people and goods from place to place has been a human activity for thousands of years. Sometimes a necessity, sometimes merely for pleasure, providing safe and efficient transportation has become an important goal for many people.

How are all these vehicles being powered? For most of our history, all transportation was directly dependent on human and animal power (and sometimes the wind). Only in the last few centuries have invention and technology allowed us to surpass these three basic modes of transport.

In modern American society, people-powered vehicles such as bicycles and skateboards are generally used for fun and exercise. They are used only occasionally for transportation. A major reason for this is that inexpensive, reliable methods of powering vehicles are now available. There is little actual need to resort to human or animal resources.

After reading this chapter, you should be able to

Explain how combustion provides heat and power.

Describe internal and external combustion devices.

Give examples of how electricity may be used to power vehicles.

Discuss the development of lighter-than-air vehicles.

Words you will need

combustion
reciprocal
rotary
external combustion
internal combustion
photovoltaic cells
dirigible

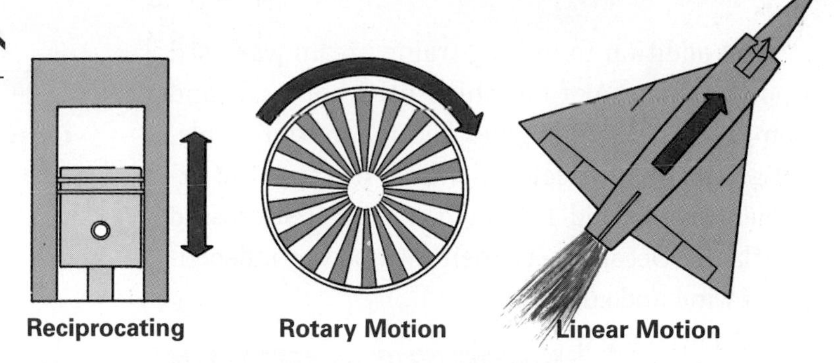

Fig. 29-1. The three basic methods of transmitting engine power are reciprocal, rotary, and straight-through.

Combustion/Heat Engines

Engines, or motors, are the devices used to power most vehicles. But where does the power come from? To a large degree, the power comes from heat. Typically, a substance is burned in a process called **combustion**, and the resulting heat is used to cause movement or motion of some type.

The system used to transmit the motion may be **reciprocal** (back and forth), **rotary** (circular), or linear ("straight-through"). Fig. 29–1. The major differences in the types of engines often hinge upon what is being burned, where it is being burned, and how the heat is being applied to cause motion. At this time, the most widely used design is the reciprocal variety, found in almost every land vehicle, as well as in many air and sea craft.

External Combustion

The earliest combustion engine was the steam engine, first developed in the 1700s. Steam engines are called **external combustion** engines because the fire or burning takes place outside the engine, in a separate firebox/boiler structure. Fig. 29–2. Various materials may be used for fuel, including wood, coal, and oil.

Steam engines revolutionized land travel, allowing a vehicle to propel itself forward for perhaps the first time. Requiring great bulk, steam engines were used primarily for railroads. Steam locomotives were popular for more than 100 years.

IMPACT

Steam-powered railways helped settle the American West. They brought people and supplies to areas that once could be reached only by horse and wagon.

Fig. 29-2. This diagram shows a reciprocating steam engine at work. Water is heated, creating steam. The steam is directed into cylinder A, where it pushes against the piston, moving it to the right. The movement of the piston causes the wheel to make half a turn. As the wheel turns, it causes the slide valve to move to the left, stopping the flow of steam into cylinder A. The steam then flows into cylinder B and forces the piston to move left. What do you think will happen next?

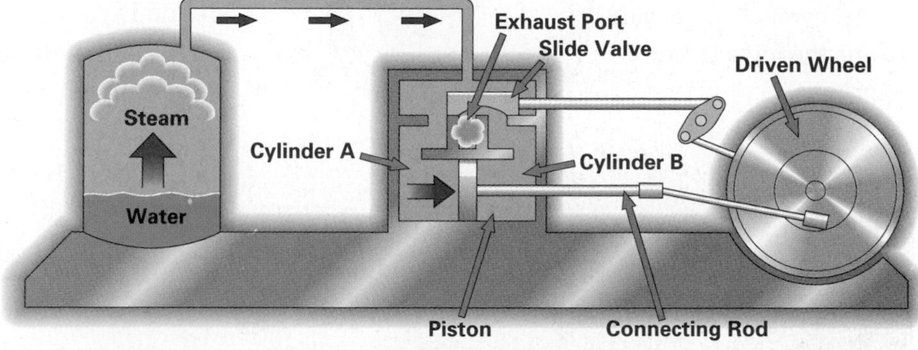

In addition to pulling trains, steam was used to power ships. At first, ships used both sail and steam, but as steam engines became more reliable, they gradually replaced wind sails. Imagine the excitement in the mid-1800s when ships first crossed the Atlantic Ocean under their own power, independent of wind and current.

Although steam engines grew very popular and were widely used for many years, they had several drawbacks. They required a great deal of maintenance, and the high-pressured steam proved to be very dangerous when an accident or defect occurred. These problems led to the decline of the steam engine as other types of engines became available.

TECHNOLOGY TRIVIA

In 1838 the Great Western, the first steamship built specifically for transatlantic service, sailed from Bristol, England, to New York in just 15 days.

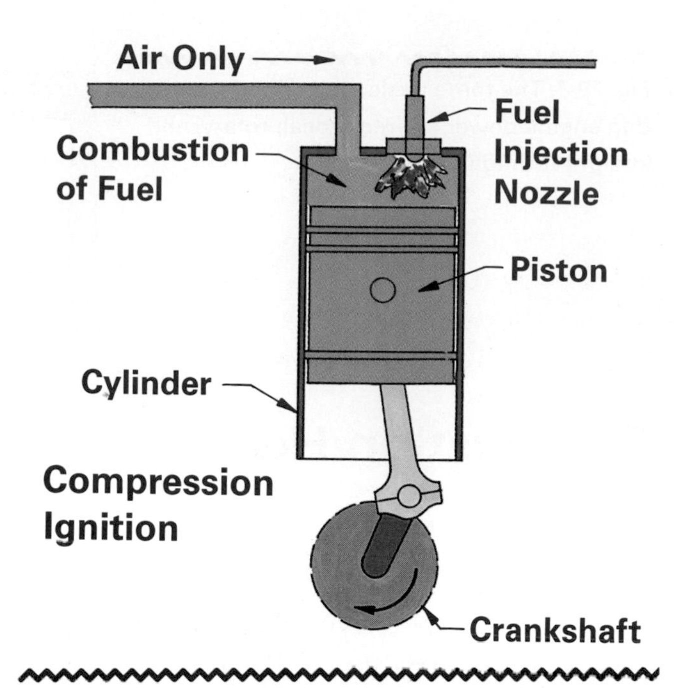

Fig. 29-4. Extremely high pressure causes the fuel-air mixture to ignite spontaneously in a diesel engine. Diesels do not have spark plugs because they do not need ignition sparks.

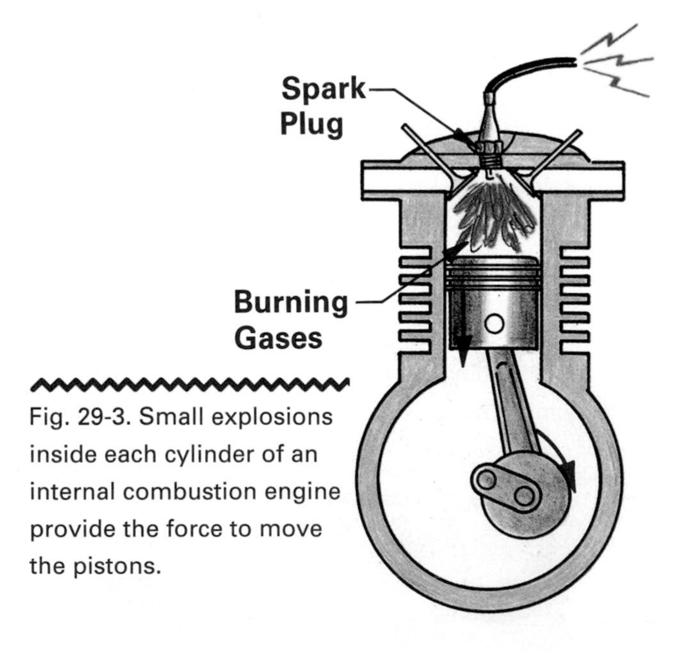

Fig. 29-3. Small explosions inside each cylinder of an internal combustion engine provide the force to move the pistons.

Internal Combustion

The second category of heat engine is the **internal combustion** type. In this design, the fuel is burned inside the engine itself. Invented in the late 1800s and refined ever since, the most common model burns a gasoline-air fuel mixture. The fuel mixture is put under pressure and ignited. The resulting combustion causes a great expanding force inside the engine. This force is used to create motion. Fig. 29–3.

A similar model of the internal combustion engine is called a *diesel* engine. Using a fuel much like gasoline, diesels put the fuel under great pressure, causing it to ignite (without using a spark plug). Fig. 29–4. Because diesels work well under conditions of long operation, diesel engines are very common in trucks, machinery, and boats.

The straight-through type of internal combustion engine is used primarily in the design of rockets and jets. The major distinction between them is that rockets carry their own fuel and oxygen with them. Thus rockets may be used for space exploration even though there is no air in space. Jets carry their own fuel, but they use the air through which they fly to get oxygen. Fig. 29–5.

▶▶▶ FOR DISCUSSION ◀◀◀

1. Why are internal and external combustion engines sometimes described as "heat engines"?
2. What might be a safety concern around steam engines?

Extension Activity

■ **Research is being done to find a clean-burning fuel to replace gasoline. Find a newspaper or magazine article that discusses one or more alternative fuels and share the information with your class.**

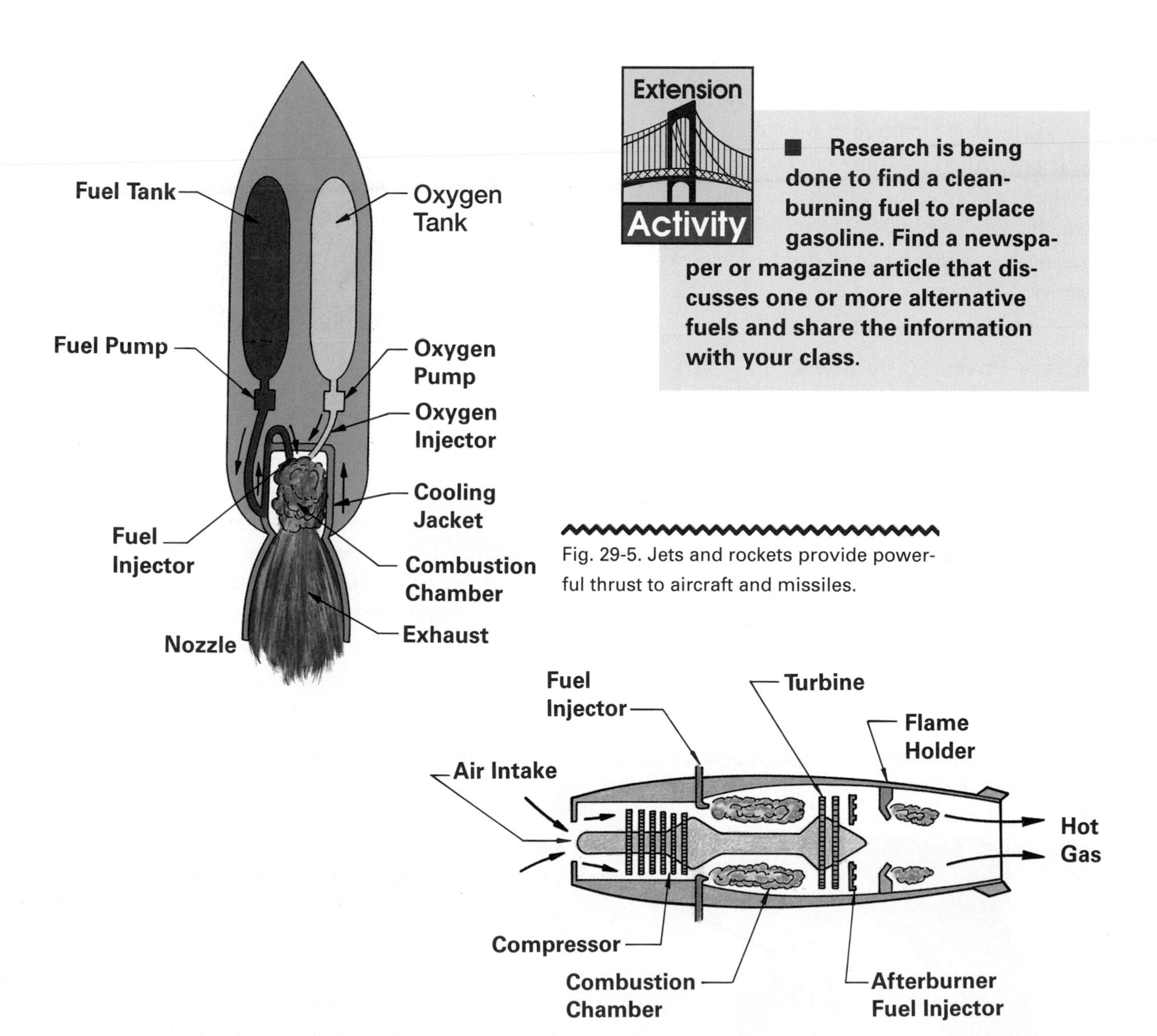

Fig. 29-5. Jets and rockets provide powerful thrust to aircraft and missiles.

Other Types of Power Sources

Several types of power sources are available in addition to internal and external combustion engines. The selection of an engine depends to a large degree on the purpose to which it will be put. Each type has its own characteristics.

Electricity

One major method of providing power is by electricity. When we think of electrical energy, we usually think of the many tools and appliances in the home that are powered by electricity. However, vehicles and large machines can also be powered in this manner.

Battery-Powered Vehicles

Moving vehicles by means of electric motors has actually been tried for more than 100 years. Since extension cords obviously wouldn't work, efforts have focused on battery power. As we enter the 1990s, research continues on developing batteries that will prove more suitable. Electric motors are quiet and non-polluting. For these reasons, interest in developing them for use in automobiles remains high. Fig. 29–6.

Fig. 29-6. Electric cars must be recharged, and this can take several hours.

Fig. 29-7. For vehicles that travel along fixed routes, electricity can be supplied along the route.

Rail and Cable Vehicles

Electricity can also be delivered to vehicles that travel on fixed routes. Trolley cars, subways, and many railroads get their power from electric lines above or below ground or along the path of travel. The electricity is used to run electric motors on the vehicle. Fig. 29–7.

■ **Construct a commuter train system for your community that includes a stop near MegaIndustries.**

Solar-Powered Vehicles

Power from the sun? Sure. "Photocells" (or **photovoltaic cells**) capture the energy from light and convert it into electricity. You may have seen this on calculators or other devices. Fig. 29–8. When used for vehicles, photocells are often combined with storage batteries to permit operation when lighting is dim. Still under development, solar-powered vehicles hold promise for the future. Fig. 29–9.

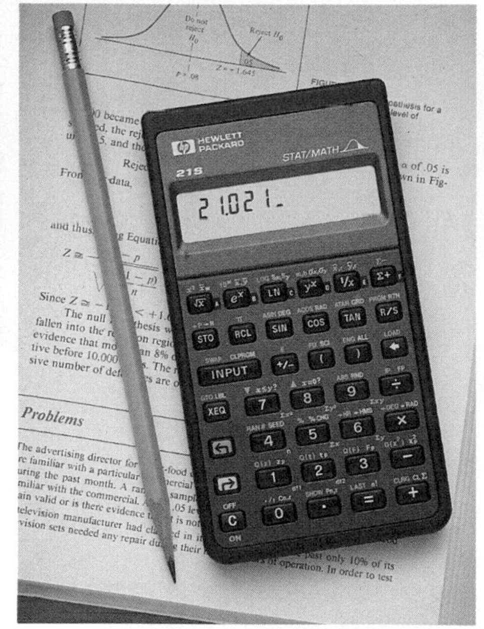

Fig. 29-8. Photocells are perfect for devices such as calculators, which do not draw much current. Not having to replace batteries is a real convenience to consumers.

Nuclear-Powered Vehicles

Getting power from splitting atoms has gone from a dream to a reality. We have also, however, learned of the potential for danger. Nevertheless, nuclear-powered ships have been operated successfully for many years. In operation, the great heat released by the atomic reaction is used to provide motion or movement of whatever mechanism is being driven. Fig. 29–10.

TECHNOLOGY TRIVIA

The first nuclear-powered submarine, the Nautilus, was built by the U. S. Navy in 1955. The first nuclear-powered surface ship, the Soviet icebreaker Lenin, was commissioned in 1959.

Fig. 29-9. Solar-powered cars are being developed. Range, speed, and battery maintenance are still problems to be solved.

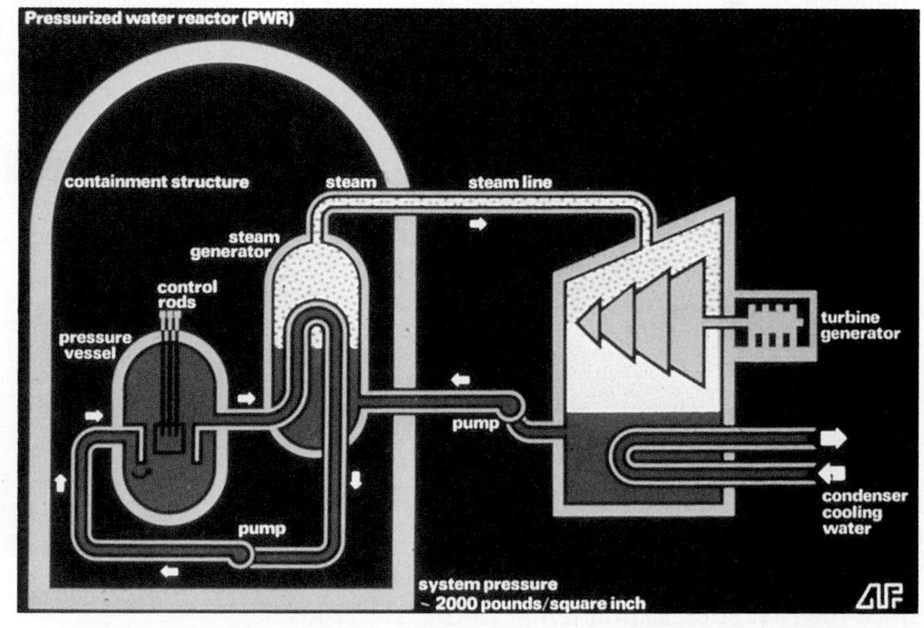

Fig. 29-10. Creating great power in a small area with a small quantity of radioactive fuel, nuclear reactors are very suitable for ships at sea. This diagram shows how a typical water reactor works.

Fig. 29-11. Using the wind to provide power for movement is a very old activity. Wind power is still used today—a welcome source of free, non-polluting energy. The boat in the foreground uses hollow wind cylinders instead of sails. (The Cousteau Society)

Wind-Powered Vehicles

Sail-driven vehicles, primarily boats and ships, are still used today. In addition, newer designs for sails have been developed. Popular as a quiet, non-polluting source of power, wind-driven sailboats continue to be used for fun and recreation, but are less used for commerce and transportation. Fig. 29–11.

■ **Design and build a model wind-powered land vehicle.**

Lighter-Than-Air Vehicles

Ballooning, or floating through the air while suspended below a balloon, has fascinated people since the first successful attempt in the late 1700s. Lifted by heated air or by a lighter-than-air gas such as helium or hydrogen, these craft provided a thrilling view of the earth below when there was no other way to see it. They are still used for sport and sight-seeing.

Dirigibles were large structures with rigid frames that used lighter-than-air principles in the first half of this century. Unfortunately, some depended on hydrogen for their lifting power. Many of these dirigibles came to a tragic end as accidents caused the flammable gas to burn out of control.

Today we use large-scale floating structures called *blimps*. They do not have rigid frames, and they use nonflammable helium gas for lift. Blimps are primarily used for advertising and as floating platforms for television cameras. Fig. 29–12.

►►► FOR DISCUSSION ◄◄◄

1. Why would environmentalists favor electric cars over gasoline-powered models?
2. What was the problem with using hydrogen in lighter-than-air craft?

Fig. 29-12. Blimps today are used as platforms from which to televise sports specials and as floating billboards.

Chapter Highlights

- Modern transportation relies on various types of power.
- Heat engines are powered by either internal or external combustion.
- The first self-propelled vehicles used steam engines for power.
- Electric motors provide quiet, non-polluting power.

Test Your Knowledge

1. What powered transport before the steam engine?
2. Why are steam engines considered to be external combustion engines?
3. Name the two most common fuels used in internal combustion engines.
4. How do diesel engines differ from ordinary gasoline engines?
5. Why must rockets used in space carry their own source of oxygen?
6. How are batteries used in electric cars?
7. Why don't electrically powered vehicles that follow fixed routes need batteries?
8. What is the function of photocells?
9. Where does the power come from in nuclear-powered sea craft?
10. Why do hot air balloons float up?

Correlations

SCIENCE

1. It costs money to convert energy into power. To demonstrate this, devise an experiment to compare the cost of a battery to the time the battery operates a mechanical toy. Do different brands of batteries give different cost to time-of-use ratios?

MATH

1. A car's fuel economy is measured in miles per gallon; that is, the number of miles the car can be driven on one gallon of gasoline. Suppose the family car can be driven 396.9 miles on one tank of gasoline. If the tank holds 18.9 gallons, how many miles did the car go on each gallon of fuel?

LANGUAGE ARTS

1. The source of power in transportation has changed drastically through the years. In a creative essay, describe the power source possibilities for transportation of the future.

SOCIAL STUDIES

1. Find out how the invention of the internal combustion engine made a change in the oil industry. Prior to the invention of this heat engine, what were the primary uses of oil in the United States?

CHAPTER 30

Impacts of Transportation Systems

Introduction

Our transportation systems affect virtually everyone on a daily basis, even though some people may not realize it. As a result, many people take today's transportation systems for granted. They tend not to appreciate its overall influence. For example, how did you get to school today? If you live far from the school, you probably took a school bus or some other mass transit carrier.

Do you know any adults who walked to work this week? It is not common for people to walk to their jobs for many reasons. The distance may be too great, weather conditions may be bad, or the people may need to carry materials to and from the workplace, for example. These workers must utilize some type of transportation system—land, air, or sea.

In this chapter, we will explore the tremendous impact that transportation has on our lives and surroundings.

After reading this chapter, you should be able to

Discuss the impact transportation has had in moving people, delivering goods and services, and providing employment.

Describe how transportation systems have contributed to our being known as a mobile society.

Give examples of effects that transportation has had on the environment.

Words you will need

mobile society
contaminants
smog
acid rain

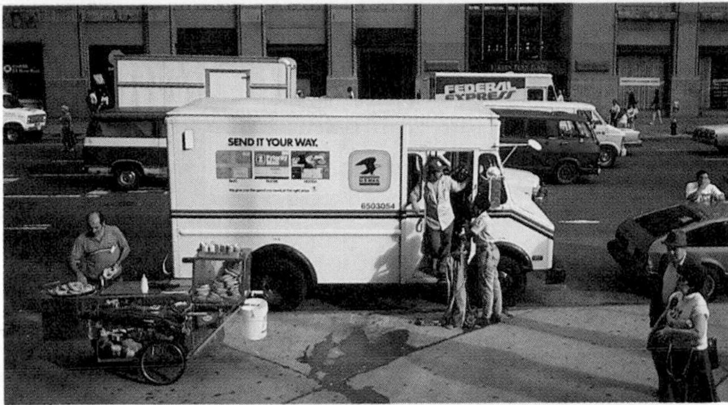

Fig. 30-1. Transportation has a real impact on people in at least three ways: transporting people, delivering goods and services, and providing employment.

Impacts on People

Generally speaking, the impacts of transportation upon people can be divided into three major categories. These three categories are: transporting people, transporting goods and services, and providing employment. Fig. 30–1.

Transporting People

In today's world, time is regarded as a valuable resource—much too valuable to squander away. It is not surprising, then, that for many people, getting someplace fast is an important goal. Whether they travel by car, train, bus, or plane, people tend to be on a schedule with little time to spare.

The importance of an adequate transportation system cannot be emphasized strongly enough. To see how significant it is, one only has to look at the chaos that results when a major transportation artery, such as a metropolitan airport, is temporarily closed down for some reason. Fig. 30–2.

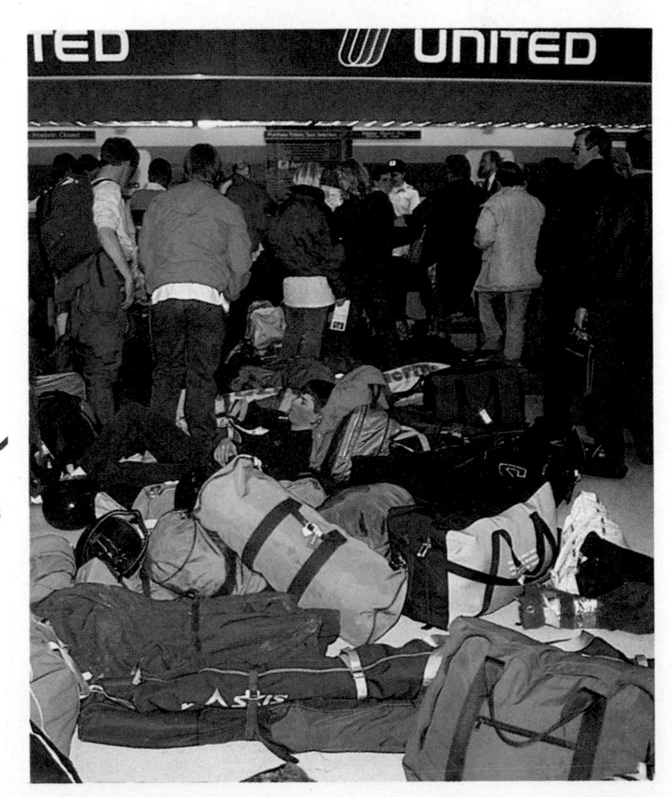

Fig. 30-2. The degree to which we depend on transportation is obvious in airport terminals and other mass transit depots.

In a similar vein, think of all the possible uses you or your family might have for transportation: employment, education, health, recreation, shopping, emergency assistance, helping a friend, going on errands, visiting, vacations—the list goes on and on. Fig. 30-3. Can you imagine the impact on your life if all of the modes of transportation you use for these purposes were restricted?

TECHNOLOGY TRIVIA

The federal government devised the system of assigning numbers to interstate highways. Even numbers are assigned when the overall direction is east and west. Odd numbers are assigned to north-and-south highways.

Fig. 30-3. Transportation systems even affect our health by providing a quick means of transporting people to health-care facilities when they are sick or injured.

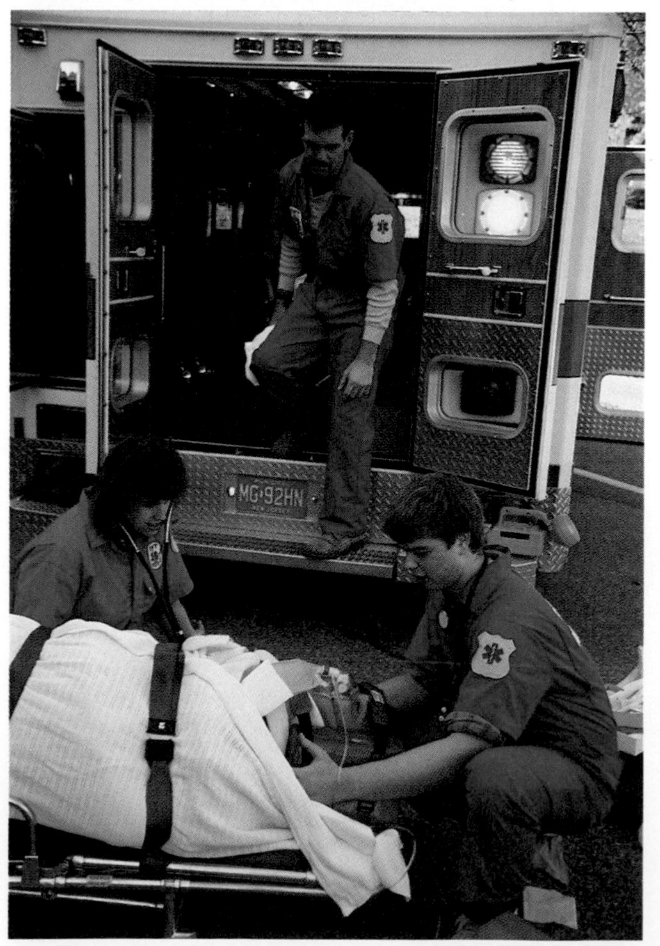

Transporting Goods and Services

Where do you go when you want to purchase something? To the nearest store, perhaps, or to your favorite shopping mall? How do you think all the products got to the stores originally? That's right, they were shipped there by some type of transportation system. Fig. 30-4. One or more means of transportation had to be used to get the products to the stores. The materials to build the stores themselves (wood, steel, concrete, and so on) had to be brought to the site from somewhere. From raw materials to finished products, from the smallest items to railroad cars themselves—all of these things had to be transported for manufacture, assembly, and sale.

The same situation exists for those who provide (and receive) services. Think for a moment about all the people and services that depend on transportation systems. Taxicabs, delivery vans, takeout food establishments, medical personnel, fire protection, business meetings, repair companies—once again the list is too long to include each possible element. It is plain to see that the impact of transportation on the delivery of goods and services is very big indeed.

Fig. 30-4. Transportation is the key to many of our delivery systems for goods and services. This truck will deliver its load of oranges to a food broker.

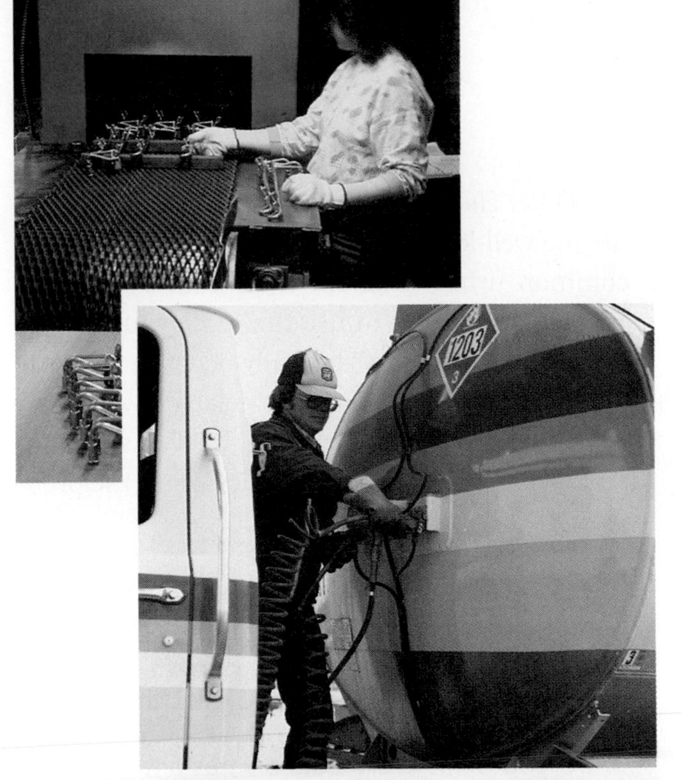

Fig. 30-5. Many jobs owe their existence to transportation systems and technology.

Providing Employment

The number of people employed within the transportation industry is quite large. Vehicles (land, sea, and air) are designed, manufactured, delivered, sold, used, maintained, and finally discarded at the end of their useful life. People are employed at every stage of this process. Engineers, factory workers, drivers, salespeople, mechanics, and junkyard operators are just some of the people involved. Other people have jobs that are related in some way to transportation systems. These people may work in materials processing, safety regulation, or in many other jobs related indirectly to transportation. Fig. 30–5.

■ Collect at least three pictures of people employed in different jobs within the transportation industry. Identify each type of occupation and write a brief description of the work tasks and responsibilities involved in that type of job. Work with other members of your class to make a scrapbook of transportation occupations.

▶▶▶ FOR DISCUSSION ◀◀◀

1. Why might time be such an important element to a traveler today?
2. Select a common store-bought product and try to determine how it was shipped there.

■ Identify and make a list of the impacts of the transportation systems in the model community.

Impacts on Society

As much as transportation systems affect people as individuals, they have also influenced society as a whole. A hundred and fifty years ago, our society was based on agriculture. Farming was a labor-intensive effort—it required many people. It was typical for people to both live and work on a farm for generations. Then came the Industrial Revolution, and with it came a transportation revolution. You may recall from earlier chapters that the Industrial Revolution was that period of time in which factories began to use mass production to produce goods at a faster rate than was previously possible.

No longer did people have to live where they worked. Farms, too, became more mechanized, needing less and less human labor. The result was a mass migration to the cities, to new jobs and new housing. As time went on, the cities became crowded, and people started moving from the cities to *suburbs* (less crowded areas surrounding cities). Transportation was what made all of that possible. Fig. 30–6.

Today we have what is sometimes called a **mobile society**. That means that people travel a lot, change jobs more frequently, and move to new homes more often. It is not unusual today for every member of a family to travel to a different place in a different vehicle on a daily basis. You may know someone who works in a different county, or even a different state, than the one in which he or she lives. People can do this because of modern transportation systems.

Other effects of our transportation systems include the well-known rush-hour traffic jams. This is a common urban problem that results from too many vehicles trying to use the same route at the same time. "Rush hours" typically occur in the early morning and late afternoon on weekdays, when people drive to and from work. Fig. 30–7. As of now, there seems to be no one best solution to the problem.

▶▶▶ FOR DISCUSSION ◀◀◀

1. Why do you think so many people stayed at the same jobs their ancestors held prior to the Industrial Revolution?
2. Do you think every country has a mobile society? Why or why not?

Fig. 30-6. An efficient transportation network has allowed city workers to reside in suburbs and commute to work in nearby cities each day.

Fig. 30-7. Each morning and evening, many cities experience rush-hour traffic jams as workers struggle to arrive at and leave their jobs in the city.

Impact on the Environment

As responsible citizens, we must consider the effect that transportation has had and continues to have on the environment. We are now sensitive to the many negative impacts of transportation on the environment, and we are trying to address the problems.

Two major environmental problems have surfaced as a result of transportation. One is the tremendous toll taken on our natural resources by people and companies who mine iron ore for steel, by those who provide fuel for steel-making plants, and by the factories themselves. Fig. 30-8. The other major concern is air pollution. Millions of vehicles spew out their poisonous exhaust, as do many of our manufacturing plants. Unfortunately, these things were not known to be problems until the last 20 years or so.

TECHNOLOGY TRIVIA

Though pollution problems have been around since ancient times, environmental pollution has become more serious and widespread since World War II. The Environmental Protection Agency (EPA) was established in 1970 to enforce pollution standards and to protect the nation's environment.

Fig. 30-8. Mining for the raw materials from which steel is made has had a major impact on the environment. Much of the resulting steel is used in transportation.

The government is trying to help solve these problems by requiring permits for certain types of mining and by regulating the permissible **contaminants** (pollutants) in tailpipe and chimney smoke. In this manner, we can minimize effects such as smog and acid rain. **Smog** is a fog-like condition worsened by airborne pollutants. **Acid rain** is precipitation which combines with sulfur dioxide, a form of air pollution. The resulting rain is acidic enough to damage plants, animals, and even nonliving things such as statues. Fig. 30–9.

Transportation systems have also caused a great change in our landscape. Thousands of miles of roads, canals, trackage, and so on have been constructed to provide access for different vehicles. Bridges, tunnels, airports, train stations, harbors—all of these and more—have been built to enhance transportation. The cumulative effect of all this construction is difficult to gauge, mostly because of its sheer size. There probably is not a town or city anywhere that hasn't been directly concerned with transportation technology in one form or another.

▶▶▶ FOR DISCUSSION ◀◀◀

1. The steel industry has been central to most transportation systems. Why do you think this is so?
2. Is pollution always associated with transportation? Explain your answer.

Fig. 30-9. Acid rain has damaged this statue. It can also harm plants and animals.

CHAPTER 30 Review

Chapter Highlights

- Almost everyone is affected by a transportation system—directly or indirectly.
- The ability to provide goods and services is based upon an efficient transportation system.
- Large numbers of people are employed in transportation-related jobs.
- Our society as a whole has been strongly influenced by transportation systems.
- Transportation has a direct and profound effect on the environment.

Test Your Knowledge

1. What are some reasons why most people don't walk to their places of employment?
2. Name the three major categories in which transportation affects people.
3. Why are so many people inconvenienced when a major traffic hub (such as a train terminal or airport) breaks down?
4. Why are many products shipped by a combination of methods (intermodal shipping)?
5. Why are so many people employed in the transportation industry?
6. Describe how the Industrial Revolution influenced the transportation revolution.
7. Why do you think people continue to travel on routes they know are prone to rush hour traffic jams?
8. Why weren't people worried about air pollution from automobiles 50 years ago?
9. Did methods of transportation before the invention of automobiles (such as horses) cause pollution? Explain.
10. Explain why almost everyone is affected in some measure by a transportation system.

Correlations

SCIENCE

1. What are the ingredients of smog? (Look in a science book or encyclopedia.) Write a one-page report explaining how smog is formed and why it is unhealthy.

MATH

1. Survey the faculty and staff at your school to find out how far each person travels to school each day. What is the average distance traveled?

LANGUAGE ARTS

1. Rush-hour traffic is a problem in most major cities. In a paragraph, describe a possible solution to this growing problem.

SOCIAL STUDIES

1. The invention and mass production of the automobile not only transformed transportation but also helped to create a mobile society. Conduct a survey within your school to find out how often the students in your school have moved. When you are finished with your survey, compute an average for the school.

CHAPTER 31

Transportation in the Future

Introduction

Personal air-jet pacs and time machines—is this the future of transportation? We really don't know for sure. Although transportation has been revolutionized during the last hundred years, the next few years seem to promise somewhat less excitement. Today's technology is focused on improving present means of transportation, not at changing it entirely.

A popular exhibit at the 1939–40 World's Fair was the *Futurama*, a preview of the "transportation of tomorrow." If this prediction had been right, we would all be zipping along on 100-mile-an-hour superhighways. Huge skyscrapers would be surrounded by a cloverleaf of highways, allowing quick access in and out. What evolved instead were cities so traffic-clogged that driving actually became a drawback in many large cities.

Will our future transportation systems be more like the *Futurama* exhibit or more like a traffic jam? This chapter will examine what is in the offing and give you a peek into the future.

After reading this chapter, you should be able to

Describe fixed and random transportation.

List four major concerns which have developed about the automobile and tell how each will be addressed in the future.

Trace the development of rail transportation from the past into the future.

Discuss the future of air transportation.

Discuss the future of marine transportation.

Words you will need

fixed transportation
random transportation
maglev trains

Fig. 31-1. Land transportation may be divided into two categories: fixed and random. The train is an example of fixed transportation; automobiles are random.

Land Transportation

Land transportation is sometimes divided into the two categories: fixed and random. **Fixed transportation** includes those vehicles that follow a fixed route, such as buses and trains. **Random transportation** refers to those vehicles that allow virtually unlimited travel with respect to direction—cars, motorcycles, bicycles, and so on. Fig. 31–1.

Until the mid-1900s, fixed transportation was the major moving force in this country. Since that time, however, random transportation has become the dominant means of travel. Of all the types of random transportation, the most popular by far has been the automobile. Although the automobile has many advantages, it has created some real problems which must be addressed in the future.

Random Transportation—The Automobile

Over the last 20 or 30 years, four major issues have arisen concerning automobiles: fuel efficiency, exhaust emissions, passenger safety, and traffic congestion. Because there are so many cars in this country, these problems have become so big that they concern us all.

In spite of these problems, the car has proved to be the most popular personal transportation product in this century. Along with air travel, the car has grown to dominate passenger travel in the United States.

Fuel Efficiency

Petroleum (from which gasoline is refined) is an expensive and increasingly limited resource. As a result, there has been a strong motivation to create more fuel-efficient engines. Both by government regulation and by consumer demand, car manufacturers have attempted to design cars which get more miles per gallon. This has been accomplished by a variety of means, including:

- using lighter materials to lower vehicle weight
- streamlining the car's shape to minimize wind resistance
- feeding the fuel into the engine via computer-controlled injectors

These efforts are on-going, and fuel efficiency is likely to continue to improve as long as people push for improvements. Fig. 31–2.

Fig. 31-2. Smaller, more efficient cars are one result of the search for answers to problems created by widespread use of automobiles.

The electric car of the future is still being pursued. Ideally, this car would be capable of rapid acceleration, moderately long trips, and quick charge-ups. For now, however, it seems the internal combustion gasoline engine is what we will be using.

There is at present a strong effort to develop electronic navigation systems for automobiles. Fig. 31–3. Its designers explain that it would conserve fuel, among other things, by minimizing the number of times drivers lose their way and roam around looking for a landmark. If such a system could be made interactive, drivers could request information on the shortest route, traffic delays, and so on.

Fig. 31-3. Electronic in-car navigation is now a reality, soon to be an option available to new-car buyers.

■ In the future, we will be exploring other planets more and more. Design and construct a model of a vehicle that could travel on the rough terrain of another planet.

Exhaust Emissions

After many years of simply not realizing that a problem was building, people gradually became aware that auto exhaust emissions were polluting the atmosphere. As more and more cars were manufactured and driven, the problem became worse and worse. Some places with high car concentrations, such as Los Angeles, have "smog alerts" to warn people who have certain medical conditions to stay indoors.

To combat this pollution, individual states (most noticeably California) and the federal government passed laws requiring auto manufacturers to meet certain emission standards. Exhaust emissions were also made part of vehicle inspections in many states, so that even older cars had to burn fuel as cleanly as possible. Since this problem continues to concern people, it will continue to be addressed in the future.

Passenger Safety

Have cars become any safer in the last 30 years? Most experts say yes. There are still a large number of injuries and fatalities, but that is primarily due to the larger number of drivers, not the cars themselves. Technology has already made numerous improvements: safety glass; collapsible steering columns; center-mounted, eye-level brake lights; and seat and shoulder belts, to mention a few. Fig. 31–4.

Many other safety improvements are now under consideration. These improvements include air bags for all passengers, automatic collision avoidance braking, electronic sensors to warn of dangerous situations, and generally more reliable cars.

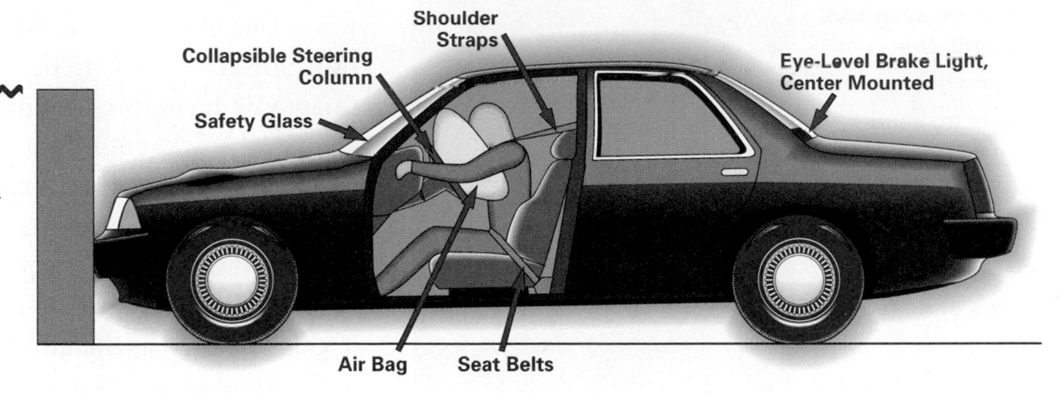

Fig. 31-4. Several safety-related improvements have been made to automobiles in recent years.

Traffic Congestion

As the use of automobiles has increased, we seem to have more and more traffic jams. In the past, the only attempt to alter this situation was aimed at building more roads. This solution has not proved to be very successful. Every day most of our cities are subjected to traffic jams that barely ease before the next jam begins building.

TECHNOLOGY TRIVIA

The world's first traffic lights were installed in London, England, in 1868. The red and green gas lights, used at night for the convenience of members of Parliament, were set in a revolving lantern perched atop a cast iron pillar some 23 feet high. A manually operated lever changed the lights and extended and lowered the signal arms.

Solving this problem will take more than simply constructing additional lanes for travel. One suggestion has been to charge higher tolls for people driving alone, or perhaps charge no fee at all for those vehicles carrying three or more passengers. A similar idea is to restrict certain lanes to cars carrying two or more people.

Other transportation experts think the solution lies in making mass transit (car/van pools, trains, and buses) more desirable so that people won't want to take their private cars. Some cities have established stop-and-ride parking lots, where car drivers can park and continue their journey via mass transit. Fig. 31–5.

Fig. 31-5. Park-and-ride parking lots encourage people to leave their personal vehicles outside of congested areas, reducing traffic jams and pollution in metropolitan areas.

Fig. 31-6. One method being used to help manage traffic flow is communication. Electronic message boards inform drivers of road conditions ahead. Some can even be programmed to warn drivers of changing rush-hour traffic conditions.

Still another plan involves using electronic two-way communication to direct drivers toward alternate routes to avoid traffic tie-ups, construction delays, and so on. At this time, some areas already have electronic overhead signs which state brief messages that can be read by people in the cars below. Fig. 31–6. It is within the capability of today's technology to expand this concept to display units inside individual cars.

Fixed Transportation

Rail transportation has declined in this country to the point that it now represents only a few percent of our transportation miles. In many areas, however, people are reconsidering the concept if not the actual trains. One idea is to have smaller, more personal railed vehicles that a few people at a time would use. These personal train cars, perhaps suspended from a monorail, could be programmed by the occupants to stop only where desired so that trips would be as quick as possible. Fig. 31–7. The fee structure would also be electronic, allowing convenient monthly billing. Such personal rapid transit systems could be electrified, by rail or battery, to minimize urban air pollution, which is now a serious problem.

One of the most widely discussed ideas for rail transportation lies in the technology of **maglev** (magnetic levitation) **trains**. By floating the train above the track, engineers have almost eliminated friction and vibration. As a result, speed can be vastly increased, and the trains themselves are more durable. Fig. 31–8. High-speed, comfortable maglev trains might recapture the public that once thought of going by train as the ultimate travel experience.

Another future possibility is that cars may be automatically guided along a chosen path by electronic means. These paths could be established along existing rights-of-way, either highway or rail. If such "smart cars" could be electronically controlled along certain routes, many urban accidents could be prevented. Congestion could also be controlled more effectively, and more efficient use could be made of existing roadways.

▶▶▶ FOR DISCUSSION ◀◀◀

1. Why weren't people concerned about car exhaust emissions many years ago?
2. Do you think cities always had some type of traffic congestion? Explain your answer.

Fig. 31-7. Can railroads be downsized and modified into the "personal rapid transit system" of the future?

Fig. 31-8. Maglev trains may be able to revitalize the mass transit rail industry. The cutaway portion of this model shows the train's interior.

Air Transportation

The future of air transportation is in many ways similar to that of the automobile. Engines spewing out poisonous fumes, noisy take-offs and landings, congestion getting into and out of the airports (both on the ground and in the air), and other problems will be addressed. It seems that the problems which have arisen will attract the most attention, and radical changes or improvements will probably not appear in the near future. The next great frontier—space—seems a long way off for ordinary passengers.

There is still talk of faster versions of supersonic aircraft, such as the *Concorde*. At the moment, however, the large investment of time, technology, and money needed to create these aircraft prevent people from actively pursuing them.

TECHNOLOGY TRIVIA

The Anglo-French Concorde made its first test flight in 1969 and entered service carrying passengers in 1976. This supersonic airliner carries up to 100 passengers at speeds of more than 1,000 mph. It crosses the Atlantic ocean in less than 3 hours.

One suggestion is to create new airports outside the urban boundaries, avoiding the problems of noise and traffic. High speed mass transit systems could be built to get people from the airports to the nearby cities.

Other efforts are being made to design more fuel-efficient and quieter aircraft. In a similar vein, better electronic landing systems may help prevent the backup of aircraft waiting to land at so many of our metropolitan airports.

■ **Describe and illustrate the transportation systems for the space-based MegaIndustries factory.**

▸▸▸ FOR DISCUSSION ◂◂◂

1. Why do airplanes sometimes have to wait before they are given permission to land?
2. Why are existing airports usually surrounded by commercial and residential structures?

IMPACT

We've gotten used to traveling fast. We can reach almost any place on earth in a matter of hours. If space travel becomes common, we will have to get used to longer journeys again. It will take days to get to the moon, and a trip to Mars could take a year.

Marine Transportation

The use of water-borne vessels for passenger transportation has dropped to just a small fraction of what it was at the beginning of this century. Although ships are still used for certain types of cargo (coal barges, oil tankers, and so on), most oceanic crossings are now being made by airplane. The last remaining refuge for large ships lies in leisure cruises for vacationers.

Partially as a result of this disinterest, there has been little advancement in seacraft in recent years. There is some speculation, however, that future cruise ships may incorporate some jet engine technology. These jets would raise the ship a bit in the water, reducing friction. This would increase the speed of the craft and smooth out the ride. Fig. 31–9.

Although boats are ideal for personal use and recreation, they have not proved popular for daily mass transit or long-range travel. Present indications show that they are not likely to be used for these purposes in the near future.

▶▶▶ FOR DISCUSSION ◀◀◀

1. Why don't as many people use boats for transportation as in the past?
2. What would trans-Atlantic flight offer a traveler that a ship could not? What would a ship offer a trans-Atlantic passenger that air travel could not?

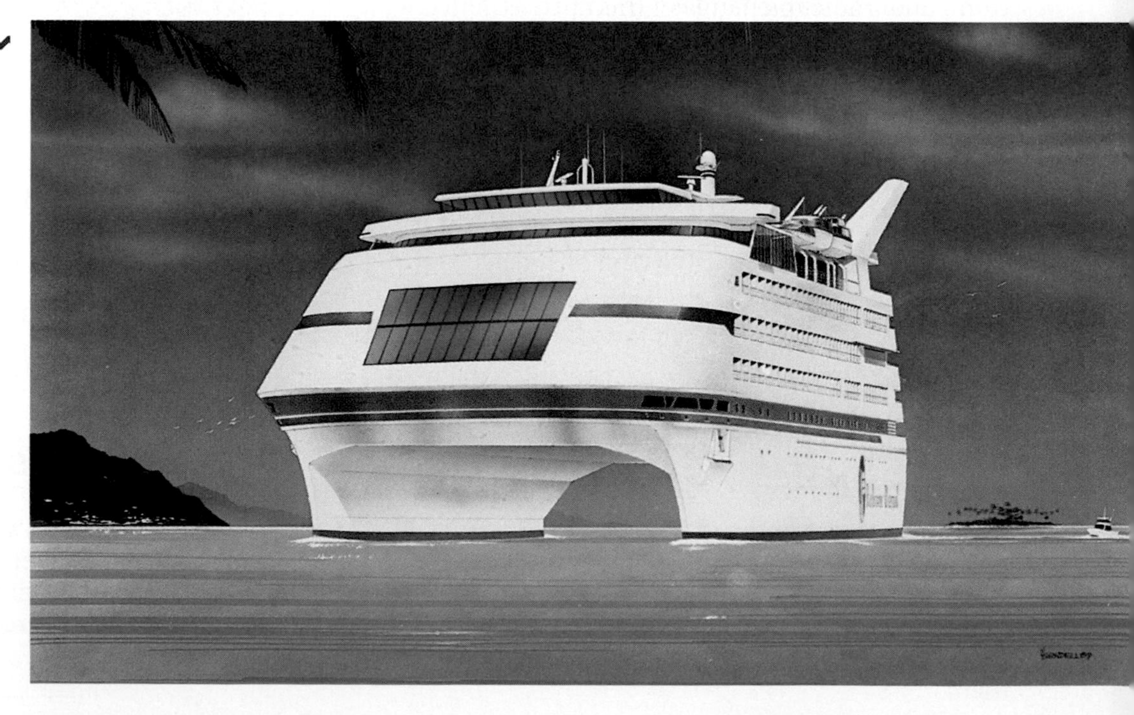

Fig. 31-9. In the future, ships may be powered by jet engines, reducing friction and increasing speed.

CHAPTER 31 Review

Chapter Highlights

- Land transportation may be classified as fixed or random.
- The automobile, a form of random transportation, is by far the most popular means of travel today.
- The increasing number of automobiles has produced four major problems: fuel consumption, exhaust emission pollution, safety, and traffic congestion.
- Maglev trains are being considered in many areas to reduce traffic congestion.

Test Your Knowledge

1. What is the purpose of such exhibits as *Futurama* in a World's Fair?
2. Why are trains considered "fixed" transportation?
3. Name four major issues that have resulted from the use of automobiles in this country.
4. Why can't we assume that there will always be sufficient petroleum available to make gasoline?
5. What is an electronic navigation system? Explain how this technology could help drivers.
6. Name at least two improvements that will probably be made to automobiles in the near future.
7. Why didn't traffic congestion decline as more roads were built?
8. Explain at least two methods people can use to reduce traffic congestion in large cities.
9. Explain what a "smart car" is.
10. Describe a personal rapid transit system.

Correlations

SCIENCE

1. Make two electromagnets. Each one requires a large iron nail, about 1 meter of bell wire with 6 cm of insulation stripped from each end, and a large flashlight battery with the terminals on top. Place the iron nail in the middle of the wire and tightly wrap the wire around the nail 100 times. Connect the ends of the wire to the poles of a battery. Observe the way the two electromagnets repel each other. Experiment by reversing the connection on one battery. Can electromagnetic force be used to lift a train above a track?

MATH

1. Your family is planning a vacation to the Dallas-Ft. Worth area. Compare the cost of flying, driving, riding the bus and traveling by train (if available). Check with a travel agent for plane, train, and, bus fares. Determine fuel costs if driving. Which method of travel would be least expensive?

LANGUAGE ARTS

1. Land transportation is sometimes divided into two categories: fixed and random. In a paragraph, compare and contrast these two categories.

SOCIAL STUDIES

1. This chapter describes some of the problems of using automobiles today. Find out what problems automobiles were causing in the 1920s, when they began to be widely used. How do those problems compare to today's?

You Can

Make a Difference

5 . . . 4 . . . 3 . . . 2 . . . 1 . . . LIFT OFF!

Near the end of training at the U.S. Space Academy in Alabama, it's time to head into space or sit at Mission Control. The trainees—who range from middle-school students to adults—must use what they have learned during the week to complete a successful flight.

Upon arriving at the Space Academy, trainees are assigned to teams. Although the "future astronauts" may be from different parts of the country, they soon learn to work together as a team to accomplish their mission. Every minute is filled with exciting program activities, including a visit to NASA's Marshall Space Flight Center.

During the early part of the week, the trainees actually get hands-on experience in the Shuttle module. In the Shuttle cockpit, they experience pitch and roll during a simulated launch and return to Earth. They also receive instructions to correct problems they might encounter on their mission. In addition to the Shuttle cockpit, the Mission Control Center in Houston has been reproduced so that trainees can track the simulated flights of their teammates. Trainees also become familiar with the full-scale Spacelab or Space Station module where they will conduct lab experiments later in the week.

Mission positions are selected on the basis of the trainee's knowledge. Each trainee trains for two different jobs from a choice of crew, science, or ground position. Since each team will conduct two flights at the end of the week, all the trainees will get a chance to use what they have learned on the ground as well as in flight. A major highlight of the week includes the study of space suits—very important as life-support systems during space walks.

Training intensifies during midweek. Trainees learn how to operate the remote manipulator arm, put a satellite into space, and handle experimental hardware. They must also understand the scientific investigations they will be expected to complete and know how to evaluate the results of the experiments.

As trainees begin their missions, computers make the flights seem real as ground control talks to the commander and pilot to help them reach the proper orbit. Payload specialists conduct real experiments in the Spacelab and Space Station. The "1-6th" Trainer allows mission specialists to experience aspects of microgravity during their space walks outside of the crew cabin. The Shuttle crew must dock with the Space Station before returning to Earth. After a debriefing, team members switch assignments to learn more about onboard activities and ground support roles.

Graduation from U.S. Space Academy takes place on the morning following the simulated flights. Families are invited to attend.

On the trip home, trainees remember the friends they have made from all over the United States. They have had a chance to use their brains and be rewarded for it. They can think about the variety of "high tech" jobs that might be available to them in the future. Most of all, they know that they can meet a challenge.

SPECIAL SECTION

PUTTING IT ALL TOGETHER

Using Systems to Solve a Problem

What have you learned about technology? If you were asked to write a composition answering this question, where would you begin? What would you include?

These are difficult questions to answer. Technology is a broad subject. You could probably write for days on hundreds of topics.

What if you were asked to help plan a space colony? Could you plan a community that provided for the basic needs of its citizens? Could you make choices about technology and give reasons for your decisions? This special section will give you the opportunity to use your knowledge and imagination to plan a community in space.

Planning a Space Community

Introduction

Suppose you were on a team whose task is to create a new community in space. What kind of shelter would you build and why? How would your energy needs be met? What impacts would your community have on the space environment?

As you work through this special section, you will help weave together all the aspects of technology that make a people-made world.

After reading this chapter, you should be able to

Discuss how communication, production, transportation, and biotechnical systems combine to meet the needs and wants of people.

Describe how the resources of technology are used to create the above systems.

Make some predictions about technology in the future.

Discuss the impacts of the above technologies on people and the environment.

Words you will need

Low Earth Orbit (LEO)
prefabricate
Manned Maneuvering Unit (MMU)
Personal Calling Number (PCN)

Fig. S-1. A space community in Low Earth Orbit could serve as a scientific laboratory and as a springboard for long-range space explorations of other planets, stars, and perhaps even other galaxies.

Your Assignment: Colony Planner

As co-planner for a space colony, you will help decide which technologies will be used to satisfy the needs of the community. Your decisions will be based on your knowledge of technology and your creative thinking skills.

Each part of this special section will describe how a basic need might be met through a proposed technology. At times, your input will be requested in the "Research and Development" feature of the section. You might be asked to sketch some solutions to problems or try to predict the impact of a technology. Before making any suggestions, be sure to consider what you have learned about technology.

Shelter and Construction Systems

Scientists have suggested that our space community be located in **Low Earth Orbit (LEO)**, about 300 miles above our planet. Fig. S-1. Engineers have provided detailed drawings of a space station that is 500 feet long. The station will consist of a frame or tower made of truss beams. The tower will serve as an anchor for the 15′ × 40′ modules. Fig. S-2. The modules will serve as housing, laboratories, factories, storage facilities, and recreation areas.

Our first decision as planners is how to build the station. We could **prefabricate** (construct beforehand) the tower and modules into small units. The units could be delivered into space by a shuttle and assembled by astronauts. Fig. S-3.

We might choose to deliver just the construction materials and have the entire station constructed in space. We might even design and build a machine that will manufacture the beams in microgravity. Please give your opinion on this subject.

Fig. S-2. The frame of the space station will be made of trusses strong enough to support the living and working modules. Why do you think trusses were chosen as the structural shape for the frame?

Fig. S-3. Prefabricated parts of a space station could be assembled by astronauts "on-site."

Living in Space

As planners, we must consider the physical and mental health of the people who will live in our community. We must provide healthy food, fresh water, clean air, and recreation for each person living in the space station.

Food and water can be transported from Earth and stored. Perhaps a better idea might be to make the space station self-sufficient so that the people could supply their own food and water. Engineers and scientists believe that some modules might be used for Controlled Environment Agriculture. Fig. S-4.

People living in the community will breathe a mixture of oxygen and nitrogen. Storage tanks containing these gases will supply the community with fresh air. In the meantime, scientists will continue to study methods of processing oxygen from certain moon rocks.

Each community member will have his or her own living space within a module. The space will include a music system, computer library, video system, exercise equipment, and sleeping space. The community members will share kitchen and bathroom facilities. Fig. S-5.

▶▶▶ RESEARCH AND DEVELOPMENT ◀◀◀

1. What factors must we consider when we select a method for constructing our station? (list at least three)
2. Make two different sketches showing how the modules and towers might be combined to form the community.
3. Using straws or toothpicks, construct a model section of the space tower.

Fig. S-4. Controlled Environment Agriculture may prove to be the most effective means of producing food for the occupants of a space station.

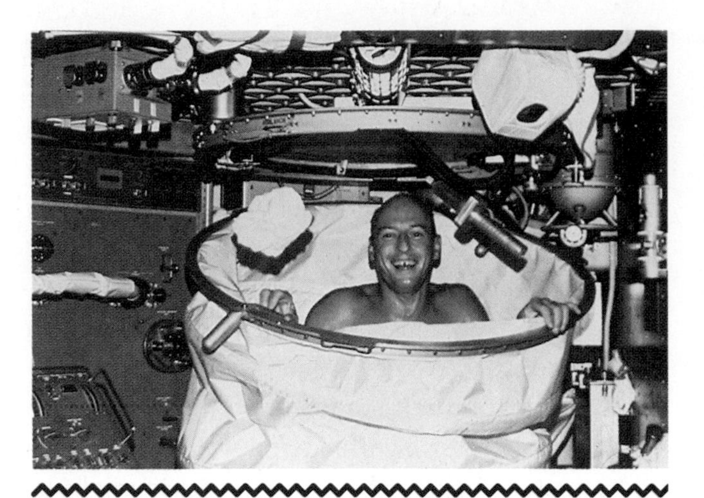

Fig. S-5. The same basic needs we meet at home must be satisfied in space. This photo shows the shower that was used in the *Skylab II* space station.

▶▶▶ RESEARCH AND DEVELOPMENT ◀◀◀

1. **Suggest how hydroponic farming and genetic engineering could be used in space-station food production.**
2. **Simple things such as eating and bathing will be made difficult by microgravity. Give three suggestions of how these problems might be overcome.**
3. **Make a series of sketches showing the layout of a typical living module.**

Transportation and Energy in Space

How will people, products, and materials be transported to the station? What modes of transportation will people use while living in the colony? What source of energy will supply power to the space community?

Shuttles, with their large cargo bay areas, will have to do much of the hauling from Earth to space. It will take many trips to bring enough materials into orbit to complete construction. An orbital warehouse will be needed to store supplies.

Should each citizen have a personal vehicle for transportation around the community? Mass transit or people movers might be a better idea. Fig. S-6.

To work outside the station, astronauts will have to wear **Manned Maneuvering Units (MMUs)**. Fig. S-7. These are jet-propelled backpacks used to fly free in space. Robot maneuvering vehicles are also being considered by engineers. These vehicles can move through space, controlled from the safety of the space station.

Providing Energy in Space

Electrical power will be supplied by 40′ by 90′ solar panels. The panels will be anchored to the tower in the same way as the modules. Large batteries will store the solar-voltaic energy for emergencies. Heat energy will be provided by fuel cells burning hydrogen fuel obtained from wastewater.

Fig. S-6. What transportation needs will occupants of a space station have? What form of transportation will be most appropriate?

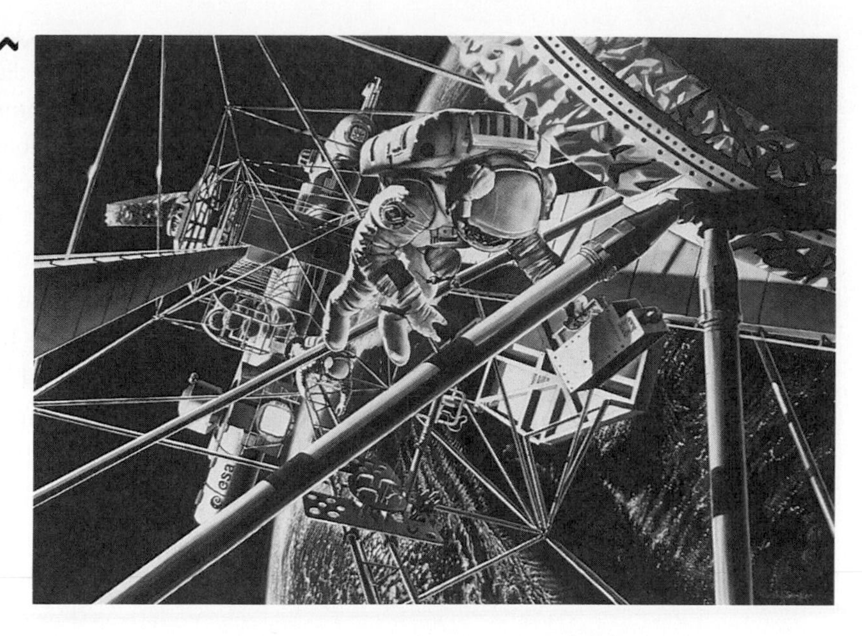

Fig. S-7. Manned Maneuvering Units allow astronauts to work outside the environmentally controlled space station. MMUs or similar suits will be needed by the astronauts so that they can perform maintenance and repair on the station.

Communication Needs in Space

A space community is much like a ship at sea. It is very important that the community stay in touch with Earth. Earth is the storehouse for resources, the most important of which will be information. Being able to communicate with Earth is vital to the success of the space community.

Anchored to the main tower will be a 100′ antenna to receive electromagnetic transmissions from Earth. Using the Tracking and Data Relay Satellite System (TDRSS), messages will be relayed by satellite from the space station to ground controllers on Earth and back. Fig.S-8.

Aboard the space station, each person will have his or her own **Personal Calling Number (PCN)**. A computer will route to each person all communications addressed to that person's PCN, no matter where the person is in the station. Videophones will also be available so that people can talk to and see family members at home on Earth.

▶▶▶ RESEARCH AND DEVELOPMENT ◀◀◀

1. Describe a system that could be used to move groups of people through the station. Be sure to consider cramped spaces, microgravity, and oxygen needs.
2. Develop a schedule of shuttle missions to bring supplies to the space station site. Make a list of what each shuttle will be carrying. List items in order of priority.
3. Describe two other sources of energy that can be used to meet the community's power needs. These must be non-polluting sources.

Fig. S-8. This diagram shows one idea about how communications with Earth could be established in a space station. Can you think of some other ideas?

▶▶▶ RESEARCH AND DEVELOPMENT ◀◀◀

1. Draw a diagram showing the position of the earth, the space station, and the locations of satellites. Position the satellites so messages can be sent and received from any point of Earth. Show how the electromagnetic waves might travel.
2. Describe how electronic mail might be used to communicate within the station and between the station and Earth.
3. Time on the videophones has to be limited to one minute per week for each person. Write a script for a conversation you are about to have with a friend on Earth. What will you say?

Working in Space

Two of the modules anchored to the tower will be space factories. The factories will conduct primary manufacturing, producing industrial materials to be used in the colony. A laboratory module will conduct experiments to further our understanding of manufacturing in space. Fig. S-9.

Space taxis will transport workers, materials, and products back and forth from the moon. Areas of the moon that are rich in natural resources will be mined and harvested.

A space energy power plant located on the moon will produce electrical energy using solar voltaic cells. The electrical energy will be sent back to Earth as electromagnetic waves and then converted back into electricity.

▶▶▶ RESEARCH AND DEVELOPMENT ◀◀◀

1. Using your knowledge of manufacturing in space, what products do you think should be created in the space factory? How could they be used?
2. Write a letter to a company inviting the company to move its manufacturing plant to the space colony. Include the benefits and problems of manufacturing in space.
3. Make a sketch of a space taxi transporting raw materials from the moon to the station.

Fig. S-9. The space station will include laboratories for research and experiments. Here is an artist's idea of what might be done. An astronaut is collecting data on the burning of the Amazon rain forest.

Protecting the Space Environment and Its People

Whether in space, below the sea, or on Earth, when we use technology to satisfy needs, we run the risk of making mistakes. We do not want to create the same problems in space that we have created on Earth. Pollution, waste, and resource depletion will not be allowed in our new environment. Fig. S-10.

Where possible, products used in the colony will be recyclable or biodegradable. All waste materials will be collected, sorted, and sent to recycling modules for processing.

Any processes that require the burning of fuels will require strict monitoring of all smoke stack emissions. Emissions harmful to the environment will be cause to close the plant.

Synthetic materials will replace natural materials where possible. A management system will be organized to determine the amount of natural materials that may safely be consumed.

▶▶▶ RESEARCH AND DEVELOPMENT ◀◀◀

1. Another member of the planning committee has recommended a floating garbage dump to be placed in orbit around the station. What is your reaction to this recommendation? Support your decision based on the impact it will have on the community.
2. Make a sketch of a container that can be used to store recyclable materials in microgravity.
3. Make a poster to advertise living in a space colony. Use graphics and text to point out the positive aspects of this lifestyle.

Further Suggestions

This concludes our preliminary plans for a space community. As a special project, you might want to create a complete set of plans for your space community. These plans could describe every aspect of your planned community in detail. For example, you could specify the dimensions and purpose of each proposed module (living quarters, warehouse, agriculture, factories, and so on).

You might even want to team up with classmates to build a complete model of your planned community. However you use your ideas, keep your plans and drawings together in a notebook that you can refer to in the future. Who knows, you may be in a position to use them someday!

Fig. S-10. When people begin to populate space, disposal of waste products will become a problem. Space "garbage" could orbit the Earth for years, posing a danger to spacecraft. By planning in advance, people may be able to limit pollution of space.

Appendix

Appendix A. Measuring with the Metric System

Products made by American production systems must compete in world markets. Most other countries use the metric system of measurement. More and more American products are being made to metric sizes.

There are many measures used in the metric system. The important ones are explained here. (See also Table 1.) A handy conversion chart can be found in Table 2.

Metric Units

Length. The base unit for measuring length is the **meter** (**m**), which is equal to about 40 inches. Another important unit is the **millimeter** (**mm**), which is about as long as a dime is thick. All metric drawings for production are made to millimeter dimensions.

Area. The unit for area is the **square meter** (**m²**), which is a little larger than the square yard. The **square millimeter** (**mm²**) is used for small area measurements.

Volume. The **cubic meter** (**m³**) is used to measure volume. It is slightly larger than the cubic yard. For measuring liquid volume, the **liter** (**L**) is used. It is a little larger than the quart. The **milliliter** (**mL**) is used for smaller volumes.

Mass or Weight. The metric base unit for mass is the **kilogram** (**kg**), which weighs a little more than two pounds. The **gram** (**g**) is used for smaller weights.

Temperature. Temperature is measured in **degrees Celsius** (**°C**). Important temperatures to know are:

100°C	Water boils
37°C	Body temperature
20°C	Average room temperature
0°C	Water freezes

TABLE 1. Metric Equivalents

Linear Measures

1 kilometer = 0.6214 mile
1 meter = 39.37 inches = 3.2808 feet = 1.0936 yards
1 centimeter = 0.3937 inch
1 millimeter = 0.03937 inch

1 mile = 1.609 kilometers
1 yard = 0.9144 meter
1 foot = 0.3048 meter = 304.8 millimeters
1 inch = 2.54 centimeters = 25.4 millimeters

Area Measures

1 square kilometer = 0.3861 square mile = 247.1 acres
1 hectare = 2.471 acres = 107,639 square feet
1 are = 0.0247 acre = 1076.4 square feet
1 square meter = 10.764 square feet
1 square centimeter = 0.155 square inch
1 square millimeter = 0.00155 square inch

1 square mile = 2.5899 square kilometers
1 acre = 0.4047 hectare = 40.47 ares
1 square yard = 0.836 square meter
1 square foot = 0.0929 square meter
1 square inch = 6.452 square centimeters

Cubic Measures

1 cubic meter = 35.315 cubic feet = 1.308 cubic yards
1 cubic meter = 264.2 U.S. gallons
1 cubic centimeter = 0.061 cubic inch
1 liter = 0.0353 cubic foot = 61.023 cubic inches
1 liter = 0.2642 U.S. gallon = 1.0567 U.S. quarts

1 cubic yard = 0.7646 cubic meter
1 cubic foot = 0.02832 cubic meter = 28.317 liters
1 cubic inch = 16.38706 cubic centimeters
1 U.S. gallon = 3.785 liters
1 U.S. quart = 0.946 liter

Weight Measures

1 metric ton = 0.9842 ton (long) = 2204.6 pounds
1 kilogram = 2.2046 pounds = 35.274 ounces
1 gram = 0.03527 ounce
1 gram = 15.432 grains

1 long ton = 1.016 metric tons = 1016 kilograms
1 pound = 0.4536 kilogram = 453.6 grams
1 ounce = 28.35 grams
1 grain = 0.0648 gram
1 calorie (kilogram calorie) = 3.968 Btu

TABLE 2. Approximate Customary-Metric Conversions

When you know:		You can find:	If you multiply by:
Length	inches	millimeters	25.4
	feet	millimeters	304.8
	yards	meters	0.9
	miles	kilometers (km)	1.6
	millimeters	inches	0.04
	meters	yards	1.1
	kilometers	miles	0.6
Area	square inches	square centimeters (cm^2)	6.5
	square feet	square meters	0.09
	square yards	square meters	0.8
	square miles	square kilometers (km^2)	2.6
	acres	square hectometers (hectares)	0.4
	square centimeters	square inches	0.16
	square meters	square yards	1.2
	square kilometers	square miles	0.4
	hectares (ha)	acres	2.5
Mass	ounces	grams	28.4
	pounds	kilograms	0.45
	tons	metric tons (t)	0.9
	grams	ounces	0.04
	kilograms	pounds	2.2
	metric tons	tons	1.1
Liquid Volume	ounces	milliliters	29.6
	pints	liters	0.47
	quarts	liters	0.95
	gallons	liters	3.8
	milliliters	ounces	0.03
	liters	pints	2.1
	liters	quarts	1.06
	liters	gallons	0.26
Temperature	degrees Fahrenheit	degrees Celsius	0.6 (after subtracting 32)
	degrees Celsius	degrees Fahrenheit	1.8 (then add 32)
Power	horsepower	kilowatts (kW)	0.75
	kilowatts	horsepower	1.34
Pressure	pounds per square inch (psi)	kilopascals (kPa)	6.9
	kPa	psi	0.15
Velocity (Speed)	miles per hour (mph)	kilometers per hour (km/h)	1.6
	km/h	mph	0.6

Metric Prefixes

The metric system is based on units of 10. Fractions are shown in decimals. Certain prefixes can be used with many units to show how they have been multiplied or divided by tens.

kilo = 1000
centi = 1/100 (0.01)
milli = 1/1000 (0.001)

For example, if you know that the distance from school to town is 1000 meters, you would say "one kilometer." Instead of saying "1/1000th of a meter," you would say "one millimeter."

Reading a Metric Rule

You read a metric rule the same way you read a customary rule. You count the markings. Look at the metric rule shown in Fig. A. It is marked in millimeters. To read it, count the number of millimeter spaces for a given length.

For example, the distance from A to B is 15 mm. The distance from A to C is 25 mm. What is the distance from A to D? A to E? A to F? Practice measuring paper clips, nuts and bolts, and other common objects. Round off your measurements to the nearest millimeter.

Designing in Metric

When designing a product to metric measurements, use a dual-reading metric rule. Think about replacement sizes, not conversions. For example, if you are designing bookshelves, a shelf width of 9 inches is about right. But now you want a metric size. Look at the dual-reading rule. You will see that 9 inches converts to about 229 mm. However, that is an odd size. It can be rounded off to 230 mm.

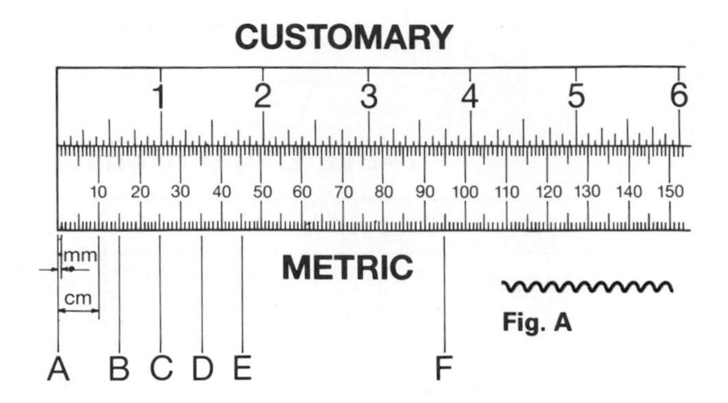

Fig. A

Because 230 mm is easier to read on your rule and more convenient to use, it's called a "rational" size. Try to select rational sizes ending in 5 or 0. Do the same for other measurements you want to use.

Changing Customary Sizes to Metric Sizes

Many product drawings are in customary inch sizes. You might want to change them to metric dimensions.

Study the customary inch-size bowl in Fig. B. Note the 5½-inch diameter. Now look at the dual reading rule. If you do not have a rule, you can multiply the inch dimension by 25.4 to convert it to millimeters. This is because there are 25.4 millimeters in one inch. It is easier to do this if you change the fraction to a decimal.

$$5.5 \times 25.4 = 139.7$$

The result can be rounded off to 140 millimeters. Change the diameter on the drawing to 140 mm. Convert the other dimensions.

If the size you need for a product is not one of the standard sizes, choose your materials in the closest standard size instead. For example, the drawing of the bowl calls for 22 gauge sheet metal. The nearest metric replacement size is 0.60 mm.

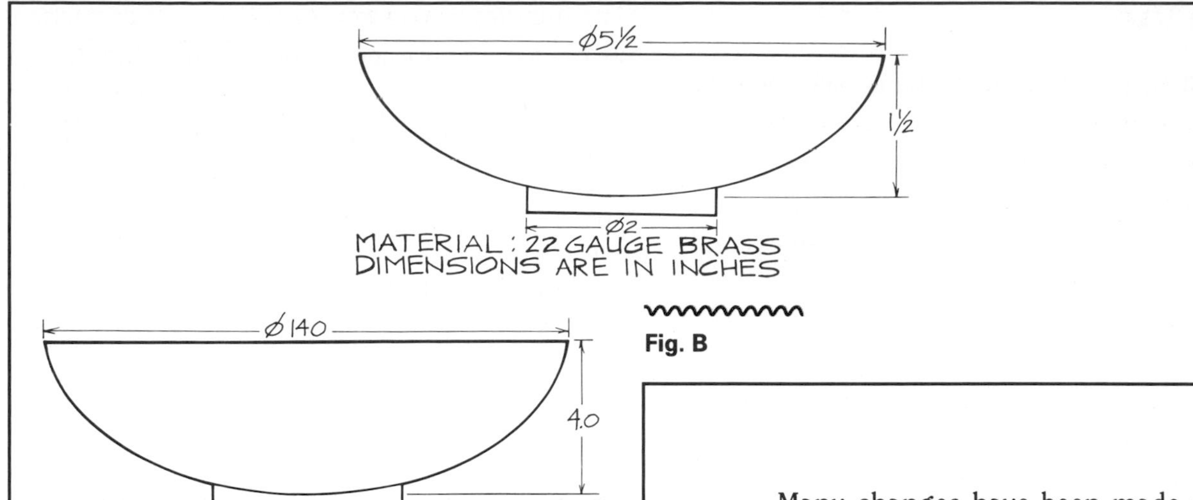

Fig. B

Appendix B. How to Conduct a Meeting

At some time during this course, your class will form a student company. Your company will then manufacture or construct a product. In Chapter 2, "Forming a Company," you will learn how a company is started and run. One thing that often takes place within any company is meetings. Meetings allow workers to communicate with one another. At meetings, most decisions are usually made.

People have used meetings as a way to get things done for many centuries. Over time they found that they needed rules for the meetings. Without rules, some people never got to speak. Others did all the talking. To have a discussion in which all could participate, the group needed a leader who would guide the discussion. This leader had to know the rules for discussions. In England, Parliament began to use these rules when conducting business. The rules became known as parliamentary procedure.

Many changes have been made since then. However, the purpose is still the same. Groups can conduct orderly meetings on the basis of democratic principles. *All* members may contribute ideas and opinions, not just a select few.

The club meetings described here are fairly formal. Most businesses are more relaxed about meetings. However, the basic procedure is the same. There is one other important difference. In some businesses, final decisions must be approved by a supervisor.

Rules of Order

Parliamentary procedure is based on the following rules:

- During a debate, group members must be fair and polite.
- Any member of the group may debate under the rules established.
- The majority (side with the most votes) decides the issues.
- The minority (side with the least votes) is free to express its opinion.
- Minority members must go along with the decision made by the majority.

Terms to Use

To participate in groups using parliamentary procedure, members must know certain terms. Only then can the meeting move ahead smoothly.

- **A quorum:** the number of members who must be present for the group to do business and make decisions.
- **Minutes:** a written record of the business covered at a meeting.
- **Majority:** at least one more than half of the members present.
- **Motion:** a suggestion by a member that certain action be taken by the group.
- **Second the motion:** a second member approves the suggestion.
- **Amendment:** a change in a document.

Conducting a Meeting

No official business can be done at any group meeting unless a quorum is present. Usually a quorum is a majority of members. It can also be a number stated in the rules of the group. When a quorum is present, the meeting usually proceeds like this:

The group leader calls the meeting to order. Then he or she presents business in the order stated in the group's rules. If an order of business has not been agreed upon, the group should use the order described below. The club meeting that follows would be typical for many groups.

The president would make these statements:

"The meeting will come to order. Marie Smith, the secretary, will read the minutes of our last meeting."

"Are there any corrections to the minutes?" (If no corrections are made, the meeting proceeds.) "The minutes are approved as read." (If corrections are needed, the president asks the secretary to correct the report.) "Are there any corrections?" (pause) "Then, the minutes stand approved as corrected."

"The next order of business is the reports of other officers." (Examples would be a secretary's report or a treasurer's report.)

"The next order of business is the reports of committees."

"Is there any unfinished business to act upon today?"

"Is there any new business to be considered today?" (New business includes communications, presenting bills for payment, future plans, setting dates for other activities, etc.)

"Is there future business or any announcements?" (After announcements are made, the president may end the meeting.)

Motions

In a motion, a member suggests that the group take a certain action.

Any member may make a motion to introduce an item of business. A motion must be seconded. After it is seconded, the motion may be discussed. Finally, a vote will be taken. The majority always rules. An example follows:

John stands and addresses the president, "Mr. President."

The president recognizes John by calling his name.

John says, "I move (make a motion) that the club buy a new ceremonial emblem before the state contest in April."

Carol, another club member, agrees, "I second the motion."

Then the president announces, "It has been moved and seconded that the club buy a new ceremonial emblem before the state contest in April."

At this point, the other members discuss the motion.

Finally, the president asks, "Is there further discussion?" After a pause, he prepares for the vote. "All in favor of the motion say 'Aye.'" He counts the favorable votes. "Those opposed say 'No.'" He counts. "The ayes have it. The motion has passed that we buy a new ceremonial emblem before the state contest."

Electing Officers

The democratic process requires that persons in leadership positions be nominated and elected by the group. Nominations are made by the members, by a committee, or by ballot.

In many organizations, new officers are nominated by a committee appointed by the president or elected by the members.

On the date set for election of officers, the president asks for a report from the nominating committee. Then, the persons nominated are named and a vote taken.

Glossary

A

acid rain A form of pollution that occurs when rain combines with sulfur dioxide and nitrogen oxide, two pollutants that are given off by automobiles, by power plants burning high-sulfur coal, and by some manufacturing processes.

adhesives Chemicals used to combine materials or join them (make them "stick together").

Agricultural Era A period of time from about A.D. 1100 to 1750, in which most people were farmers and farming technology advanced greatly.

alloy A mixture of two or more metals to create a new, human-made metal.

architect A person trained to design structures.

artificial intelligence Computer hardware and software systems that can evaluate data and respond to questions people ask about a particular subject.

atomic energy The energy stored in the nucleus of atoms.

automated system Any system in which machines control machines.

automation Using computers instead of—or in addition to—people to produce products in a manufacturing system.

B

beam Horizontal support used in construction systems; most often used with columns to support weight in a structure.

bending A type of forming by which materials such as plastics, metals, woods, and composites are shaped.

bill of materials A concise list of the sizes and characteristics of the parts shown on a drawing that will be needed for construction of the object shown in the drawing.

binary code The basic "language" computers use to accomplish all their work; binary code is based on the binary number system, which includes only 0's and 1's.

biomedical engineering The application of technology to the field of medicine. Biomedical engineering produces new equipment and techniques to help physicians diagnose and treat disease.

bionics Artificial body parts manufactured to take the place of human body parts that no longer function properly.

bioprocessing Work done by microorganisms in a biotechnical system.

biotechnical systems Technologies that involve living organisms.

biotechnology The use of living cells to help create new products.

board of directors Elected members of a corporation who set policy and goals for the corporation.

brainstorming A method of gathering information in which one or more people suggest ideas without trying to perfect them first. The purpose of brainstorming is to gather as many ideas as possible.

bulletin board system (BBS) An electronic system, run by computer, that allows people to send and receive electronic mail. The host computer (the central computer that holds the electronic "mailboxes") is also referred to as a bulletin board system.

C

capital A company's assets, which may include money, property, buildings, equipment, and all the goods a company owns.

cash flow A comparison of a business's income and expenses.

casting A technique in which a material is poured into a mold and takes the shape of the mold. When the mold is removed, the material retains the shape of the mold.

cellular phone network A type of portable phone that allows people to make calls from their cars or other places where no telephone lines exist.

central processing unit (CPU) The "brain" of the computer, which carries out the user's instructions.

ceramics Natural materials made from minerals called *silicates*. Sand, clay, and quartz are examples of silicates that are used to make ceramics.

chemical energy Energy stored in chemicals. Matches and firecrackers are examples of items that have chemical energy.

coaxial cable A type of cable in which an electrically conductive wire, such as copper, is surrounded by another conductor and wrapped in an insulating material.

cohesion The use of solvents or cements to melt two materials and cause them to flow together. In cohesive bonds, the molecules of the materials being joined mix together, so cohesive bonds are stronger than adhesive bonds.

column Vertical supports used in construction systems; most often used with beams to support weight in a structure.

combining A technique that allows people to fasten materials together using nails, screws, or other fasteners.

combining processes Processes used to join materials. The combining process used for a given manufacturing job depends on the materials being assembled and how strong the joint must be.

combustion The process of burning a fuel or other substance to produce heat energy.

communication technology The family of technology that helps people gather, store, and share important information.

composite A material formed from combining or mixing two or more different kinds of materials.

compression The force that tends to push two parts of an object toward each other. Gravity is usually the cause of this force. Compression increases as height or depth increases.

computer-aided drafting (CAD) Computer systems that allow drafters and artists to make complicated, accurate drawings using computers.

computer-aided manufacturing (CAM) A system that organizes manufacturing using computer numerical control.

computer modeling Creating a pictorial image of an object electronically. Some software allows people to test and experiment with an object before it is built, greatly decreasing the amount of time and money that must be spent on product development.

computer numerical control (CNC) A system in which computers are used to operate machines.

conditioning processes Processes used to change the inner structure of a material.

conserve To waste less in order to reduce pollution and make natural resources last longer.

construction The process of building structures such as buildings, tunnels, dams, roads, and bridges. Construction is done on-site. See also *manufacturing*.

consumer research A method of finding out what potential buyers want in a product.

consumers The people who buy or use a product.

contaminants Pollutants.

contractor In a construction project, the person in charge of building a structure.

controlled environment agriculture (CEA) A biotechnical system that makes it possible to control environmental conditions to produce agricultural products under artificial conditions.

corporation A company that has many part-owners called *stockholders*. Corporations are the most common form of large business.

cottage industry Craftspeople working out of their homes and small shops. Cottage industries were common before the Industrial Revolution.

credit history Credit files kept on people regarding their financial history. A credit history may include information about loans, mortgages, credit cards, or legal judgments a person may have had in the past.

D

data Facts. Data has little meaning until it is organized into information.

database A collection of information organized around a topic. Most of today's databases are stored on computers for easy access.

data processing system Any system used to organize raw data into information.

dead load The weight of a structure; static load.

demodulation The process of separating an electronic signal from the carrier signal with which it was combined for transmission through telephone lines.

design criteria Goals that the solution to a design problem must fulfill.

dimensioning The inclusion of numbers on a technical drawing, showing accurate sizes, shapes, and locations.

dirigible Large structures with rigid frames that used lighter-than-air principles to transport people through the air during the first half of this century.

dividend Money received by stockholders for each share of stock they own.

drafting A method of communicating detailed information about a proposed object or device.

E

electromagnetic pulses Electronic signals that can travel at the speed of light; used for long-distance communication.

electronic banking A method of banking using touch-tone phones and computers; electronic banking allows people to do their banking during hours when the bank is normally closed.

electronic cottage A common name for any arrangement in which people work at home, using computers, modems, and fax machines to communicate with central offices.

electronic mail (E-mail) Messages people send and receive using computers. Messages are stored in an electronic "bulletin board" stored on a host computer.

energy The ability to do work and create movement.

entrepreneur A person who starts his or her own business.

Environmental Protection Agency (EPA) The office of the federal government that creates and enforces regulations to help protect our air, land, and water.

ergonomics A type of designing in which the designer studies how to match the product to the human user; human factors engineering.

erosion Loss of top soil due to rains that wash away earth. Erosion often happens where land is not anchored by trees or other plants with strong roots.

external combustion A type of combustion in which the fire, or burning, takes place outside the engine, in a separate structure.

extrusion A process in which a soft material is squeezed through an opening called a *die*. The material takes the shape of the die.

F

factory system The use of factories and mass production techniques to produce products faster and less expensively.

feedback Information about the output of a system.

ferrous A word that describes metals that contain iron.

fiber-optic cable A cable made of thin glass or plastic fibers through which light signals can be transmitted. Fiber-optic cable weighs less than coaxial cable and is less likely to be affected by electrical interference. It also has a greater information-carrying capacity.

field-test To use a product under realistic circumstances to see how well it works.

finance The management of money.

fixed transportation Methods of transportation, such as railroads, in which the vehicles follow a fixed path.

force The push or pull that gives energy to an object.

forging The process of shaping metal by heating it and then hammering it into shape.

forming Changing the shape of materials such as clay and metal without removing any materials.

forming processes Processes used to shape liquids and soft materials into finished products without removing or adding material. Common forming processes include casting, pressing, forging, molding, bending, and extruding.

fossil fuels Fuels created when heat and pressure act on decaying plants and animals over millions of years.

foundation In construction, the part of a structure that is in contact with the ground; sometimes called the substructure.

framing In construction, building the frame of a structure.

futurist Someone who tries to predict the future.

G

generator A device that changes mechanical energy into electrical energy.

genetic engineering A type of biotechnology that includes the study of genes and how they can be changed to create desired traits in living things.

graphic communication Written communication, or communication meant to be seen in some form.

gross national product (GNP) An index of the production and consumption of a country's goods; used to determine the health of the country's economy.

H

hardwoods Woods that come from trees that have broad leaves, such as maple and oak trees.

high-definition television (HDTV) A television system that can form pictures made up of 1000 or more lines per screen, forming a cleaner, sharper image than is currently available on regular televisions.

hydroelectric plants Power plants in which the mechanical energy of falling or flowing water is used to provide electricity.

hydrofoil A specialized watercraft that can skim through the water at high speeds.

hydroponics A biotechnical system used to grow plants in materials other than soil.

I

impact Effect; the connection between one action and later actions that are influenced by that action.

indirect employment Secondary employment; jobs created by a company, even though the company does not hire these people directly.

Industrial Era A period of time that began around 1750 in England; characterized by large industrial growth and many new technical inventions.

Information Era The period of time in which we live; so-called because of the huge amounts of information we use each day to create products and services to meet our wants and needs.

information technology Technology that provides a means of handling (collecting, organizing, storing, sending, and processing) data and information.

innovation A modification to a product that already exists.

input Most widely used in reference to data and commands given to a computer system, but in a broad sense, anything put into a system.

insulate To provide a barrier against heat, cold, and/or sound.

interest A fee charged by banks or other lenders for the use of their money.

intermodal A method of transporting goods using more than one mode of transportation.

internal combustion A type of combustion in which the fuel burns inside the engine.

isometric A fairly natural-looking drawing in which angles are about 30 degrees from the horizontal.

K

kinetic energy Energy put into motion.

L

law of conservation of energy The physical law that states that energy can be changed from one form to another, but it cannot be created or destroyed.

layout The technique of putting lines, angles, and circles on materials to show where the materials should be cut or bent to form a product.

live load Weight on a structure that may vary, such as the weight of snow on a building or traffic on a bridge.

Low Earth Orbit (LEO) A relatively low orbit around the earth, located about 300 miles above the earth.

M

machine vision Sensors in machines that can scan manufactured parts as they are produced, searching for mistakes.

macroengineering Large-scale engineering projects that affect the environment.

maglev trains Trains that are propelled by magnetic levitation. Maglev trains are faster and more durable than traditional trains.

management A system of people that provides organization for a business; the role of management is to plan, organize, and direct the activities of a business.

Manned Maneuvering Unit (MMU) Jet-propelled backpacks used to fly freely in space.

manufacturing The process of creating products in a factory.

market research Gathering information about competition for a new product or service.

mass production The process of producing large quantities of products in factories.

material processing A series of steps or operations used to change materials into finished products.

measuring tools Type of tools used to determine the sizes of objects.

mechanical advantage The number of times a machine increases the force applied.

mechanical energy The energy found in moving things.

megaphone A device that helps project sound in a chosen direction.

microgravity A condition of almost no gravity.

mobile society A society in which people travel a lot, change jobs frequently, and move to new homes often.

mock-up A model that looks like the real product but does not necessarily work or contain the actual materials or mechanisms of the real product.

modem A device that allows electronic messages to be sent by modulating an electronic signal with a "carrier" signal so that it can be transmitted over telephone lines, then demodulating the signal (removing it from the carrier signal) so that it can be interpreted by the receiver.

modular homes Structures that are prefabricated off-site and trucked to the assembly point.

modulation The process of combining an electronic signal with a "carrier" signal so that it can be transmitted through telephone lines.

molding The process of shaping materials such as plastics by techniques such as injection molding and vacuum forming.

molecule The smallest part of a substance that still has the properties of that substance.

monorail A railed vehicle that uses a single rail.

Morse code A communication system in which various combinations of long and short pulses stand for letters of the alphabet and the numerals 0–9.

movable metal type Individual pieces of metal type that are combined to print many copies of a text.

multiview Drawings that allow you to show one or more sides of an item at a time; orthographic projection.

N

natural resources Materials that are found in nature.

network A system that links many computers to a central controlling computer and/or to each other.

nonrenewable resources Resources of which we have only a limited supply, such as fossil fuels and nuclear fuels.

nuclear fission The splitting of the nucleus of an atom into smaller nuclei.

O

oblique A type of drawing that can be drawn quickly but appears distorted to the eye. The front view of an oblique drawing is always drawn without any changes.

Occupational Safety and Health Administration (OSHA) A federal office that creates and enforces regulations designed to improve working conditions.

offset printing A type of printing in which the image to be printed is transferred from an inked form to a rubber blanket and then to paper.

optical Relating to the eye or vision; optical tools and machines such as microscopes and telescopes allow people to view things they could not ordinarily see.

optical telegraphs Sophisticated visual systems such as the Chappe semaphore and the Murray shutter semaphore in which telescopes and mechanical devices were used to transmit messages over distances of several miles.

oral communication Spoken communication.

output The information a computer supplies after it processes the data that has been input. In a broader definition, output is the result of the process in any type of system.

P

pager A device that beeps when its number is dialed, getting the attention of the person carrying it.

Personal Calling Number (PCN) A telephone number assigned to each person, instead of to each telephone. When a person's PCN is dialed, the person will be able to answer the call, regardless of where the person is at the time.

perspective The least distorted type of pictorial drawing.

photography A technique used to capture images on light-sensitive film.

photovoltaic cells Devices that capture the energy from light and convert it into electricity.

pictorial A type of drawing that gives the most realistic view of what an object looks like.

picturephone A type of telephone that enables people to see and hear each other.

plasticity The ability of a material to be formed into shape easily and to stay in that shape.

polymer chemists Scientists who work with plastics.

polymers Plastics, usually made from carbon obtained from petroleum or natural gas.

potential energy Energy at rest, or stored energy.

prefabricate To construct beforehand.

pressing A technique similar to casting, except that a plunger is used to force the material into the mold; often used to form powdered materials.

primary manufacturing The conversion of raw materials into industrial materials.

printing press A machine that presses movable metal type against paper to create a printed image.

process section The part of a system that does the actual work and achieves the desired result.

production The manufacture or construction of products.

production technology Technology that provides us with the manufactured and constructed products we use each day.

profit A business's income after expenses are deducted.

program A set of computer instructions.

project management A method of keeping track of the pace of a construction project, delivery of materials, number of employees, and so on.

prototype The first working model of a device.

public works Structures built for the "public good." For example, roads and government buildings are public works.

pultrusion A manufacturing process that allows composites to be made from continuous fibers, yielding a much stronger final product.

purchase power A person's ability to buy products; determined by the person's income and the state of the economy.

R

random transportation Methods of transportation that allow vehicles to travel in almost any direction.

raw materials Materials such as wood, oil, cotton, and iron ore that are used to create products.

reciprocal A system that uses back-and-forth movements to transmit motion to the axles of a vehicle.

recycle To reuse products that have already been used once in order to reduce pollution and to conserve natural resources.

renewable resources Materials that are replaceable after they are removed from nature. Trees, plants, and animals are examples of renewable resources.

resources All the things needed to accomplish a goal or get a job done.

rotary A system that uses circular movements to transmit motion to the axles of a vehicle.

rural Outside of the cities and suburbs; in an area of low population density.

S

sanitary landfill A large hole in the ground where garbage and waste products are buried between layers of clay that help keep pollutants from contaminating groundwater.

satellite communication Communication in which signals are beamed to satellites in the sky and then bounced back to a different location on the earth.

scale A reduction or enlargement in which an object is drawn to the correct proportions, but not the correct size. Used to allow people to draw or model very large or very small objects.

science The study of our natural world.

screen printing A printing method in which ink is squeezed through a "screen" onto an area defined by a stencil.

secondary manufacturing Manufacturing processes that use industrial products to produce finished products.

semaphore signaling A communication system that consists of two flags that can be read as they are brought to each of several possible arrangements; used mostly on ships and along railroad tracks.

separating Refers to tools and machines that cut or remove part of a material.

separating processes Processes in which one piece is removed from another. Examples of separating processes include shearing, sawing, drilling, planing, milling, turning, and shaping.

service industry Industry in which services rather than objects are the final products.

shear A force that exerts pressure on material in opposite directions.

shearing A process in which a knife-like blade is used to slice materials; no loss of materials takes place during shearing.

signal A message sent from one point or person to another.

signal processing A technology in which voice or any other signal is modified into a form that can be transmitted over long distances.

site In construction, a place on which to build.

smart houses Houses in which computers monitor and adjust all the electrical systems, such as lighting, heating, and security systems. The occupants of the house can program the computer to customize the adjustments.

smog A fog-like condition that is worsened by airborne pollutants.

softwoods Woods that come from coniferous trees (trees that have needles and cones rather than leaves).

soil sample A small amount of soil that is tested; for example, soil may be tested to determine how much weight the soil can support.

solar voltaic energy Electrical energy derived from the energy of the sun.

stock A small piece, or share, of a corporation.

stockholders People who invest their money by purchasing stock in the company.

structural plastics Plastics that are immune to corrosion and electrical interference; these plastics are inexpensive and are also good insulators.

structure Any arrangement of parts, either rigid or flexible.

subsystems Smaller systems that make up large systems.

suburban Around the outskirts of a city or urban area.

subway A railroad that is run underground because buildings block the right-of-way above the ground.

survey A technique that determines the exact property and building lines for a construction project.

synthesized speech Speech that has been created electronically.

synthetic materials Materials not found in nature.

system A group of parts working together to achieve a goal.

system diagram A chart that makes it easier to understand how a complex system works.

T

technological system A system designed to help people accomplish goals and produce desired outcomes.

technology The means by which we try to improve our people-made world.

techostress The pressure that people face from technology and technological developments.

telecommunication Communication over long distances.

telecommuting "Commuting" to work electronically, through the use of modems and fax machines, without leaving one's home.

teleconferencing A technology that allows several people, all in different locations, to speak together.

telegraph A device that uses a switch called a telegraph key to open and close a circuit in a predetermined pattern that can be interpreted by the receiver.

tension The force that exerts a pull from the center toward the ends of an object.

terminal A computer station.

thermography Computerized mapping using infrared photography; a method by which heat loss from a building can be analyzed.

thermoplastics Plastics that can be repeatedly reheated and reshaped.

thermoset plastics Plastics that, once formed, cannot be reheated and reshaped.

torque A twisting force.

toxic Poisonous.

transmit To send.

transportation technology Technology that applies to the movement of goods and people from one place to another.

trend The choices most people seem to be making at a given time or in a given place.

truss A structural shape that provides great weight-to-strength ratio and excellent stability.

U

Underwriters' Laboratories (UL) A group that tests products to ensure that they work properly and safely.

urban Within a city or highly populated area.

V

videoconferencing A conference call in which participants can see and hear each other.

voice mail A combination of electronic mail and the telephone. If the person you want to talk to is unavailable, you can leave a message in the person's electronic "mailbox."

W

wigwag An organized system of red and white flags that was used during the Civil War to transmit messages.

Z

zoning Local building codes that specify what types of structures can be built on a piece of property.

Photo Credits

Academy of Awards/Roger B. Bean, 157
ACF Industries, 98
Adept Technology Inc., 95
Alaska Division of Tourism, 22
A.O. Smith/Scott Witte, 22
Apple Computer, Inc.: 178; J. Greenleigh, 241
Argonne National Laboratory, 343
Arnold & Brown, 22, 28, 56, 61, 69, 78, 79, 96, 99, 112, 209
ART RESOURCES, *(Pieter Brueghel the Elder, Tower of Babel, Vienna),* 147
AT&T Archives, 191
Ball Corporation, 255
Ball Seed Company, 231
Roger B. Bean, 22, 28, 43, 57, 65, 67, 78, 121, 138, 139, 182, 191, 219, 281, 284, 285, 286, 289, 290, 291, 295, 317, 342
Keith M. Berry, 124, 150
Bethlehem Steel Corporation, 216
The Bettman Archive, 13, 27, 28, 315
Bettmann Newsphotos, 29, 30
Black & Decker, 57, 58, 59, 62
Boise Cascade, 76
British Airways, 20
Brown & Sharpe Grinding Machines, Inc., 59
Carlson, James, 28
Carnival Cruise Lines, 293
Caterpillar Inc., 95
Certainteed Corporation, 285
Chrysler Corporation, 28, 126
Cincinnati Milacron, 58
Circle Design/Carol Spengel, 12, 38, 40, 41, 109, 114, 115, 116, 140, 142, 143, 144, 198, 206, 207, 228, 229, 264, 281, 308, 309, 310, 312, 342, 348, 354, 355
Cleveland Cliffs, Inc., 335
Colt Industries, 22
Compression Labs Inc., 180
Comstock: 120; Robert Houser, 154, 244; Russ Kinne, 99
The Cousteau Society, 328
Coventry Creative Graphics, 68, 97, 237, 250, 292.
CS&A, 45, 71, 80, 85, 106, 122, 126, 148, 149, 189, 205, 215, 223, 227, 238, 239, 243, 247, 254, 260, 279
David Sarnoff Research Center, 194
Howard T. Davis, 54, 55, 56, 57, 58, 70, 90, 199, 200, 202, 215, 216, 217, 218, 219, 220, 221, 222, 265, 323, 324, 325
Deere & Company, Moline, 19, 22
Delta International Machinery Corp., 59, 220
Design Associates, 291
Detroit Edison, 326
DoAll Company, 220
Dover Elevators, 228
Dow Chemical Company, 217
Dresser Industries, Inc., 58
Emerson Electric Company, 96
Epson America, Inc., 178
E-Systems, Inc., Dallas, Texas, 75
Fiberflex, Inc., 302
FINA, 185
Foxfire/Jason Kilby, 196, 197
David R. Frazier/Photolibrary, Inc.: 8, 19, 91, 95, 136, 169, 170, 209, 210, 230, 304, 320, 326, 331, 334; K-Mart, 245
Bob Gangloff, 13, 125, 133, 157, 158, 159, 161, 168, 169, 174, 175, 176, 177, 190, 194
Ann Garvin, 28, 32, 33, 43, 66, 95, 96, 97, 104, 183, 247, 280, 289, 291, 292
General Motors Corp., 121, 327
General Motors/Pontiac, 293
Georgia Pacific Corp., 209, 211
Goes Lithography Company, Chicago, 97
Graphic Associates/Eric Fanning, 209
Jeff Isaac Greenberg, 255, 293, 320, 331, 332
Marshall R. Greenberg, 212, 224, 234, 249, 256, 283, 316
Habitat for Humanity/Julie Lopez, 306, 307
Handy & Harmon, 96, 333
Harris Corporation, 184, 331
Henry Ford Museum & Greenfield Village, 253
Hewlett Packard, 20, 136, 327
Hughes Aircraft Company, 20
Illinois Bell/John Firucan, 185
Index Stock International Inc., 83, 94
JET Equipment & Tools, Tacoma, WA, 58
K's Merchandise/Woodruff High School/Roger B. Bean, 44
Knight-Ridder, Inc./Chuck Mason, 178
L.A. Visitors and Convention Center, 28
Laser Alignment, Inc., 301
L. B. Foster Company, Pittsburg, 282
Lockheed Missiles & Space Co. Inc., 8, 349
Manual High School/Roger B. Bean, 125, 220, 222
Manual High School Rams/Roger B. Bean, 165
Kathy L. McCloskey, 227
McDonnell Douglas Corporation, 259
Bob McElwee, 160, 236
Midwest Dental Products Corp., 58
Ted Mishima, 22
Morton Water Treatment Plant, Roger B. Bean, 294
NASA: 8, 22, 24, 28, 186, 259, 298, 318, 319, 350, 351, 352; Ames Research Center, 351; Harold Smelcer, 298, 353, 354
National Machinery Co., 216
Nikon, 78
Noah Herman Sons, House of the 1990s/Roger B. Bean, 182
North Wind Picture Archives, 13, 30, 48, 151, 155, 156, 181, 207, 208, 275
Numonics Corporation, 78
Parker Hannifin Corporation, 184
Patent Reproduction Company/Sarah E. Callahan, 46
City of Peoria, IL/Roger B. Bean, 48
Peoria Public Library/Roger B. Bean, 123
David Perry, Rockwell International, 322
Judith Peterson/Arnold & Brown, 205
Phelps, Brent, 43, 49, 78, 79, 104, 194, 209, 245, 291
Philips & DuPont Optical Company, 79
PhotoEdit: 22, 50, 60, 340; Robert Brenner/ The Lighthouse, Inc., 95; Paul Conklin, 209, 249; Mary Kate Denny, 122; Tony Freeman, 22; Hewlett Packard, 248; Felicia Martinez, 76; Alan Oddie, 50
Photo Researchers Inc.: 298; James Holmes/ Thomson Laboratories, 7; Will & Deni McIntyre, 172; Hank Morgan, 257, NASA, 338, 350; David Parker, 4; Dr. Morley Read, Science Photo Library, 113
PHOTOTAKE/Bob Masini, 233
Picker International, Inc., 28
Plymouth, 22, 339
Potlatch Corporation/Tom Tracy, 209
Purcell, Elizabeth, 27, 28, 34, 67, 92, 240, 273, 275
P&W Builders, Inc./Roger B. Bean, 280, 282, 283, 284, 285, 286, 291
Quaker State Corporation, 333
Questar, 6
Randy Feucht Construction Co., Inc./Ann Garvin, 282
Rauma Yards, Finland for Diamond Cruise, Ltd., 344
Reynolds Metals Company, 22, 241
Cloyd Richards, 129, 130, 131, 132, 134, 135, 266, 275, 276
Mark Romine, 5, 128, 140, 159, 253
Doug Schermer, 47, 53, 68, 69, 70, 84, 85, 87, 103
Scotti Bureau of Marketing Research/Roger B. Bean, 75
Sears Roebuck Co., 59
Smart House/L.P., 303
Sonoco Products Co./Terry Parke, 219
Space Biospheres Ventures/Scott McMullen, 299
Space Camp®, 346, 347
StarServe, 263, 264
Jeff Stoecker, 14, 15, 16, 17, 65, 76, 77, 84, 86, 88, 105, 106, 107, 108, 165, 166, 167, 173, 190, 192, 258, 317, 323, 341
Strippitt/Di-Acro-Houdaille, Inc., 242
Superstock, Inc.: 146, 215, 341; B. Apton, 164; G. Fritz, 7 (Four By Five); D. Junor, 278; G. Martin, 4; C. Orrico, 6, 18, 52, 214, 220 (Four By Five); PI Prod., 7, PR Prod., 74, 198, 252; Rivera Collection, 5, 26; M. Roessler, 82.
Tandon Computer Corporation, 79
Tensar Structures, 271
Thompson's Food Basket/Arnold & Brown, 28, 182
Thunder & Colt Airships, 329
The Timken Company/Great Lakes Lithograph Co., 316
TJI® Joist, 283
Tony Stone Worldwide: 308; Tim Brown, 42; Rainer Grosskopf, 314; Brownie Harris, 226; David Jacobs, 5; John Moss, 102; Peter Pearson, 270; Ed Pritchard, 330; Andrew Sachs, 45, 188; Stephen Scudd, 249; Vince Streano, 64; Charles Thatcher, 6
Underwriters Laboratory Inc.®, 49
U.S. Council for Energy Awareness, 327
U.S. Department of the Interior, 336
U.S. Environmental Protection Agency, 112
USDA Forest Service/Pacific Northwest Region, 112
USX Corporation, 292
VeriFone, 186
VersaCad, 160
Washington Grade & Jr. High Schools, 36, 37
Washington Intermediate School, 138, 139
Westlight®: Mark Adams, 335; M. Angelo, 38; Craig Aurness, 210, 264, 294, 303; Cradoc Bagshaw, 89; Digital Art, 114; Annie Griffiths Belt, 249; Kenneth Garrett, 317; Walter Hodges, 95, 233; Robert Landau, 272; Larry Lee, 92, 210, 274, 294; Ian Lloyd, 231; Lawrence Manning, 348; C. Moore, 288; Charles O'Rear, 61, 89, 332, 339; Optical Artists, 12; P.L.I. Ltd., 204; Jim Pickerell, 253; Richard Price, 83, 99; Diana Rasche, 232; Jim Richardson, 295; Bill Ross, 210, 237; David Seawell, 83; William James Warren, 89; Bob Waterman, 294; Ron Watts, 205; Nik Wheeler, 320
Woodruff High School: Roger B. Bean, 182, 294; Roger B. Bean/P&W Builders, Inc., 291
Wyffels, Terry, 23
David Zalaznik/Journal Star, 228
Zehr, Duane R., 36, 37, 104, 170, 182, 318, 333

Index

T

U

V

W

X, Y, Z